The Internet of Things:
Do-It-Yourself Projects with Arduino, Raspberry Pi, and BeagleBone Black

Donald Norris

New York Chicago San Francisco
Athens London Madrid Mexico City
Milan New Delhi Singapore Sydney Toronto

Cataloging-in-Publication Data is on file with the Library of Congress

McGraw-Hill Education books are available at special quantity discounts to use as premiums and sales promotions, or for use in corporate training programs. To contact a representative, please visit the Contact Us pages at www.mhprofessional.com.

The Internet of Things: Do-It-Yourself Projects with Arduino, Raspberry Pi, and BeagleBone Black

1234567890 DOC DOC 1098765

ISBN 978-0-07-183520-6
MHID 0-07-183520-2

Sponsoring Editor
Roger Stewart

Editorial Supervisor
Patty Mon

Project Manager
Harleen Chopra, Cenveo® Publisher Services

Acquisitions Coordinator
Amy Stonebraker

Copy Editor
Nancy Rapoport

Proofreader
Claire Splan

Indexer
Ted Laux

Production Supervisor
George Anderson

Composition
Cenveo Publisher Services

Illustration
Cenveo Publisher Services

Art Director, Cover
Jeff Weeks

Cover Designer
Jeff Weeks

This book is dedicated to my wonderful parents, Harry and Esther Norris,
who provided the love and support to enable me to become the person that I am today.
I would also like to dedicate the book to my inspiring high school physics teacher, Mr. Bluhm,
who instilled in me a love for science, especially physics, that still burns within me.

About the Author

Donald Norris has a degree in electrical engineering and an MBA with a specialization in production management. He is currently teaching both undergraduate and graduate IT courses at Southern New Hampshire University. He has also created and taught several robotics courses there. He has over 30 years of teaching experience as an adjunct professor at a variety of colleges and universities.

Mr. Norris retired from civilian government service with the U.S. Navy, where he specialized in acoustics related to nuclear submarines and associated advanced digital signal processing. Since then, he has spent more than 20 years as a professional software developer using C, C#, C++, Python, and Java, as well as 5 years as a certified IT security consultant.

Mr. Norris started a consultancy, Norris Embedded Software Solutions (dba NESS LLC, www.nesslc.net), that specializes in developing application solutions using microprocessors and microcontrollers. He likes to think of himself as a perpetual hobbyist and geek and is always trying out new approaches and out-of-the-box experiments. He is a licensed private pilot, active member of the Civil Air Patrol, photography buff, amateur radio operator, and avid runner.

He is also the author of two other McGraw-Hill Tab books *Raspberry Pi Projects for the Evil Genius* and *Build Your Own Quadcopter*.

Contents at a Glance

Contents

Acknowledgments

Project books such as this one are never created in a vacuum. They are a product of the author's insight, creativeness, and most important, an integration of the vast resources available in the open-source community. I have tried to be true to the spirit of open source and acknowledge all the fine contributions to the technology that I have incorporated into this book. I humbly apologize if I've missed someone's contribution.

I also wish to acknowledge the fine support and encouragement that I have received and continue to receive from my colleagues at Southern New Hampshire University. This book's content is also presented in a new undergraduate IT course, which reflects SNHU's innovative spirit.

Introduction

I suppose the genesis of this book was from long-term interest in connecting computers to the Internet. Back in the early 1990s, I experimented with a variety of single-board computers that would connect, in a fashion, to the Internet and serve up relatively simple web pages. They didn't have the sophistication or capabilities of today's single-board systems exemplified by the Raspberry Pi or the BeagleBone Black boards. But they worked and provided useful platforms to experiment using simple computing devices in lieu of full-scale PC-based servers or desktop clients. Roll forward almost twenty years and there are now quite a few highly capable and functional systems, which can easily fulfill the promise of the "Internet of Things," or IoT for short. I describe the IoT in the first chapter in yet another attempt at defining a somewhat nebulous phrase, which truly means different things to different people. I also took a somewhat different approach with this IoT book in that I used three separate platforms to implement the various demonstration projects. My initial thought was to demonstrate how one platform could have strengths in one area while another would be better suited to another area. However, what I found was that the platforms had much more in common than they were different. In fact, the Raspberry Pi and the BeagleBone Black boards are just about identical from a software prospective. Let me now present a glimpse into what awaits you in this book.

In Chapter 1, I present a high-level view of what constitutes the IoT. I also introduce the Raspberry Pi (Pi) board, which is the first one of the three development platforms used in the book's projects. The LAMP framework is also introduced, which allows you to create a comprehensive data acquisition system that can not only take sensor measurements, but allows them to be stored in a relational database and later makes that data available via a web interface.

I show you how to build and remotely access a home temperature measurement system in Chapter 2. This system uses the LAMP framework, which was introduced in Chapter 1, as its basis. Analog-to-digital conversion appropriate for the Pi is also discussed and demonstrated. A program written with Python controls the system to help simplify this project. Again, all the temperature system data is available via a web page.

As a computer science/information technology instructor, I have found that many of my beginning students really do not have a good understanding of the nature of OO and why they should even have to learn it. Chapter 3 contains an introduction to object-oriented (OO) programming using the Java language, which should help you understand the program presented in Chapter 4. This chapter's content will also help you gain more insight into what makes up OO and how to properly use it.

Chapter 4 explores the principles and concepts discussed in the previous chapter, and applies them in the construction and operation of a home weather station, which uses the Pi as a controller. The Pi also is programmed using the Java language, which was just recently "ported" over to the Pi by the software developers at Oracle. The latest software Linux Wheezy distributions, which run on the Pi, all contain a fully functional Java runtime. I also discuss how to program the general-purpose input/output (GPIO) pins to implement the weather station interface. Programming this interface is only made possible by the use of a very clever Java library named Pi4J.

Chapter 5 covers three projects involving a webcam and the Pi camera. The purpose of these projects is to demonstrate how to implement remote video viewing using the Pi as a controller. The first project uses a generic USB webcam along with a comprehensive Linux software suite named Motion. The Motion software package provides for literally a plug-and-play situation where the webcam video can be viewed in real time via a standard web browser in a matter of a few minutes; no soldering or construction required. The second chapter project uses the Raspberry Pi camera accessory with its standard driver software to again implement a remote video viewer. The last project in this series again uses the Pi camera, but this time I use the Motion software in lieu of the standard driver software. This demonstrates Motion's flexibility and capability to interface to a variety of video devices.

The next chapter introduces the Arduino board, which is the next development platform used in this book. I briefly discuss the Arduino board features, as I realize that most of my readers are already very familiar with its operation and probably have one or two in their workshop. This chapter's demonstration project is a garage door opener, which may be actuated via a web browser. I also show you how to use a smartphone to control the garage door. In addition, some security in the form of a password is added to this project as you probably don't want strangers operating your garage door.

Chapter 7 covers an Arduino-controlled home irrigation system. This system builds on some of the concepts discussed in Chapter 6 for the garage door opener and also shows you how to incorporate a wireless sensor into the overall control scheme. This system allows a homeowner to remotely activate a specific irrigation zone using only a web browser. It also further expands the homeowner's options by reporting the current soil moisture content via a web page so that the user can make a decision on whether or not to turn on an irrigation zone.

Chapter 8 focuses on both remote activation of lights or other similar devices and the capability of locating these controlled devices anywhere in the home without using wires. Arduino boards are used in multiple locations for this project. Some boards control wireless XBee nodes, which allow for the flexible placement of lights within the home. Another Arduino board implements the main controller, which connects to the Internet to enable remote light activation via a web page. There is also a four-channel key fob RF device used in this system that allows a homeowner to quickly and directly activate up to four lights without the need to use the Internet.

I next introduce the BeagleBone Black (BBB) boards in Chapter 9. This is the third and last development platform used in the book projects. This chapter's project is a simple demonstration that displays only a line of text on an LCD, which has been sent from a web browser. The chapter focus is to discuss the BBB and compare it to the Raspberry Pi, which seems to be its principal "competitor." The BBB used in this project used the same Debian, Wheezy, Linux distribution as was used in earlier Pi projects. This demonstrates that at least these two boards are more similar than different. The BBB does incorporate some features such as analog-to-digital conversion (ADC), which are not present in the Pi and must be externally added if needed. In addition, the BBB's standard clock rate is 1 GHz while the Pi's normal clock is set at 700 MHz. The Pi may be overclocked to approximately 1 GHz, if desired, but that does increase power consumption and heat generation. The Pi does run cooler and consumes less power than the BBB, which are important considerations for portable, battery-operated applications.

Connecting the BBB to a cloud service is the chief topic in Chapter 10. I used the same temperature monitoring system, which was shown with the Pi in Chapter 2. However, in this project, the data is streamed real-time to a cloud-based service named Xively. In the Pi project, the data was stored in a local MySQL database. The BBB sensor data is streamed to Xively for storage and later retrieval as desired. Xively also provides several web interfaces that make it easy for users to both examine and manipulate sensor data as needed. The Xively developer version is free with unlimited data storage, which should suffice for most experimenters and hobbyists.

The final chapter deals with machine-to-machine (M2M) communications, which happens when two or more fully automated computer systems interchange data without any human involvement. This chapter's project uses the same temperature measurement system used in previous projects. Transferring data also requires a protocol to be used, which will ensure that data is successfully sent and received. This project uses the open-source MQTT protocol, which is an excellent, lightweight data protocol currently used by Facebook and several national wireless carriers for sending alert messages. This demonstration project uses a single channel BBB temperature measurement system to indirectly send data to a Pi system. The Pi system accesses the data from any one of a number of MQTT broker websites, which are freely available to handle MQTT message traffic.

I hope these ten chapters open your desire to experiment and further explore this exciting and ever-expanding area.

Don Norris

1
CHAPTER
Introduction to the Internet of Things

This book offers useful projects that you can build and then experiment with, using the Internet to both receive data from and/or provide control commands to devices. The "Internet of Things" (IoT) is a phrase that was first used in 1999 by Kevin Ashton while he was working at MIT's Media Center. He meant it to represent the concept of computers and machines with sensors, which connect to the Internet to report status and accept control commands. The IoT, in reality, has been around for a long time, but it didn't have a name. Machine-to-machine (M2M) communications has been in existence for many decades, often using dedicated networks that eventually converged over to the Internet. IoT is also referred to with different names, such as Ubiquitous Computing and the Internet of Everything. No matter what the name, IoT is here to stay and is progressively affecting more people in their everyday activities as time progresses.

Many books are in print and in digital media that discuss the overall ramifications of IoT upon society and where it is leading all of us. There are also books published that claim to guide you on how to make a fortune by monetizing your clever IOT project. This is not one of those books, as I mentioned in the foreword. My focus is on creating useful projects that take advantage of the tremendous communication capabilities provided by an Internet connection. My approach also differs from other IoT authors by using three separate hardware platforms, which provide project control. I should note that the Arduino platform uses three slightly different implementations for Internet connectivity, which I classified as one platform. Using different platforms was a deliberate and purposeful decision on my part to show you what is involved in creating projects using different development infrastructures yet still establishing a working Internet connection. You will likely appreciate one approach over the others. These three hardware and software development platforms are listed in Table 1-1.

Creating a project that is equipped with sensors and is capable of both sending and receiving data via the Internet is a bit challenging, especially to those readers who are attempting to do so for the first time. Let's start this journey with a discussion of hardware as that seems easiest for most folks to handle and is absolutely required for these projects.

Hardware Platform	OS	Language(s)
Raspberry Pi	Linux—Raspbian (Debian) distribution	Python, Java
Arduino	WIN7—for development	.NET framework Processing
Beaglebone Black	Linux—Wheezy (Debian) distribution	Python

Table 1-1. IoT Project Hardware and Software Development Choices

Raspberry Pi Platform

The Raspberry Pi has been in existence for almost two years at the time of this writing. Over two million Pi platforms have been produced since it was introduced, which is not too shabby considering that the creator, Dr. Eben Upton, originally thought about 10,000 would be sold. I won't go into extensive detail about the origins, history, and structure of the Pi, as I have already covered that subject in extensive detail in my recent book *Raspberry Pi Projects for the Evil Genius*. However, I will reiterate some key Pi concepts that are critical to your success in building the Pi projects, and it is always convenient to have the data immediately available and in one place. The Model B Raspberry Pi is the platform I strongly suggest for the Pi projects in this book (see Figure 1-1).

Figure 1-1 Model B, Raspberry Pi board

A cheaper model, A, is available but it does not have an onboard Ethernet port and only half the memory of the B model. Interestingly, neither one of these two constraints would prevent you from using the A model; however, you would need to provide a wireless USB adapter for Internet connectivity, and the diminished memory would certainly slow down the Pi applications while they were running.

All the projects in this book, except for the first one, involve using some type of digital input and/or output to interface with sensors and actuators. All the different platforms used in the projects refer to these input/outputs as *general purpose input output* (GPIO). Each platform's GPIO has somewhat different specifications as to the maximum voltage and current that can be handled, and I will strive to keep that very clear so as to avoid any possible damage to the project boards. Unfortunately, irreversible damage happens if you exceed the maximum voltage or current GPIO rating to a particular board, which will render it useless or non-operative.

Raspberry Pi GPIO

The Model B, rev 2 Raspberry Pi uses a multi-pin connector designated as P1 for its GPIO. This connector is shown in Figure 1-2 with the first two beginning and ending pin numbers annotated for orientation and reference.

This multi-pin connector will be the gateway through which the Pi will interface with real-world devices. As you are probably aware, there must be software drivers loaded that provide the logical interface between the control program, operating system (OS), and the GPIO pins. The particular type of driver depends primarily upon the programming language used to develop the control program. I will be using both Python and Java to develop control programs so there will be a separate set of drivers loaded to accommodate each

Figure 1-2 GPIO P1

development environment. However, many GPIO pins in the P1 connector have multiple functions that extend beyond simple digital input and/or output. Figure 1-3 shows the functions associated with each of the P1 pins for the Model B, rev 2 Pi.

I will not review these pin functions at this time but will discuss them as they become relevant to a project. Incidentally, none of the projects connect directly with the P1 pins but instead rely on the use of a Pi Cobbler, which is plugged into a solderless breadboard. Figure 1-4 shows the Pi Cobbler adapter plugged into a solderless breadboard with the 26-conductor ribbon cable plugged into the Pi's P1 connector.

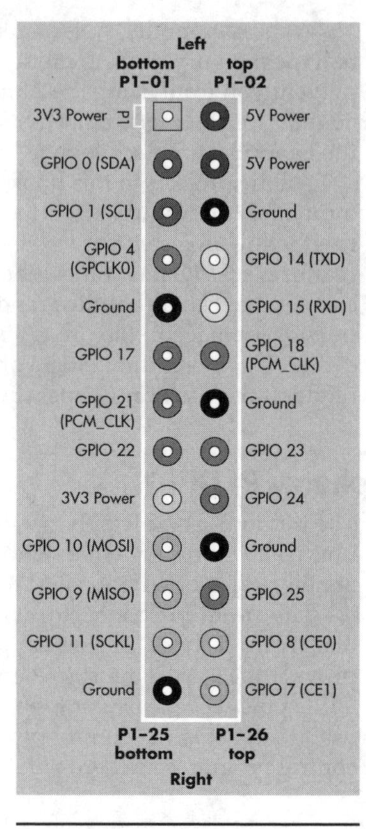

FIGURE 1-3 P1 pin functions

The Pi Cobbler is available from a variety of suppliers such as Adafruit Industries and MCM Electronics. You will have to assemble it by soldering a connector to the printed circuit board (PCB), which is not too difficult, and this task allows you to practice your soldering skills. Just don't add too much solder to the connector pins as they are close together and it is easy to form a solder bridge, which might be disastrous to the Pi when you connect the Pi Cobbler to it.

Although there are jumper wires shown connecting components on the solderless breadboard, I prefer to use manufactured jumper wires, as shown in Figure 1-5. These jumpers are very sturdy and can easily be inserted into the breadboard without the bending or crinkling that affect ordinary precut wires. Inexpensive jumper wire kits are also typically available from the same Pi Cobbler suppliers.

FIGURE 1-4 Pi Cobbler

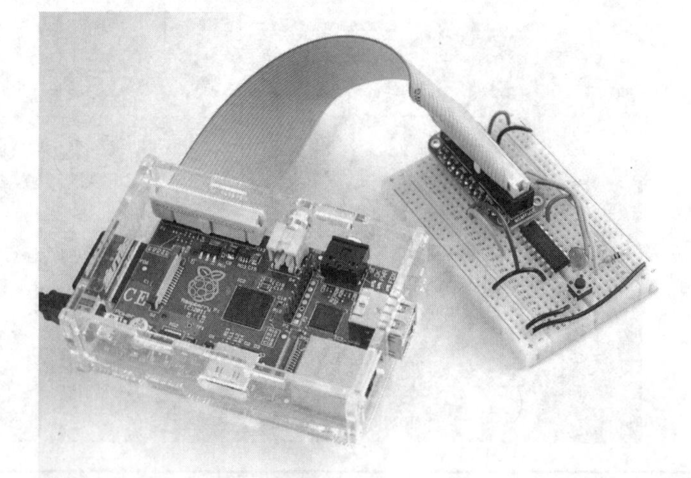

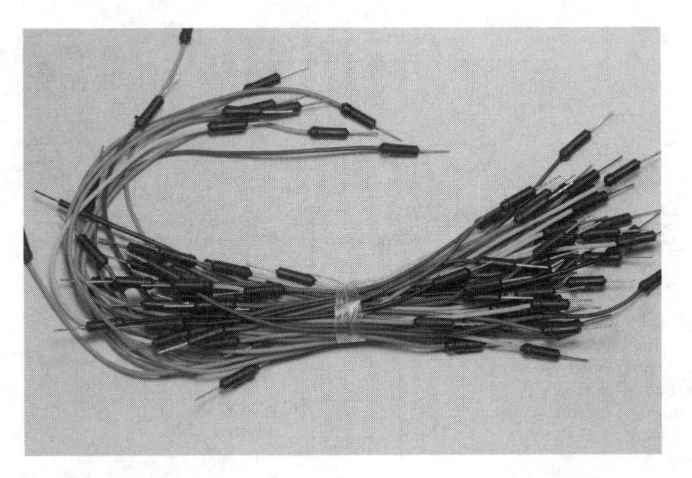

Figure 1-5 Manufactured jumper wires

Establishing a Raspberry Pi Development Station

There are several ways to set up a Raspberry Pi development station, each with its own pros and cons. I will cover two approaches that will likely fulfill most users' needs.

Standalone Setup

The first approach is what I call a standalone setup where you connect a keyboard, monitor, and mouse to the Pi. You will also need a powered USB hub and either a wireless Wi-Fi adapter or Ethernet patch cable that you can plug directly into your router. Figure 1-6 is a block diagram showing all the components needed for a standalone workstation.

The Pi has both composite and HDMI video outputs. Most readers will elect to use the HDMI output as that provides a much superior video display as compared to the analog composite video output. You will need a HDMI to VGA converter module in case your monitor does not have an HDMI input. These converters are relatively inexpensive with a typical one available from Adafruit shown in Figure 1-7.

The Pi power supply is also worth discussing. I used a "wall wart" 5V 1A supply, which is more than adequate for providing sufficient current to the Pi as long as you do not attempt to power any external USB devices from either one of the two onboard USB ports. From my experience in using the Pi now for two years, I have found the board to be a bit sensitive to the quality and level of the 5V supply. Strange and frustrating events happen if the power supply droops to 4.75V or less, which is only a 5 percent drop. Often, simply swapping the power supply clears up mysterious and intermittent operational issues, which can lead to unproductive and "hair-tearing" development sessions. In Figure 1-7, I have included a note that mentions you can also power the Pi directly from the hub using a micro-USB/USB cable as long as the hub power supply is rated for a minimum of 2.5A. I have used the Pluggable series of powered hubs to do this in the past, one of which is shown in Figure 1-8.

Any USB keyboard and mouse combination will suffice for user input. However, I did find the wireless Logitech K400 keyboard/mouse device to be a very handy and flexible combination. There were no issues with the Pi detecting this device and installing the

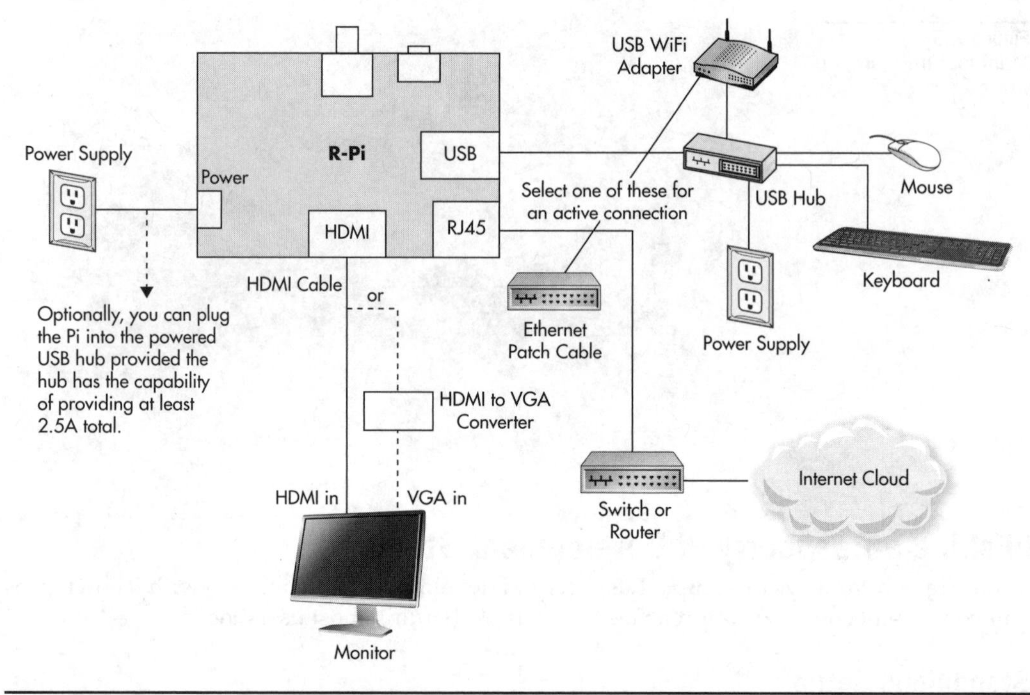

Figure 1-6 Raspberry Pi standalone workstation block diagram

proper driver. The K400 is inexpensive and is shown in Figure 1-9. I highly recommend this keyboard/mouse unit.

I would like to mention the wireless Wi-Fi adapter that I have successfully used for a number of projects. It is the EDIMAX EW-7811Un and is shown in Figure 1-10. It is very inexpensive and seems to perform quite well for the relatively low-bandwidth projects I have used it in.

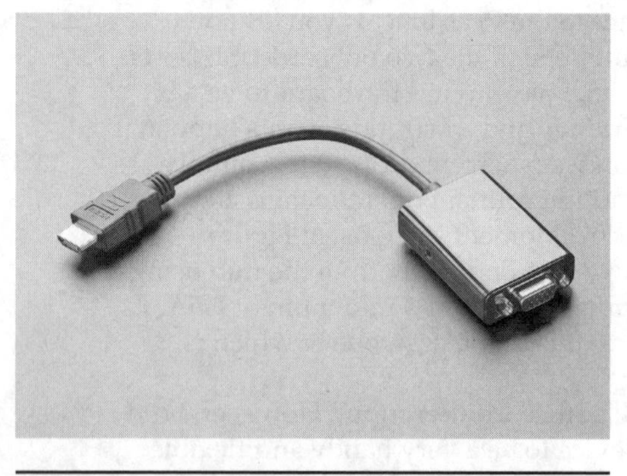

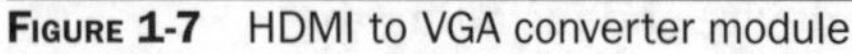

Figure 1-7 HDMI to VGA converter module

Figure 1-8 Plugable powered USB hub

Figure 1-9 Logitech K400 wireless keyboard/mouse unit

Figure 1-10 EDIMAX model EW-7811Un USB Wi-Fi adapter

You should note that it is rated at a maximum of 150 MBps, which is somewhat lower than other more expensive brands. However, none of the book projects require a very high bandwidth so why spend the money for performance you will not require?

Headless Setup

The second approach is not a gruesome Pi decapitation as the name suggests but a network-centric configuration to remotely control a Pi. For this approach you will need only a networked Pi and another computer. It doesn't matter if the Pi is connected wired or wirelessly to your network. All you really need is the Internet protocol (IP) address that your router assigns to the Pi when it discovers it upon initial startup. Note that no keyboard, mouse, monitor, or powered hub is required for this setup. Just a Pi, a power supply, and either an Ethernet cable or a wireless Wi-Fi adapter are needed. Figure 1-11 is a block diagram showing all the headless components and their interconnections.

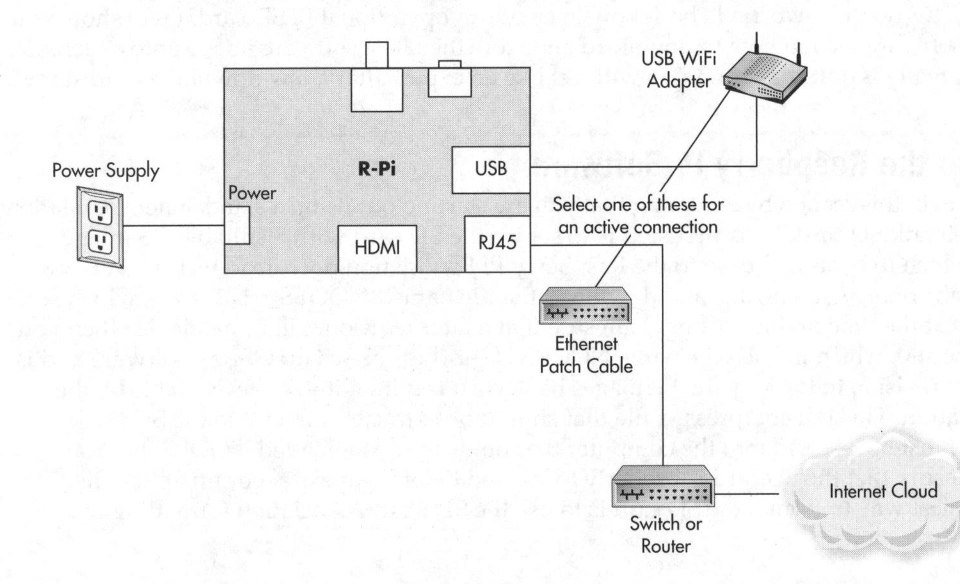

Figure 1-11 Raspberry Pi headless workstation block diagram

TABLE 1-2 SD Card Class Designations

Class	Minimum Performance
Class 2	2 MB/sec
Class 4	4 MB/sec
Class 6	6 MB/sec
Class 10	10 MB/sec

The secret to the simplicity of the headless setup is the software running both on the Pi and the computer used to communicate with the Pi. This software will be one of the items discussed in the following software section.

The last hardware item to be discussed is the SD card that stores the software that the Pi needs to function. A standard 4GB SD card is the minimum required for Pi operations, but I feel strongly that you should use at least an 8 or 16GB card to have space for all of the book projects without having to delete any of them. It is fairly easy to add software whose memory requirements can quickly add up to the point where Pi operations could be adversely affected. However, don't be deterred if you purchased a Raspberry Pi starter kit that came with a pre-built image 4GB SD card. It will be sufficient for all the book projects, but you might have to delete some early project files to ensure there is space for the later projects.

SD cards are also rated for speed with a Class number. Table 1-2 shows the various classes and associated minimum data transfer speeds.

Using a higher Class number of SD card in the Pi allows for much better performance. Just be mindful that SD cards with high Class numbers are more expensive than ones with lower numbers. However, the cost differential diminishes as time progresses. I strongly suggest you purchase at least a Class 4 or higher; anything less and you will be disappointed in how slow your Pi responds.

Finally, don't be worried about how to create an operational Pi SD card. I will show you in the software section how to download and store the latest software image onto a blank SD card. It really is quite easy and you will feel like an expert after a few downloads and stores.

Setting Up the Raspberry Pi Software

I will begin this section by assuming that you are starting out using a standalone workstation with a blank SD card. Your first step is to set up the SD card with a suitable OS image from which to boot the Pi. Go to the Raspberry Pi Foundation download website at www.raspberrypi.org/downloads and download the file named NOOBS_v1_3_4.zip, which is current at the time of this writing. I am sure that a later revision will be available when you visit the site, which is fine. The name NOOBS is short for New Out Of Box Software and is a recent revision to the way the Pi images have been traditionally made available by the Foundation. This is a compressed file that should be extracted directly to the SD card, which must be inserted into the computer that holds the downloaded NOOBS file. You must ensure that the SD card is properly formatted before you extract or unzip the file. The easiest way to format the SD card is to use the SD Card Association formatting tool,

SDFormatterv4.zip, which may also be downloaded from the same Raspberry Pi Foundation website mentioned previously. Of course, the formatting tool must also be extracted before use.

The freshly formatted and NOOBS-loaded SD card has been designed to boot the Pi into a clever menu that allows you to select one of four operating systems. To boot the Pi, first ensure that the workstation is set up per the diagram shown in Figure 1-6, without the power supply attached to the Pi board. It is okay to power up the USB hub, provided that the hub is not directly powering the Pi.

Next, insert the NOOBS SD card into the Pi and then connect the power supply to the Pi. If this has been properly done, you will see the NOOBS menu selection displayed.

The NOOBS revision menu selection has eight choices, as detailed in Table 1-3.

Menu Selection Name	Description
Raspbian (*recommended*)	This is the recommended Linux distribution you should initially use. It is based upon the Debian "Wheezy" distribution and has the LXDE desktop built-in.
Arch	This distribution is recommended for experienced users or those who wish to learn Linux using a text-based interface. No built-in desktop.
OpenELEC	This is the Ubuntu 13.04 distribution customized for the Pi. Users should try this one after using Raspbian. Usually comes with Gnome desktop.
Pidora	A fully functional distribution based on Fedora 18, using the Xfce desktop. *Note: Future open-source support for Pidora is in question. It will not likely be updated.*
RISC OS	A British developed and maintained OS especially designed for ARM processors such as the one that runs the Raspberry Pi. It is compact (memory wise) and very efficient. *Note: It is not Windows or Linux but its own open-source OS.*
RaspBMC	A distribution that is media-centric in that it blends Raspbian with the XBOX media center. Fairly mature now with many of the initial problems resolved. Turns the Pi into a very nice media center supporting most media formats.
Raspbian (Boot to Scratch)	This Raspbian variant is set to boot directly into the Scratch application. Useful for users who wish to focus on learning and using the Scratch app. Readers of this book will not find this OS useful.
Data Partition	In my opinion, an odd selection to add to the menu. It sets up a 512MB partition to the SD card. Using the gparted application is much more useful and general purpose. NOT recommended.

Table 1-3 Initial NOOBS Selection Menu

I strongly recommend you select the first menu item, which is to install the latest Raspbian distribution. The top menu Install button will become active after you click on any selection. Simply click on Install to commence the install process.

A dialog box asking you to confirm that the pending install will delete any existing data from the SD card will be shown next. This is the last time you can avoid the serious mistake of overwriting an SD card that you didn't intend to use. Click OK, assuming everything is proper and that you are indeed using the desired SD card.

Next, there will be a series of screens displayed, beginning with a Welcome message that also contains a progress bar indicating how much of the installation has been completed. The install will take a while depending on the size of the distribution and the data transfer speed of the SD card you are using. The initial installation portion has been completed when the Raspbian banner is displayed. The screen contains, in part, this statement: "… based on Linux and optimized for the Raspberry Pi." That is all true, but it really should have expressed that it is based on the Debian Linux distribution as there are significant differences between Linux distributions, as you can see in Table 1-3.

The next screen that appears in the installation sequence is very important. Figure 1-12 shows both the username and password that you will need when you attempt to run the Raspbian OS on the Pi. Every Raspbian distribution that is downloaded from the Foundation's website has the same username and password. Obviously, this is not a very secure situation if you connect the Pi to the Internet. Never fear, however: Later on, I will show you how to change both the username and password to establish much better security for your Pi installation.

You will next see the raspi-config introduction screen. The raspi-config main menu is automatically shown the first time you boot up the Raspbian OS. Its purpose is to easily allow you to configure your OS to match your needs and requirements. I will discuss the raspi-config application in detail.

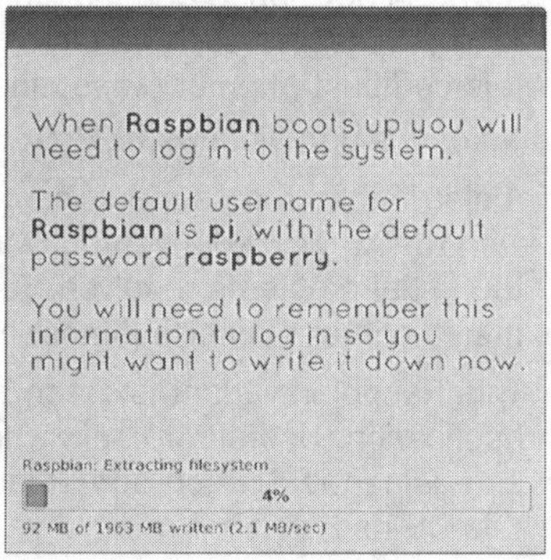

When **Raspbian** boots up you will need to log in to the system.
The default username for **Raspbian** is **pi**, with the default password **raspberry**.
You will need to remember this information to log in so you might want to write it down now.
Raspbian: Extracting filesystem
4%
92 MB of 1963 MB written (2.1 MB/sec)

FIGURE 1-12 Default username and password screen

The last display screen shown indicates that the Raspbian OS has been successfully installed. Notice the following near the bottom center of the screen display:

"For recovery mode, hold Shift"

What this means is that you can get back to the NOOBS opening selection menu by holding down the keyboard's SHIFT key while powering on the Pi. At this point, you can reinstall an old OS or select a new one. This is very useful if and when you corrupt your existing OS, which is likely to happen with all the experimenting you will be doing. Now, you must be very aware that any data files, which are stored on the NOOBS SD card, will be deleted when a reinstallation happens. This is why it is very important to copy and store any and all data files either to a network drive or to removable media such as a thumb drive. Neglecting to do frequent backups will cause you distress when you realize you have corrupted the OS and consequently lost all your data. Also realize that any applications that you might have loaded above and beyond the core Raspbian installation will be lost. This is not a problem as you can reload and reconfigure using the same procedures you followed earlier to initially install them. The data, however, is another story, and it will likely remain gone unless you have done the backups as suggested.

Clicking OK on this last screen will reboot the Pi and eventually bring you to a command line prompt where you will enter the username (pi) and the password (raspberry). The raspi-config menu screen should now appear.

Table 1-4 shows all the raspi-menu selections arranged by menu number along with the description and my recommendation as to what you should do with a particular selection. I believe you should initially follow my recommendation. You can always change at a later time.

There are also seven Advanced Options menu selections, as described in Table 1-5.

Click the Finish button after you have entered all the raspi-config menu selections. You should be returned to the command line.

Next enter the following command to check if you have successfully installed the Desktop GUI.

```
startx ↵
```

Note *I will use the* ↵ symbol from now on to denote the ENTER key.

The Desktop screen should appear, as shown in Figure 1-13, if the OS installed correctly.

This screen is the LXDE Desktop, which is the default Raspbian OS GUI interface. LXDE is short for Lightweight X11 Desktop Environment and is built upon the X-Windowing system. X-Windows has nothing to do with Microsoft Windows but is instead based on a windows framework created at MIT during the mid-80s. X-Windows is independent of any particular OS, which means developers must create appropriate interface software for it to function with a specific OS.

raspi-config menu #	Description	Recommendation
1	Expand Filesystem	Do nothing. The SD card will already have been expanded because of the NOOBS installation.
2	Change User Password	Leave it as "raspberry" for now.
3	Enable Boot to Desktop/ Scratch	Do nothing. You will need to do most of the configurations at the command line.
4	Internationalization Options	Personalize your language and set your time zone.
5	Enable Camera	Enable the camera option if you have one or are planning on buying the Pi camera.
6	Add to Rastrack	It's your choice if you wish to register your Pi at this private website. No advantage either way regarding the book projects.
7	Overclock	Not recommended. Overclocking adds additional heat stress to the Pi cpu and is not needed for any of the book projects.
8	Advanced Options	A series of additional choices that are discussed in the text that follows.
9	About raspi-config	A simple credits screen.

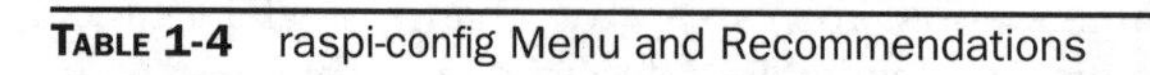

Table 1-4 raspi-config Menu and Recommendations

raspi-config Advanced Options Menu #	Description	Recommendation
A1	Overscan	Removes black bars from display. Use it if you need it.
A2	Hostname	Do nothing. It sets a network hostname for the Pi. I will show you how to do this from the command line.
A3	Memory Split	Do nothing. It changes the amount of memory allocated to the graphics processor unit (GPU). The default split is fine.
A4	SSH	Enable SSH. You will need it for headless operation.
A5	SPI	Enable SPI. SPI is short for the serial peripheral interface bus, which is used in a book project.
A6	Audio	Do nothing. There are three modes: 0—Auto (default) 1—Force audio to headphone jack 2—Force audio to HDMI connector
A7	Update	Do nothing. I will show you how to both update and upgrade from the command line.

Table 1-5 raspi-config Advanced Options Menu and Recommendations

Figure 1-13 Desktop GUI

Clicking the LXDE icon button located at the lower-left corner of the screen triggers a menu with four choices:

- Shutdown
- Reboot
- Logout
- Cancel

Shutdown turns off the Pi as the name implies. Reboot causes the Pi to cycle through a complete restart and presents you with a command line login prompt after it is done. Logout stops the GUI and brings you right back to a command line prompt. There is no reboot or resetting involved with this command. The Cancel command brings you back to the GUI screen.

You will now have a complete Raspbian OS up and running if you have successfully followed all the previous steps. Before proceeding to any more advanced instructions, I would like to show you how to set up the Pi using a complete Raspbian OS image that may be downloaded from the Foundation website.

Setting Up the Raspberry Pi OS Using an Image File

This section shows you how to set up a Pi with a raw image file. This was the only way you could create an OS prior to the NOOBS software introduction. It is important to understand this procedure because it allows you to load any OS image and not be limited to the ones contained in NOOBS.

The first step is to download the desired image file from the Foundation's download website. This is the same one mentioned earlier when you downloaded the NOOBS software. The image software is located further down in the website listing from the NOOBS section. At the time of this writing, the current Raspbian image is listed as 2014-01-07-wheezy-raspbian.zip. It will need to be unzipped or extracted before being further processed.

You cannot simply unzip the file onto an SD card. It won't work as the image must be transferred in a very specific manner for it to properly boot and function as an OS. There is a free open-source program named Win32DiskImager that you would use on a Windows computer to transfer the unzipped image to a formatted SD card. This program is available from the sourceforge site at http://sourceforge.net/projects/win32diskimager/files/latest/download. The program download is in a zipped format that must be extracted to a convenient location prior to use. Figure 1-14 is a screen capture of the Win32 Disk Imager program in action downloading the latest Raspbian image to a Class 10 SD card.

Notice the over 17MB/sec transfer rate shown in the figure. You will quickly appreciate using high-speed SD cards as they allow read and write operations to occur an average of two to three times faster than the much more common Class 4 SD cards.

All you need do next is put the newly imaged SD card into the unpowered Pi and apply power to start the boot process. This is what I did and I saw absolutely nothing on the monitor screen. This was certainly discouraging as I was sure that I had done everything as described in the Foundation instructions. It turns out that this raw Raspbian image caused the Pi to default to the analog video output instead of using the HDMI output to which my monitor was attached. This was not the case with the NOOBS installation, which apparently defaults to the HDMI video output. In any case, it is fairly easy to remedy this situation. Figure 1-15 is a listing of all the files that are installed on the SD card after the Win32 Disk Imager finishes executing. Note that this screenshot is from the laptop that I used to create the SD card and not from the Pi.

Shown in the list near the bottom is a file named config.txt, although the .txt extension is not shown in the file name list because of my Windows folder configuration. This file must be edited in order for the video display to appear on the HDMI video output. Figure 1-16 shows this file's content using the Notepad editor.

You will need to uncomment the following line:

```
#hdmi_force_hotplug=1
```

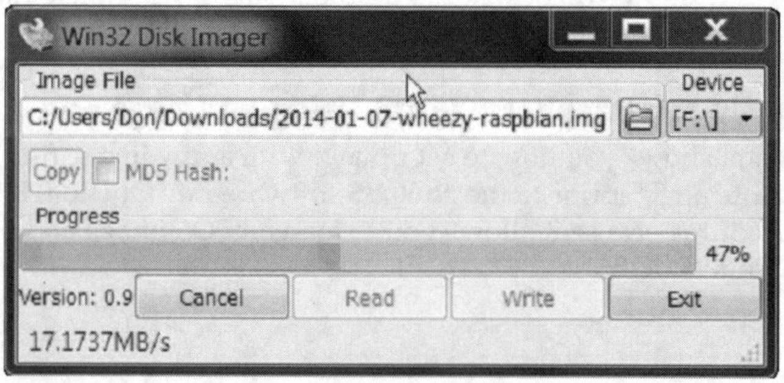

FIGURE 1-14 Win32 Disk Imager program executing

Name	Date modified	Type	Size
start	1/7/2014 7:50 PM	CodeWarrior ELF ...	2,456 KB
start_cd	1/7/2014 7:50 PM	CodeWarrior ELF ...	469 KB
start_x	1/7/2014 7:50 PM	CodeWarrior ELF ...	3,414 KB
fixup	1/7/2014 7:50 PM	DAT File	6 KB
fixup_cd	1/7/2014 7:50 PM	DAT File	3 KB
fixup_x	1/7/2014 7:50 PM	DAT File	9 KB
bootcode	1/7/2014 7:50 PM	FDT3 Data File	18 KB
kernel.img	1/7/2014 7:50 PM	IMG File	3,043 KB
kernel_emergency.img	1/7/2014 7:50 PM	IMG File	9,561 KB
LICENSE.oracle	9/25/2013 10:57 PM	ORACLE File	19 KB
cmdline	1/7/2014 9:13 PM	Text Document	1 KB
config	2/15/2014 10:25 PM	Text Document	2 KB
issue	1/7/2014 11:34 PM	Text Document	1 KB

Figure 1-15 Raspbian raw image file listing

To uncomment the line, you simply need to delete the # symbol from the line's beginning, and then save the file and exit Notepad. The SD card should now be all set to display the boot sequence from the HDMI port.

config - Notepad

File Edit Format View Help

```
# uncomment if you get no picture on HDMI for a default "safe" mode
#hdmi_safe=1

# uncomment this if your display has a black border of unused pixels visible
# and your display can output without overscan
#disable_overscan=1

# uncomment the following to adjust overscan. Use positive numbers if console
# goes off screen, and negative if there is too much border
#overscan_left=16
#overscan_right=16
#overscan_top=16
#overscan_bottom=16

# uncomment to force a console size. By default it will be display's size minus
# overscan.
#framebuffer_width=1280
#framebuffer_height=720

# uncomment if hdmi display is not detected and composite is being output
#hdmi_force_hotplug=1

# uncomment to force a specific HDMI mode (this will force VGA)
#hdmi_group=1
#hdmi_mode=1

# uncomment to force a HDMI mode rather than DVI. This can make audio work in
# DMT (computer monitor) modes
#hdmi_drive=2

# uncomment to increase signal to HDMI, if you have interference, blanking, or
# no display
#config_hdmi_boost=4
```

Figure 1-16 config.txt file content

Booting a raw image will bring you to the raspi-config screen, as was the case for the NOOBS installation. All the recommendations made for that installation hold true for this one, with the addition of expanding the filesystem. The NOOBS installation does it automatically but not for a more manual installation like this one.

Updating and Upgrading the Raspbian Image

The NOOBS software and raw image OS should be updated and upgraded to have the latest software revisions and patches in place. The update should be done first by entering the following at the command line prompt:

```
sudo apt-get update  ↵
```

This will normally take several minutes depending upon how "out-of-date" the OS image was at the time of installation versus the number of updates issued from the OS image publication date. I want to explain this command a bit further for those readers without much Linux command experience.

- **sudo** This instructs the OS that it should execute the following commands as if an administrator issued them. Linux is constructed by privilege layers with the admin at the top with the fewest restrictions.
- **apt-get** At this point you should have a fully functional and updated Raspberry Pi running the Raspbian Linux distribution after completing either the NOOBS or raw image installation. This must be in place before proceeding with any of the following IoT Raspberry Pi projects.

Headless Configuration

Unfortunately, it is one of those catch-22 situations (apologies to younger readers who don't know what that means—Google it!) where you need a fully configured SD card in order to run a headless configuration. But you can't configure it without a standalone workstation, as described earlier, or you can buy a pre-imaged SD card. That would be my strong recommendation if you know beforehand that you want to run headless.

A headless configuration was shown in Figure 1-11, where you need only to connect to a Pi in a network using either an Ethernet cable or a wireless Wi-Fi adapter. The network router will automatically provide an IP address to the Pi using what is known as the DHCP protocol, which is normally the default setup in most home or business wireless routers. What you need to do is attach another computer to your network that runs a program that can connect to the Pi and remotely run it. For Windows computers, that program is named PuTTY and is freely available for download at http://www.chiark.greenend.org.uk/~sgtatham/putty/download.html.

PuTTY uses the SSH protocol to communicate with the Pi, which must be enabled in order for the Pi to allow the communication link to function. You will need to determine the Pi's IP address in order to establish this link. The following procedure is usually successful in determining the Pi's IP address:

1. Open a browser session on the computer that you wish to use to control the Pi.
2. Go to the admin IP address for the router that is the DHCP server for your network. Often it is at 192.168.0.1.

3. Enter the username and password to get to the control web page. These are normally shown in the instructions that came with the router, but they are also readily available by doing an Internet search for your specific router model.
4. Click on Attached Devices or some similar menu selection that displays the IP addresses of all devices attached to the network whether through wires or wireless.
5. Look for the entry labeled Raspberry Pi. That's the IP address you will need for PuTTY.

It is a simple matter to connect to the Pi through your computer once you have the Pi's IP address. I will show you another way to determine or confirm the Pi's IP address at the start of the LAMP project discussion.

Start PuTTY and enter the IP address in the Host Name (or IP address) text block. Leave all the other selections and text blocks alone. Figure 1-17 shows the initial PuTTY screen with my Pi's IP address entered into the Host Name text block.

Your Pi's IP address will likely be different than the one I entered into the Host Name block. Also check that the port number is set at 22, which is the default for SSH. You should see the Pi's command line opening screen after you click the Open button located at the bottom of the PuTTY opening screen. Figure 1-18 shows the Pi login screen being delivered to the remote computer from the Pi via SSH.

Just enter the default username "pi" and the default password "raspberry," and you will see the normal command line prompt appear, as shown in Figure 1-19.

FIGURE 1-17 Initial PuTTY screen

192.168.0.13 - PuTTY
login as:

FIGURE 1-18 Opening Pi login screen over the network

pi@raspberrypi: ~
login as: pi
pi@192.168.0.13's password:
Linux raspberrypi 3.10.25+ #622 PREEMPT Fri Jan 3 18:41:00 GMT 2014 armv6l
The programs included with the Debian GNU/Linux system are free software;
the exact distribution terms for each program are described in the
individual files in /usr/share/doc/*/copyright.
Debian GNU/Linux comes with ABSOLUTELY NO WARRANTY, to the extent
permitted by applicable law.
Last login: Mon Feb 17 17:53:17 2014
pi@raspberrypi ~ $

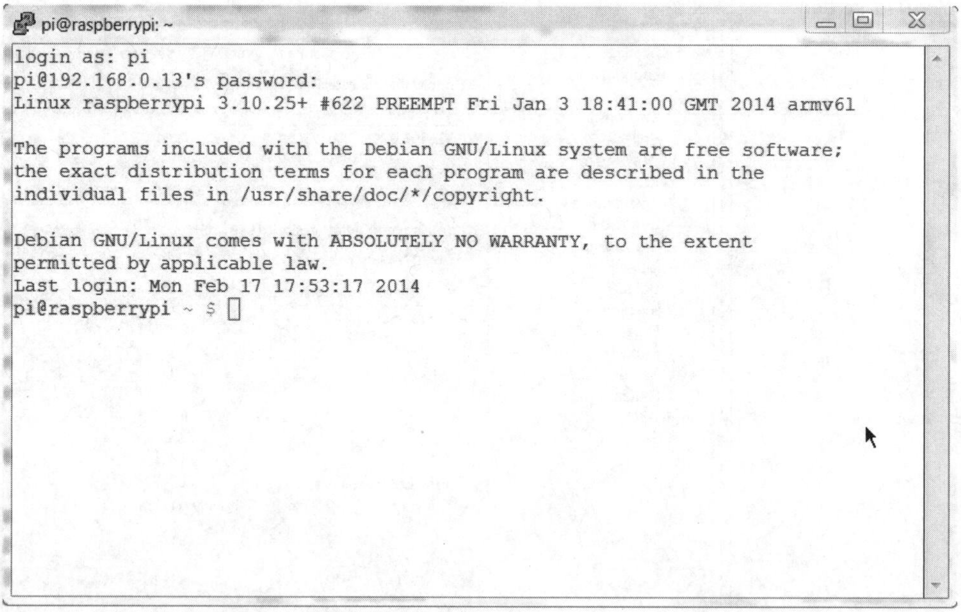

FIGURE 1-19 Pi command line prompt

You will now be able to interact with the Pi in exactly the same way as if you were sitting in front of a standalone workstation. The one major limitation with the SSH protocol is that it is text only and you cannot open a GUI desktop. This would be fine for most operations but it would prevent you from running any program with graphics, which in my opinion is a big constraint. However, there is a great solution to this situation, which I discuss in the next section.

Headless Operation with Graphics

Linux has a wonderful program suite named xrdp, which stands for X11 Remote Desktop Protocol. I first introduced the X11 server in the LDXE desktop GUI discussion. This is the same server engine used in this software suite. xrdp also contains a virtual networking connection (VNC) server name tightvncserver, which functions in a similar manner to SSH except it handles both text and graphics. Type the following command to install xrdp into the Pi:

```
sudo apt-get install xrdp ↵
```

This program suite takes only a few minutes to install and takes up about 11MB of file space. You start the VNC server by entering the following at the command line:

```
vncserver  ↵
```

Every time you start the VNC server, you will see the following line:

```
New 'X' desktop is raspberrypi:x
```

where the lowercase *x* represents a number. The first time you start, it should be a 1. You need to remember this number as it is an important parameter when you run the Windows client on the remote computer. Also, at the first startup, you will be prompted to enter a password that can be up to eight characters in length. You will need to input this password when you authenticate the remote computer with the Pi. That is all that is required on the Pi or server side; it is now time to focus on the Windows or client side.

You will need to download a free VNC suite from http://tightvnc.com/download.php.

This download includes both server and client VNC packages, but only the client package is needed for this configuration. The website has two Windows installers (.msi files), one for 32-bit machines and another for 64-bit machines. Select the appropriate one for your computer and install it.

Go into the Start menu, choose Program Files, and find the TightVNC folder. Click it, and then double-click the TightVNC Viewer menu item. You should see the screen shown in Figure 1-20.

Enter your Pi's IP address in the Remote Host text box, as you see in the figure. Also append a colon with the number that you saw when you started the Pi's VNC server. In this case, I added ":2". Yours will be different. Then click the Connect button next to the text box. If everything goes smoothly, you should see the Pi's VNC server authentication dialog box appear, as shown in Figure 1-21.

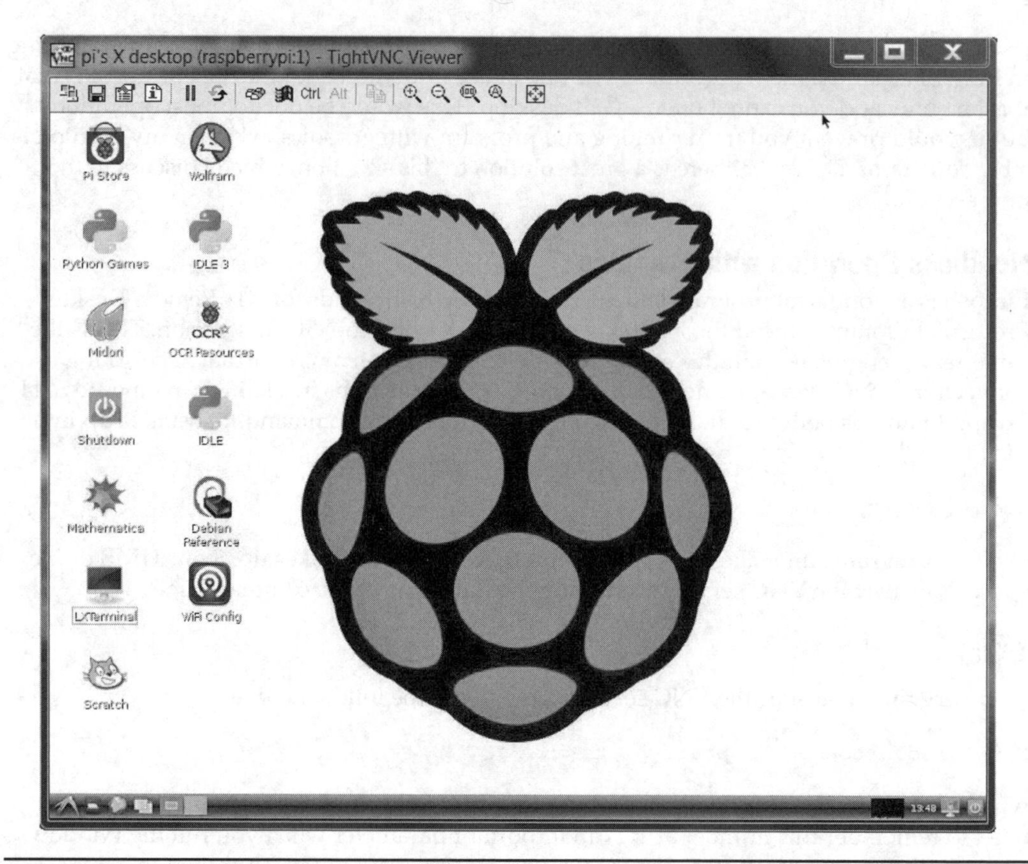

Figure 1-20 Opening screen for the TightVNC Viewer

Enter the password you created when you first configured the Pi's VNC server. Click OK, and you should see the classic Raspbian LXDE GUI desktop, as shown in Figure 1-22.

There is absolutely no difference in using this desktop GUI compared to the standalone desktop GUI. I also launched a terminal window shown in Figure 1-23 to demonstrate that everything responds as it should even though it is a remote desktop connection.

It really is very cool technology when you consider what has taken place. I am remotely controlling the GUI desktop of an extremely inexpensive Linux computer using a completely separate Windows computer, all with free, open-source software. I guess it's just the geek in me surfacing to appreciate this setup. I hope you also appreciate it. Now, on to the LAMP project.

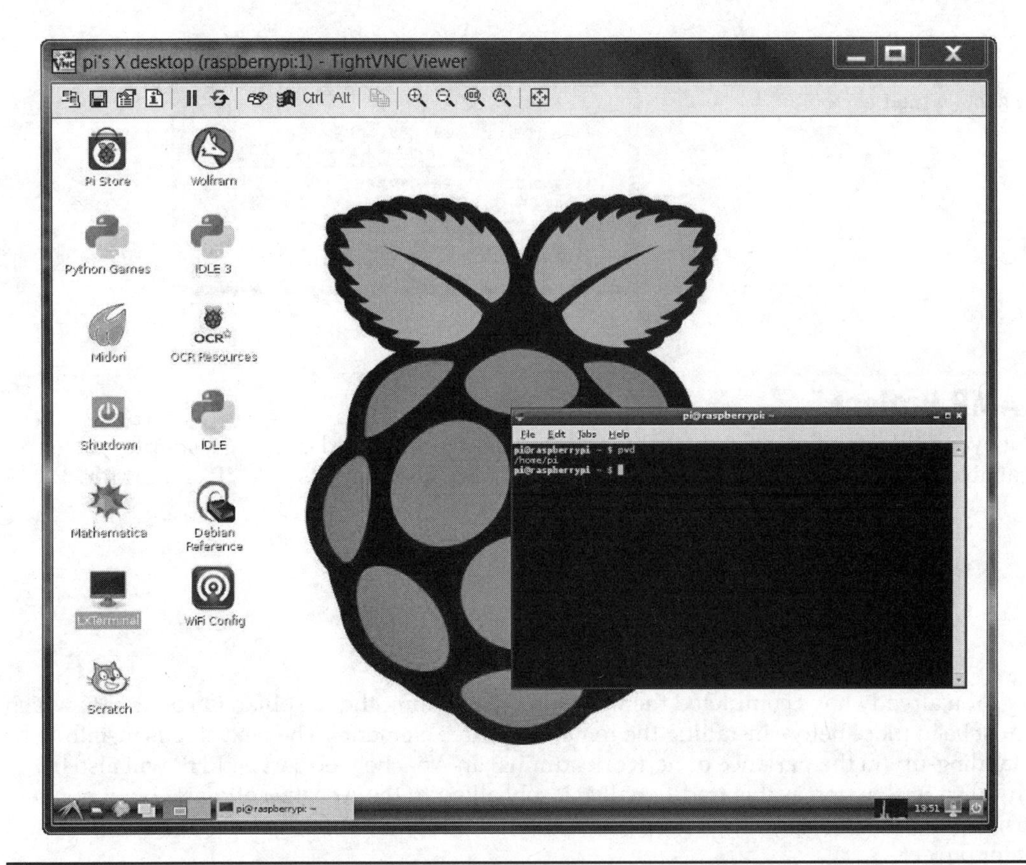

FIGURE 1-21 The Raspberry Pi VNC authentication dialog box

FIGURE 1-22 Raspbian LXDE desktop served by the TightVNC connection

New TightVNC Connection

Connection

Remote Host: 192.168.0.13:2

Connect

Enter a name or an IP address. To specify a port number, append it after two colons (for example,

Options...

Reverse Connections

Listening mode allows people to attach your viewer to their desktops. Viewer will wait for incoming

Listening mode

TightVNC Viewer

TightVNC is cross-platform remote control software.

Its source code is available to everyone, either freely (GNU GPL license) or commercially (with no GPL restrictions).

Version info... Licensing Configure...

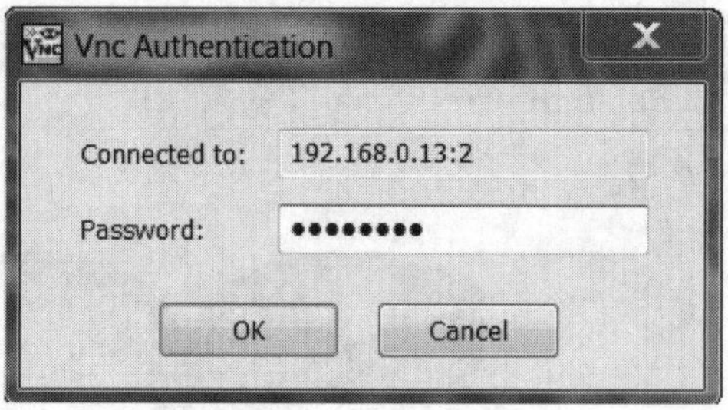

FIGURE 1-23
Launching a terminal window in the Desktop

Vnc Authentication
Connected to: 192.168.0.13:2
Password: ••••••••
OK Cancel

The LAMP Project

No, you will not be constructing a living room lamp but instead creating a simple, yet very capable, web server with a database using the LAMP components. LAMP is short for:

- Linux
- Apache
- MySQL
- Perl/PHP/Python

You already have completed the first step by installing the Raspbian OS on the Pi, which must be in place before installing the remaining three elements. The next step is installing or standing-up (in the parlance of network admins) an Apache web server. PHP will also be installed in this step as it is traditionally closely allied to the Apache software. I will not be using the Perl scripting language, however. I will use Python but separately install it in a later project.

Apache Web Server and the PHP Scripting Language

The Apache web server is by far the most popular open-source web server in existence. It is very mature, having been created and updated for almost 18 years as of the time of this writing. Its formal name is Apache HTTP Server and its logical name as far as the Linux OS is concerned is httpd. The "d" in the logical name stands for daemon, which is a background task in the OS lexicon. The latest version of Apache is 2.4.7, which is why it will be referred to as "apache2" during the install process.

Currently, Apache serves well over 100 million websites worldwide, which accounts for about 55 percent of all active Internet websites. This makes this software the most popular web server ever used. By some accounts, Apache has been the prime reason why the World Wide Web has been so popular.

The PHP web scripting language will also be installed in this step as it is closely integrated with the Apache web server software. Just a bit of history regarding PHP would be helpful in understanding what it is and how it relates to Apache. PHP is a web development scripting language that is hosted on the web server that it supports. PHP originally stood for "Personal Home Page," but that has been superseded by a fancier phrase, "PHP:Hypertext Preprocessor," which is a humorous backronym. In any case, PHP is mainly used to create dynamic web pages that the web server generates in real time based on client requests.

I will use PHP in the "Hello World" project but will use Java and JavaScript to generate dynamic web pages for the Pi projects.

Well, enough with the Apache and PHP promotions. It is time to install them on the Pi.

But first you must ensure the Pi software is the most current with all modifications, new versions, and patches installed. At the command line prompt, type the following:

```
sudo apt-get update ↵
```

Let me take a moment to further explain these commands to readers who may not be experienced with Linux commands:

- **`sudo`** As previously discussed, this command instructs the Raspbian OS to execute the commands that follow it as an administrator. Linux is structured to provide a different set of privileges to different user levels. The update must be done at the highest level or administrator level.
- **`apt-get`** As previously discussed, this portion of the command sequence refers to the advanced packaging tool (`apt`), which is a user interface that works in conjunction with core libraries that are contained in repositories located in servers worldwide. The `apt-get` must be used with at least one other command to execute the desired function. `apt-get` is really a suite of programs each designed to either install or remove software per user requirements.
- **`update`** This is the command modifier to `apt-get` that carries out an update to the local OS list of installed software. It does not install any software per se but updates the local list to match the current version(s) listed in the server repositories. The actual software is installed using the following command:

  ```
  sudo apt-get upgrade ↵
  ```

- **`upgrade`** This command modifier to `apt-get` goes out to the worldwide servers and downloads and installs the new versions along with any and all dependent software. An upgrade may take a long time (greater than an hour) to complete depending upon how out of date the OS is along with any applications that might have been installed since the initial OS installation. `Upgrade` depends upon the `update` operation to be successful, which is why you always update first and then upgrade.

It is finally time to install Apache and PHP assuming all the updates and upgrades have been completed. Type the following in at the command line:

```
sudo apt-get install apache2 php5 libapache2-mod-php5 ↵
```

The software install takes several minutes and uses a little over 21MB. It is also prudent to restart the web server after the install. You do this by entering the following at the command line:

```
sudo service apache2 restart  ↵
```

You will next need to use the Pi browser to confirm that Apache installed correctly. To do this, you will need to determine the IP address that the Pi was issued when it logged on to your network. I used a wireless adapter, so wlan0 is the device name that is associated with my IP address. It will be "eth0" if you used an Ethernet patch cable to connect the Pi to your router. In any case, type the following in at the command line to determine the IP address:

```
sudo ifconfig ↵
```

Figure 1-24 is a screenshot of the results after I entered the command.

As you can see, the local IP address shown in the wlan0 section is 192.168.0.13. Your results should differ somewhat depending on the router you use and the number of devices connected to your network.

I next started an instance of the Pi's Midori web browser and entered the IP address shown previously. It is purely optional to enter http:// before the actual address as most browsers correctly interpret only the numeric entry as a proper URL. I saw Figure 1-25 appear in the browser confirming that the Apache web server was working.

Not too impressive, but it is a start. You should replace the test file that produced the Figure 1-25 display in order to test if PHP is working. I use the nano editor as it is quick and effective and is already part of the Raspbian distribution. The file you edit is named index .html and is located in the /var/www directory. I usually like to change to the target directory before I edit as that way I am sure that any changes will be stored in that directory in lieu of another if I forget to add the path. Assuming you start in your home directory, which is /home/pi, all you need to do is issue the following command:

```
cd /var/www ↵
```

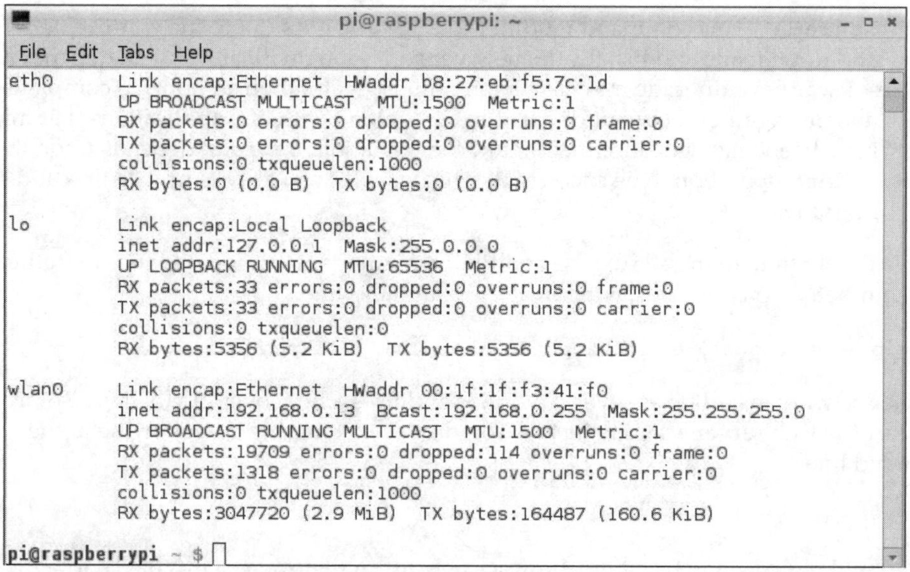

```
pi@raspberrypi: ~
File  Edit  Tabs  Help
eth0      Link encap:Ethernet  HWaddr b8:27:eb:f5:7c:1d
          UP BROADCAST MULTICAST  MTU:1500  Metric:1
          RX packets:0 errors:0 dropped:0 overruns:0 frame:0
          TX packets:0 errors:0 dropped:0 overruns:0 carrier:0
          collisions:0 txqueuelen:1000
          RX bytes:0 (0.0 B)  TX bytes:0 (0.0 B)

lo        Link encap:Local Loopback
          inet addr:127.0.0.1  Mask:255.0.0.0
          UP LOOPBACK RUNNING  MTU:65536  Metric:1
          RX packets:33 errors:0 dropped:0 overruns:0 frame:0
          TX packets:33 errors:0 dropped:0 overruns:0 carrier:0
          collisions:0 txqueuelen:0
          RX bytes:5356 (5.2 KiB)  TX bytes:5356 (5.2 KiB)

wlan0     Link encap:Ethernet  HWaddr 00:1f:1f:f3:41:f0
          inet addr:192.168.0.13  Bcast:192.168.0.255  Mask:255.255.255.0
          UP BROADCAST RUNNING MULTICAST  MTU:1500  Metric:1
          RX packets:19709 errors:0 dropped:114 overruns:0 frame:0
          TX packets:1318 errors:0 dropped:0 overruns:0 carrier:0
          collisions:0 txqueuelen:1000
          RX bytes:3047720 (2.9 MiB)  TX bytes:164487 (160.6 KiB)

pi@raspberrypi ~ $
```

FIGURE 1-24 Screenshot resulting from the ifconfig command

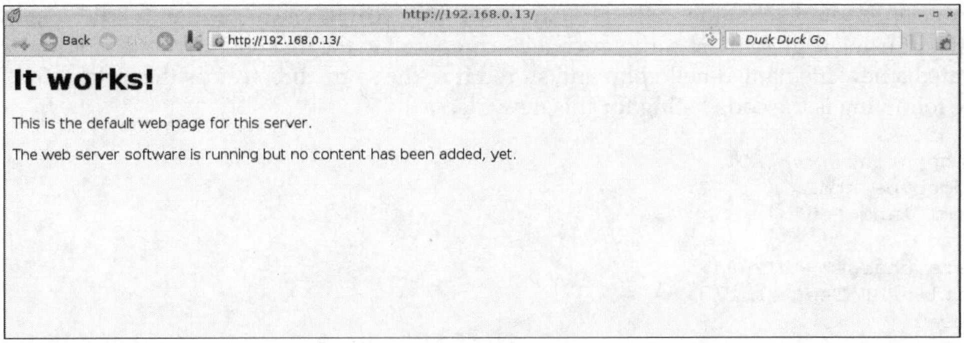

Figure 1-25 First web page served by Apache

and you will be at the proper Apache directory that contains the default web page file, index.html. Start the nano editor with this command:

```
sudo nano index.html ↵
```

I deleted everything in the file and added the following:

```
<html>
<head>
<title>PHP Test</title>
</head>
<?php echo <p>Hello World</p>
</body>
</html>
```

After you have entered the code, save it by simultaneously pressing the CTRL and O keys. Simply press the ENTER key when the nano editor asks if you wish to save it as index.html. Press the CTRL and X keys together to exit the nano editor. Figure 1-26 shows the web page that results from visiting this web page using the Midori browser as you did before.

Figure 1-26 PHP Test web page

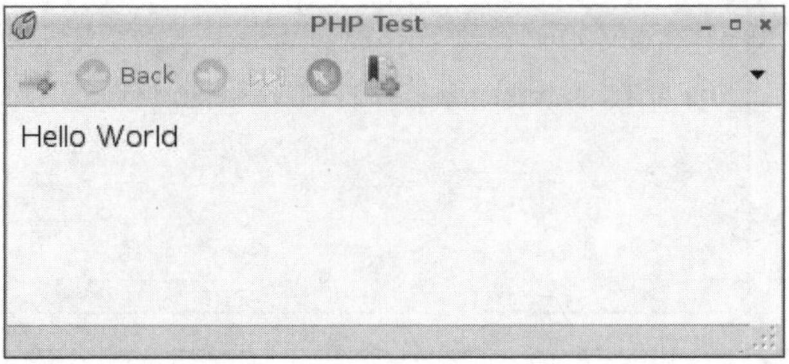

It clearly shows that the PHP software is working with no issues evident. I did one more PHP test using slightly more complex code as compared to the first one. However, for this test I created a new file named hello.php and stored it in the same directory as the index.html file. The following is the code listing for this new file:

```
<?php print  <<< EOT
<!doctype html>
<html lang="en">
<head>
<meta charset="UTF-8">
<title>PHP Test</title>
</head>
<body>
<h1>PHP TEST WAS SUCCESSFUL</h1>
<p>This test confirms PHP is working on the web server</p>
<p>The Apache web server and PHP are good to go</p>
</body>
</html>
EOT;
?>
```

When you visit this web page, you must specifically request it or else the default index .html file will be displayed. Enter the following in the browser URL line: http://192.168.0.13/hello.php.

Obviously, substitute your own IP address for the one shown in the preceding line. Figure 1-27 shows the results for this operation.

The remaining step in creating a complete LAMP project is to install and test the MySQL database, which I do in the next section.

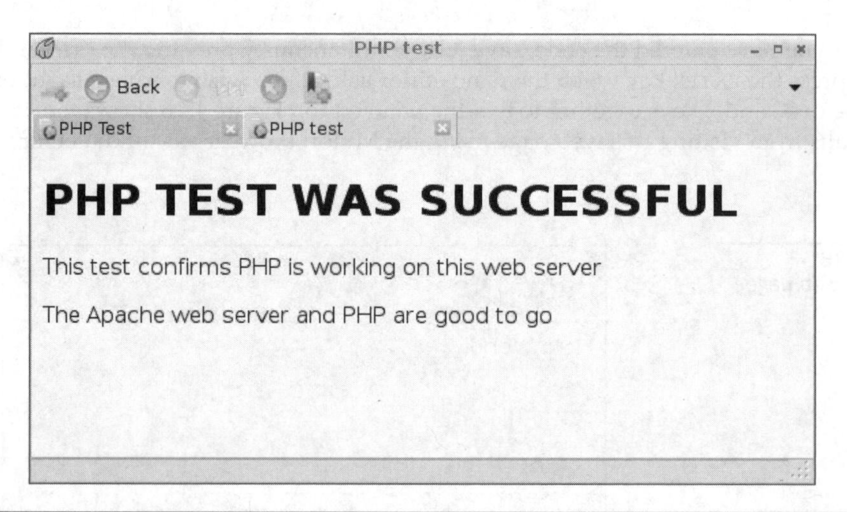

Figure 1-27 hello.php test results

MySQL Database Installation

The fourth and final component for your installation is the MySQL database software. MySQL is an open-source, full-feature relational database that is an essential part of any meaningful IoT project. It really makes no sense to acquire data unless you store it for some purposeful action. I realize that real-time data can be pushed onto the Web but somehow it should be stored for later retrieval and/or analysis. Storing it at the generation site makes sense to me if you do not have a 100 percent reliable and continuous web connection. This is an interesting and continually debated topic—whether or not sensor-driven websites should have any organic or built-in data storage. My take on it is that MySQL doesn't cost anything other than your time to install and maintain it and it does not really occupy a lot of memory, so why not increase your options and overall website reliability by using it.

MySQL has been around for a fair number of years, having been created in 1995, about the same time Apache and PHP came into existence. The installation is very straightforward; you begin by typing the following command:

```
sudo apt-get install mysql-server mysql-client php5-mysql ↵
```

It will take about 8 minutes and require about 95MB of file space to completely install both the MySQL server and client programs along with some PHP support programs. Near the end of the installation you will be prompted to enter a root password for the MySQL server. I highly recommend you do so as it adds another level of security to the LAMP project.

Testing the MySQL installation is somewhat involved because you need to create and partially populate a database to evaluate if the installation was successful. Before I show you these steps, I wish to point out that covering all the essential elements and background for a relational database would take a separate book all to itself. I will show you the necessary commands for creating and populating a MySQL database without going in-depth explaining the theory. I urge all interested readers to either take a formal course in database technology or read one of many fine books that explain relational databases to gain a comprehensive education.

I will begin by creating a simple MySQL database named test. You must first start the MySQL program at the root level by entering the following at a command line prompt:

```
mysql  -u root -p ↵
```

You will immediately be prompted to enter the password that you entered during the install process. If all goes well, you should see the cursor waiting at the "mysql > " prompt. You are now in a MySQL shell program that receives, interprets, and finally executes your commands. For your reference, I have put a summary of some of the more important commands on the website for this book.

I do want to provide some basic background on how a database is structured before demonstrating how to store data into it. A relational database shares some resemblance to a spreadsheet, which readers are most likely familiar with. It is arranged in columns called fields, and rows named records. The rows are indivisible, unlike the spreadsheet, meaning a row or record is treated as a single unit. The fields in each record are individually addressable but can exist outside of the record. Also, records are unique; there cannot be any duplicate records in a formally constructed or normalized database to use the appropriate database terminology. I have created a simple example in the text below to help clarify these concepts.

Suppose you have created a distributed, temperature-monitoring system from which you wish to log a series of temperature sensor readings along with the date, time, and sensor location. A sample log entry would consist of the following:

Date	Time	Location	Temperature

This would constitute a sample record that would be part of table structure that, in part, makes up a database. All databases have one or more tables and each table has multiple records. Tables must be created and named prior to use. Also, the record elements such as Date and Time are table fields and, as such, must have names to allow data to be stored in their respective fields. For this example, I decided to use the names as specified in this book's Table 1-6.

NOTE *I will be very careful in trying to distinguish between my book Tables and database tables using capital "Tables" for the book and lowercase "tables" when referring to the database variety.*

Table 1-6 is all that's needed to start creating the database, table, and fields. I have created the database schema using database parlance. Admittedly, it is very simple but it's all that's needed for this example.

The following command should be entered at the mysql prompt to create the database:

```
CREATE DATABASE test; ↵
```

The semicolon is very important for MySQL commands as it indicates the end of the command sequence. Neglecting to add it will cause an error or prevent the desired command from being executed. I also want to mention that commands can extend over several lines. You will notice that you are in line continuation mode when the `mysql>` prompt changes to an indented → symbol. Just remember to always end the command sequence with a semicolon or `\g` (backslash g).

You should also have noted that the `CREATE DATABASE` command is capitalized. This format signifies that it is an SQL command. Traditionally, all SQL commands, whether issued in the command shell or done programmatically, are capitalized. The MySQL program doesn't distinguish between capitalized and lowercase commands but other SQL programs are more strict. I will follow the standard format and capitalize any SQL command.

The next command instructs MySQL to use the database that was just created:

```
USE test; ↵
```

TABLE 1-6 Example Database Names and Types

Name	Description	Data Type
test	Database name	N/A
tempData	table name	N/A
tdate	Date field	DATE
ttime	Time field	TIME
tloc	Location field	TEXT
temperature	Temperature field	NUMERIC

There could be several databases that were already created in the MySQL "library." It is important that the desired database be used, which is the reason for this command.

Next, the table to store the temperature data needs to be created. The table and all the associated fields for that table are created using this single command:

```
CREATE TABLE tempData (ttdate DATE, ttime TIME,

tloc TEXT, temperature NUMERIC); ↵
```

Use the following command if you wish to check if the table and fields were properly created:

```
SHOW TABLES; ↵
```

Figure 1-28 shows the results of this command.

This figure shows not only the field names and data types associated with the fields but also whether or not Null values will be accepted for a field entry when a record is inputted. There must be a `YES` in the field's `NULL` column to allow for a missing data value or the record will not be entered into the table. Whether or not to accept partial data records is an important decision, which should be made on a case-by-case basis depending on the nature of the data logging that is desired. The figure also indicates if a default value should be entered in case there is no actual value present in the input record. Usually this is not needed as such a situation is more easily handled programmatically.

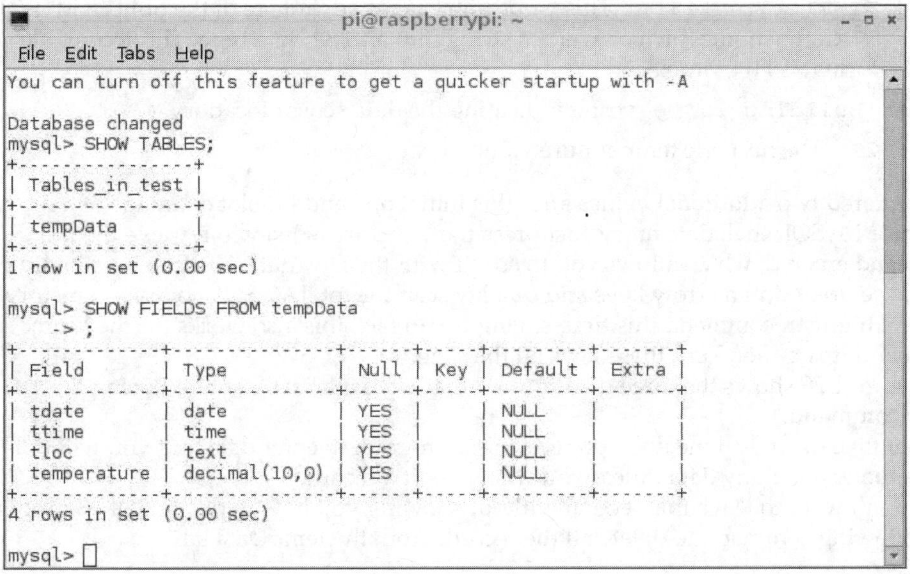

```
pi@raspberrypi: ~
File Edit Tabs Help
You can turn off this feature to get a quicker startup with -A

Database changed
mysql> SHOW TABLES;
+-----------------+
| Tables_in_test |
+-----------------+
| tempData        |
+-----------------+
1 row in set (0.00 sec)

mysql> SHOW FIELDS FROM tempData
    -> ;
+-------------+---------------+------+-----+---------+-------+
| Field       | Type          | Null | Key | Default | Extra |
+-------------+---------------+------+-----+---------+-------+
| tdate       | date          | YES  |     | NULL    |       |
| ttime       | time          | YES  |     | NULL    |       |
| tloc        | text          | YES  |     | NULL    |       |
| temperature | decimal(10,0) | YES  |     | NULL    |       |
+-------------+---------------+------+-----+---------+-------+
4 rows in set (0.00 sec)

mysql> 
```

FIGURE 1-28 SHOW TABLES command screenshot

I next used INSERT statements to manually enter three records into the tempData table. The following is an example of one of these INSERT statements:

```
INSERT INTO tempData (tdate, ttime, tloc, temperature)

VALUES (DATE('2014-02-20'), TIME('13:42:50'), 'Building 10', 21);
```

Observing the length of this statement makes one appreciate the usefulness of the line continuation prompts that automatically appear in the MySQL shell. The statement really isn't all that complex when you break it into the major parts, as I have done here:

- **INSERT INTO** The SQL command to insert what follows into a record for the specified table.
- **tempData** The specified table. Remember that I earlier issued the command `USE test;` to specify the database that contains the `tempData` table.
- **(tdate, ttime, tloc, temperature)** The fields that you wish to provide data for or "populate." You may skip one or more fields if Null values are permitted, as discussed previously.
- **VALUES** The SQL expression that indicates that the data follows. It is important to provide a matching number of data values and types to those specified after the table name. Don't enter TEXT if you specified a NUMERIC type.
- **DATE('2014-02-20')** This value entry takes advantage of the built-in MySQL `DATE` function, which converts a string into a `DATE` data type. There are two acceptable string formats, YYYY-MM-DD and YY-MM-DD.
- **TIME('13:40:45')** This value entry takes advantage of the built-in MySQL `TIME` function, which converts a string into a `TIME` data type. The acceptable string format is HH:MM:SS.
- **'Building 10'** A string indicating the data sensor location.
- **21** The numeric temperature value in °C.

I entered two additional values after this initial one and I took advantage of a real time saver for MySQL shell data entry. Just press the ↑ (up arrow) key to retrieve the last command entered, which allows you to edit it with the new data. You can repeatedly press either the up or down arrow keys and quickly scan the total MySQL command history buffer. I highly recommend this time-saving tip. In fact, this also works for the Linux command line buffer. I use these keys all the time.

Figure 1-29 shows the three data records that were manually entered using the `INSERT INTO` command.

But take heart—all the book projects use a program to enter data and you will not have to manually enter any data unless you wish to edit a record.

I don't want to leave this section without showing you how to delete data. Simply use the following command to delete all the records from the tempData table:

```
DELETE FROM tempData;
```

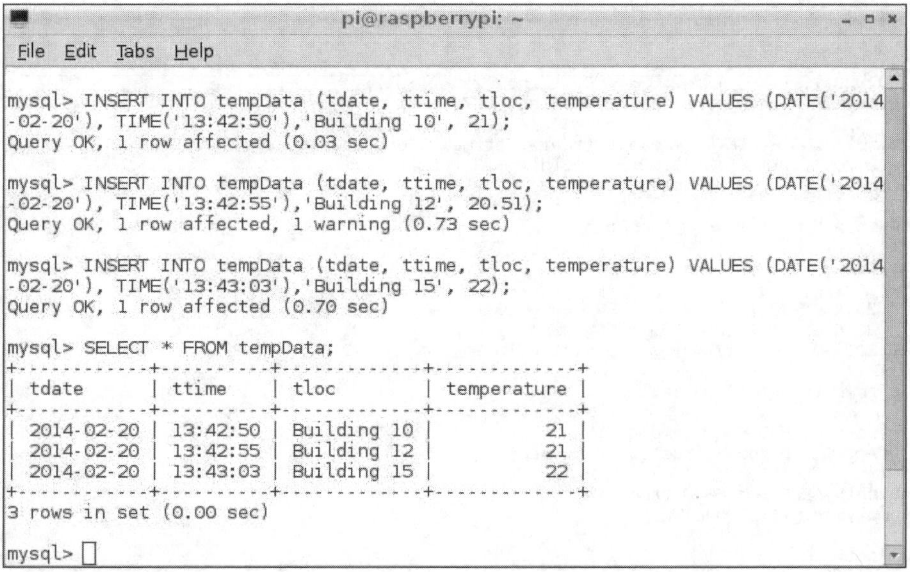

```
pi@raspberrypi: ~
File Edit Tabs Help

mysql> INSERT INTO tempData (tdate, ttime, tloc, temperature) VALUES (DATE('2014
-02-20'), TIME('13:42:50'),'Building 10', 21);
Query OK, 1 row affected (0.03 sec)

mysql> INSERT INTO tempData (tdate, ttime, tloc, temperature) VALUES (DATE('2014
-02-20'), TIME('13:42:55'),'Building 12', 20.51);
Query OK, 1 row affected, 1 warning (0.73 sec)

mysql> INSERT INTO tempData (tdate, ttime, tloc, temperature) VALUES (DATE('2014
-02-20'), TIME('13:43:03'),'Building 15', 22);
Query OK, 1 row affected (0.70 sec)

mysql> SELECT * FROM tempData;
+------------+----------+-------------+-------------+
| tdate      | ttime    | tloc        | temperature |
+------------+----------+-------------+-------------+
| 2014-02-20 | 13:42:50 | Building 10 |          21 |
| 2014-02-20 | 13:42:55 | Building 12 |          21 |
| 2014-02-20 | 13:43:03 | Building 15 |          22 |
+------------+----------+-------------+-------------+
3 rows in set (0.00 sec)

mysql> 
```

FIGURE 1-29 tempData table with three records

Be cautious as this is a powerful command that will irretrievably delete all the data records. There is no recourse unless you have separately backed up the data. Also, realize that the tempData table schema is not affected by this command. All the field names and associated specifications remain intact. Figure 1-30 shows the results of removing all the data from the tempData table.

If you wish to remove only certain select records, you can use what is known as a "where_clause". For the preceding example, suppose I wanted to delete only the records from Building 12. I would rewrite the `DELETE` command as follows:

```
DELETE FROM tempData WHERE tloc='Building 12';
```

I applied this command to the tempData table after I restored the three records using the buffer key hint I gave previously. Figure 1-31 shows the result where the record containing the `tloc` field value of "Building 12" was deleted. Actually, all records that contained that value would have been deleted if there had been multiple records with that field value.

The where_clause is a very valuable tool that can be applied to most of the MySQL commands. If you study Figure 1-31 carefully, you will see where I used the clause to simply display the desired record with a `SELECT` command. Applying selective operations to table records is a key database manipulative element that you should find very handy, especially with large datasets.

Adding a New User to a MySQL Database

Up until now, I have been logging into the MySQL as "root" using the following command:

```
mysql  -u root -p ↵
```

```
pi@raspberrypi: ~
File Edit Tabs Help
-02-20'), TIME('13:42:55'),'Building 12', 20.51);
Query OK, 1 row affected, 1 warning (0.73 sec)

mysql> INSERT INTO tempData (tdate, ttime, tloc, temperature) VALUES (DATE('2014
-02-20'), TIME('13:43:03'),'Building 15', 22);
Query OK, 1 row affected (0.70 sec)

mysql> SELECT * FROM tempData;
+------------+----------+-------------+-------------+
| tdate      | ttime    | tloc        | temperature |
+------------+----------+-------------+-------------+
| 2014-02-20 | 13:42:50 | Building 10 |          21 |
| 2014-02-20 | 13:42:55 | Building 12 |          21 |
| 2014-02-20 | 13:43:03 | Building 15 |          22 |
+------------+----------+-------------+-------------+
3 rows in set (0.00 sec)

mysql> DELETE FROM tempData;
Query OK, 3 rows affected (0.02 sec)

mysql> SELECT * FROM tempData;
Empty set (0.01 sec)

mysql> 
```

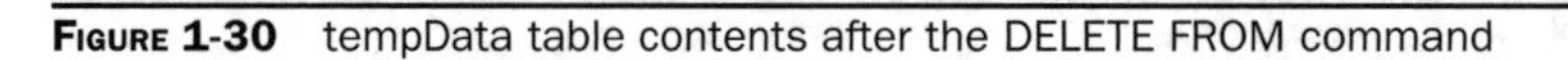

Figure 1-30 tempData table contents after the DELETE FROM command

```
pi@raspberrypi: ~
File Edit Tabs Help
Database changed
mysql> SELECT * FROM tempData;
+------------+----------+-------------+-------------+
| tdate      | ttime    | tloc        | temperature |
+------------+----------+-------------+-------------+
| 2014-02-20 | 13:42:50 | Building 10 |          21 |
| 2014-02-20 | 13:42:55 | Building 12 |          21 |
| 2014-02-20 | 13:43:03 | Building 15 |          22 |
+------------+----------+-------------+-------------+
3 rows in set (0.01 sec)

mysql> DELETE FROM  tempData WHERE tloc='Building 12';
Query OK, 1 row affected (0.03 sec)

mysql> SELECT * FROM tempData;
+------------+----------+-------------+-------------+
| tdate      | ttime    | tloc        | temperature |
+------------+----------+-------------+-------------+
| 2014-02-20 | 13:42:50 | Building 10 |          21 |
| 2014-02-20 | 13:43:03 | Building 15 |          22 |
+------------+----------+-------------+-------------+
2 rows in set (0.01 sec)

mysql> 
```

Figure 1-31 tempData table after the DELETE FROM with a where_clause command

This is not a very good practice especially if the database will be made available to others. I will now show you how easy it is to create new users and have them attached to specific databases. This is a very nice way to allow different users to access only the databases they need to access. Now you have the ability to maintain many databases and select which users should have access to each one.

You must first log in as root to have the privileges to create new users. Use the login command shown in the preceding code. Suppose there is a new user named tester1 who needs access to the test database. Enter the following command to create the user in MySQL and allow access only to the test database:

```
CREATE USER 'tester1' IDENTIFIED BY 'password'; ↵
```

Of course, I have purposefully used the worst password known—"password"—but this is only to keep it simple. You must next grant this new user privileges in order to perform operations on the target database. You do this with the following command:

```
GRANT ALL PRIVILEGES ON test.* TO 'tester1'; ↵
```

Notice the command element `test.*`, which tells MySQL that the user tester1 only has access to all components of the test database and no others. However, the user can perform all MySQL commands without restriction on that database.

Next you should issue the following command:

```
FLUSH PRIVILEGES; ↵
```

This forces the MySQL grant table, which holds all user privileges, to reload, thus saving the new privileges that you just established for the user tester1. Next, quit MySQL so you may log in as tester1.

```
quit; ↵
```

Now log in to MySQL as tester1 with this command:

```
mysql -u tester1 -p ↵
```

You will now be asked for tester1's password, which I just gave you. You will be at the `mysql >` prompt from which you can enter all the commands, as you did for user "root."

Being a naturally curious individual, I decided to confirm whether or not the tester1 user was only confined to the test database and had no access to any other one. I quit the MySQL application and logged in as root. This time I created a null database named test1. A null database, as the name implies, has no content. I then quit the application, logged in again as the tester1 user, and tried the `USE test1;` command to try to switch from the test database to the newly instantiated test1 database. Figure 1-32 shows the resultant error message confirming that MySQL will not allow a non-root user into any database other than one it is registered to.

This last error check ends this section and chapter. In the next chapter, I demonstrate how to remotely connect with a database as well as many other interesting and related topics.

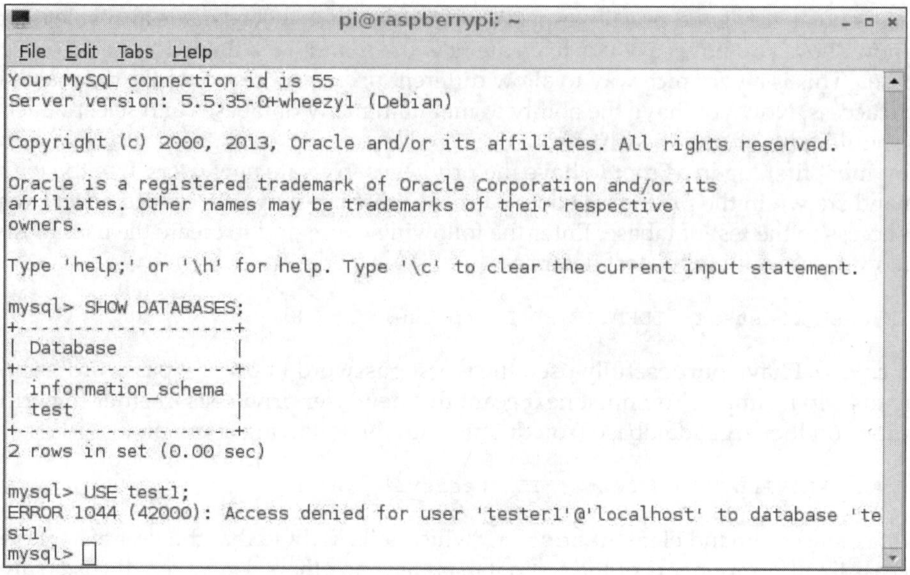

```
pi@raspberrypi: ~
File  Edit  Tabs  Help
Your MySQL connection id is 55
Server version: 5.5.35-0+wheezy1 (Debian)

Copyright (c) 2000, 2013, Oracle and/or its affiliates. All rights reserved.

Oracle is a registered trademark of Oracle Corporation and/or its
affiliates. Other names may be trademarks of their respective
owners.

Type 'help;' or '\h' for help. Type '\c' to clear the current input statement.

mysql> SHOW DATABASES;
+--------------------+
| Database           |
+--------------------+
| information_schema |
| test               |
+--------------------+
2 rows in set (0.00 sec)

mysql> USE test1;
ERROR 1044 (42000): Access denied for user 'tester1'@'localhost' to database 'te
st1'
mysql> 
```

FIGURE 1-32 tester 1 error message

Summary

This chapter began with a brief overview of what makes up the Internet of Things, or IoT. I listed the platforms and software that I use in the book projects to demonstrate how to build simple, but effective, IoT projects.

The next section introduced the Raspberry Pi (hereafter referred to as the Pi) as the first platform to host an IoT project. I very briefly reviewed what makes up the Pi and went into some depth about the general purpose Input/Output (GPIO) pins as that will be the way real-world sensors connect with the Pi.

The next part of the chapter dealt with how to set up a standalone Pi workstation. The key to Pi operations is a properly configured SD card that stores the operating system (OS) as well as all applications and data. I first showed you how to use NOOBS (New Out Of Box Software) to set up the SD card. NOOBS is a relatively new means of providing several Linux distributions to the Pi community. I followed that with a discussion on how to implement the older and more traditional method of using a raw image to create the SD card.

The next section dealt with headless operation where the Pi is controlled over a network. There are two variants on headless operations, one that provides only text or command line operations and another that supports a full graphical user interface (GUI). I first showed you how to use the PuTTY application on a client Windows machine. You learned that SSH is the underlying communication protocol that PuTTY uses along with the SSH server software installed on the Pi. This configuration is the text-only means to implement Pi remote control. SSH is a non-graphical application that allows you to control the Pi over a network using only a command line interface. I next showed you how to install the virtual network communication (VNC) software on both the Pi and the client Windows machine. The actual software suite is known as TightVNC, and it provides a complete, remote Pi graphical desktop.

The chapter discussion next moved to the LAMP project. LAMP is short for Linux, Apache, MySQL, and Perl/PHP/Python. I used the LAMP project as an introduction to how an IoT platform can both host a website and provide data storage using a highly capable relational database named MySQL.

The Linux portion of LAMP was already covered in the beginning of the chapter. I showed how to install the Apache web server along with the PHP programming language. I created two PHP scripts and executed them to demonstrate that both Apache and PHP worked as they should.

The MySQL relational database was installed next. I provided some basic background on what makes a database and how to properly create one and populate it with some sample data. I also covered various commands that manipulate database tables, along with the records that constitute the tables.

I ended the chapter with a discussion on how to add a new user to a MySQL database, which will be necessary for the projects that follow.

2 CHAPTER

Home Temperature Monitoring System

The project in this chapter walks you through the steps in building a home temperature monitoring system that is controlled by a Pi, which can be accessed from anywhere there is an active Internet connection. I will also show you two ways to create the worldwide access, each with its own advantages and disadvantages.

There are many components to this project, including building the sensor network, programming the Pi, standing up a MySQL database, and establishing the Internet connectivity. I will separately discuss each component and provide clear steps on how to proceed; I will also cover some of the theory behind the technology used in that step. Let's begin with the hardware.

Temperature Sensor Network

Figure 2-1 shows a block diagram of the system with three sensors that are connected using wires to an interface block that in turn connects to the Pi.

Wires were used in the first system design to simplify both the hardware and software designs. This was a deliberate decision on my part in order to focus on creating a successful sensor system without being concerned with the potential difficulties that often arise when using wireless sensors. Don't be too concerned that I am not using wireless sensors as I do show you how to use this type of sensor connector in later projects, and you can use that information to modify this system if you so desire.

TMP36 Temperature Sensor

The basic temperature sensor I will use in this project is an Analog Devices model TMP36, shown in Figure 2-2. It is housed a in a standard TO-92 plastic form factor that is also common to most transistors. The TMP36 is far more complex than a simple transistor in that it contains circuits to both sense ambient temperature and convert that temperature to an analog voltage. The functional block diagram is shown in Figure 2-3.

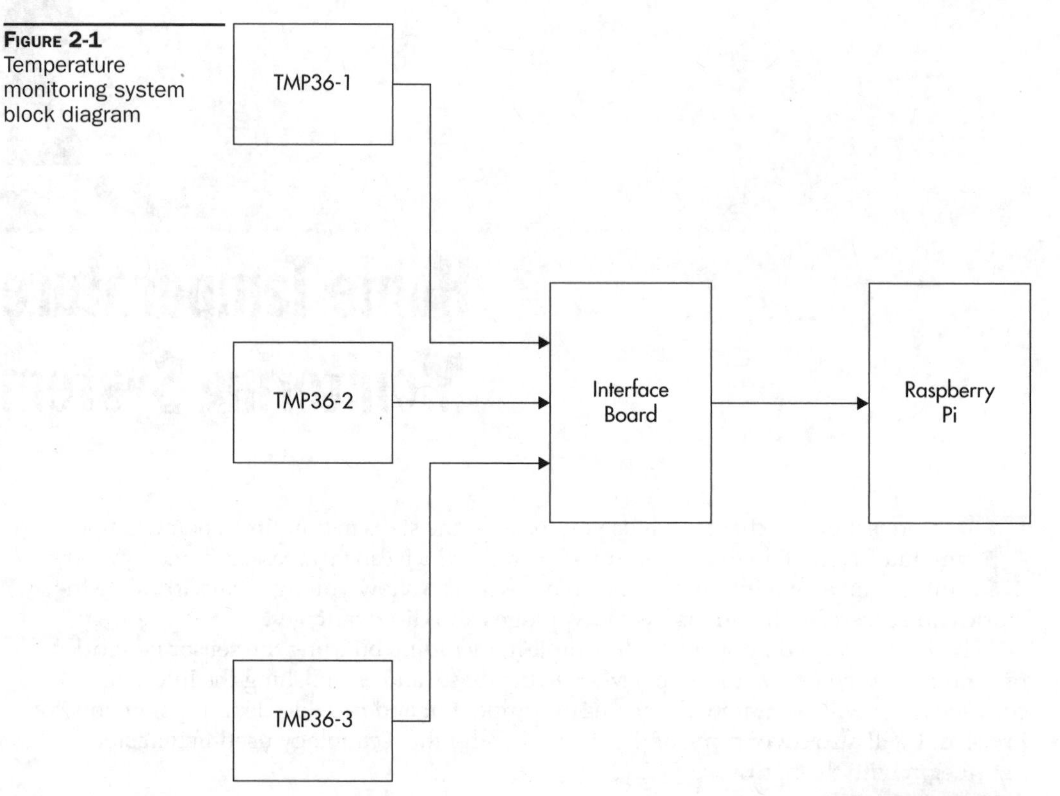

FIGURE 2-1 Temperature monitoring system block diagram

The TMP36 has only three leads, which are shown in a bottom view in Figure 2-4.

Table 2-1 provides details concerning these three leads, including important limitations.

The voltage representing the temperature is dependent upon the TMP36 supply voltage, which must be considered when converting the V_{OUT} voltage to the equivalent real-world temperature. I do account for this in the software that converts the V_{OUT} voltage to an actual temperature. See Figure 2-5 for a graph of the V_{OUT} voltage versus temperature using a 3V supply voltage.

The actual temperature measurement range for the TMP36 is –40 to +125°C with a typical accuracy of +/–2°C and 0.5°C linearity. All in all, not too shabby specifications considering the cost of the TMP36 is typically less than $2 USD. The TMP36 range, accuracy, and linearity are well suited for a home temperature monitoring system.

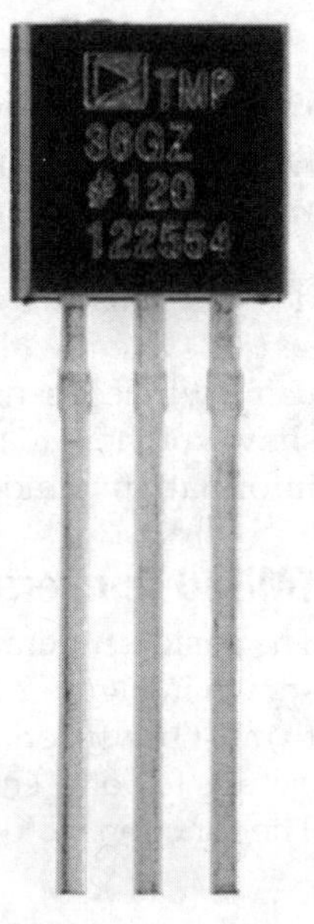

FIGURE 2-2 Analog Devices model TMP36 temperature sensor

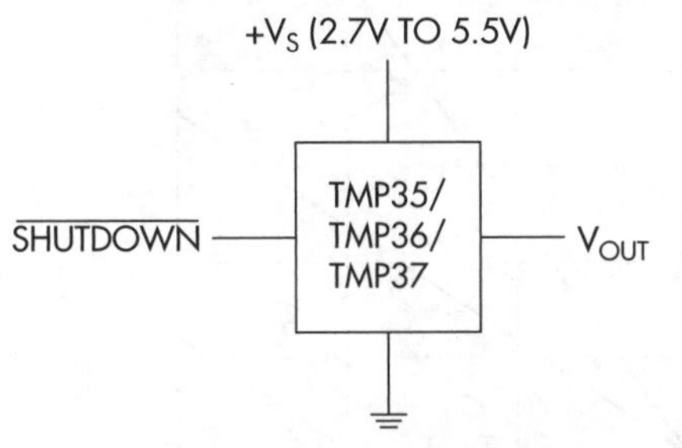

FIGURE 2-3 Model TMP36 functional block diagram

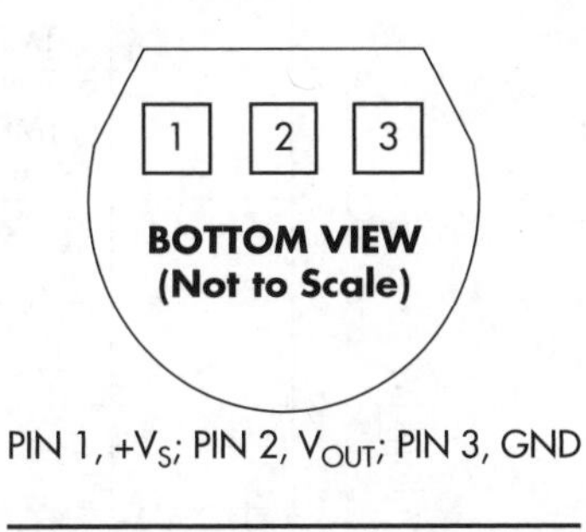

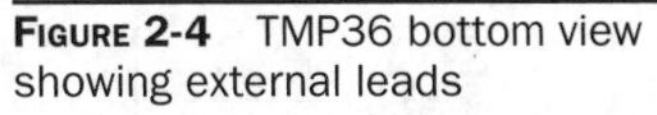

FIGURE 2-4 TMP36 bottom view showing external leads

Analog to Digital Conversion

The Pi does not contain any means by which analog signals can be processed, as most readers will already know. This means that some type of analog voltage to digital converter (ADC) must be used before the Pi can deal with the temperature signals.

I used a Microchip model MCP3008, which is described in the Microchip datasheet as a 10 bit, SAR ADC with SPI data output. This means the MCP3008 uses a Successive Approximation Register (SAR) technique to create a 10-bit digital result that in turn is output in a serial data stream using the serial peripheral interface (SPI) protocol, which is discussed after the sidebar. The very inexpensive MCP3008 ADC chip has impressive specifications despite its very low cost. Figure 2-6 shows the package form and pin-out for this chip.

The MCP3008 chip, as used in this project, is in a dual-in-line package (DIP), which means that either a custom printed circuit board (PCB) or a solderless breadboard must be used for it to interface with the Pi. I discuss how to connect the Pi to MCP3008 after the sidebar. I encourage you to read the sidebar if you are interested in how the MCP3008 accomplishes the analog-to-digital conversion process.

Pin Number	Description	Remarks
1	$+V_S$	Supply voltage. Ranges from 2.7 to 5.5V.
2	V_{OUT}	The analog voltage representing the temperature. The maximum voltage depends upon the supply voltage.
3	GND	Common reference used by both the supply and V_{OUT} pins.

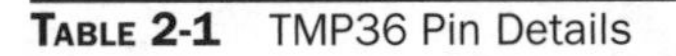
TABLE 2-1 TMP36 Pin Details

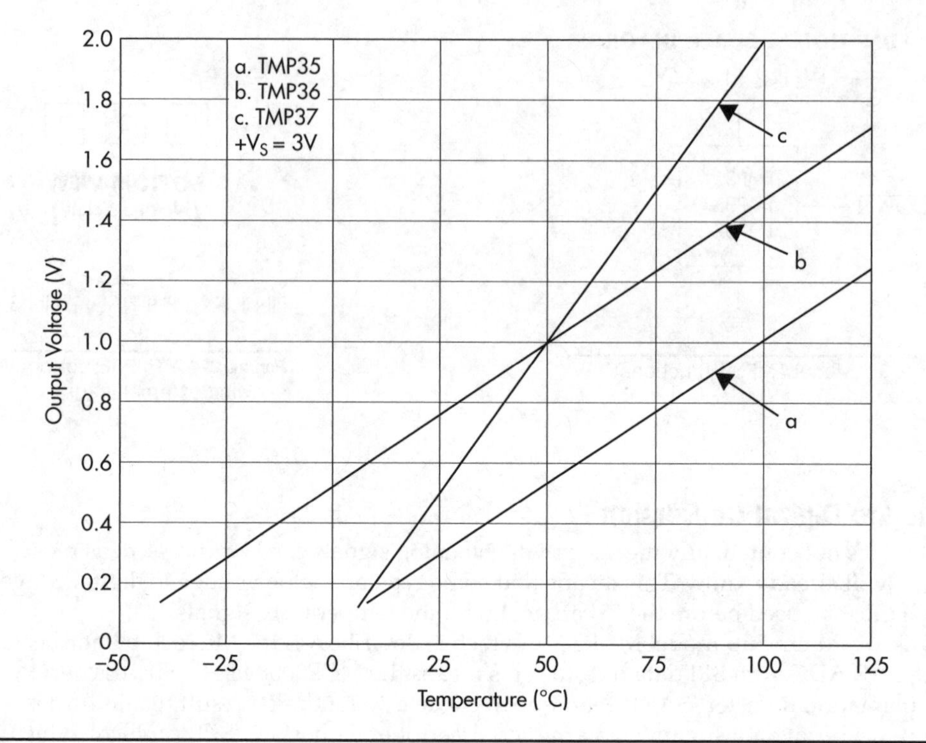

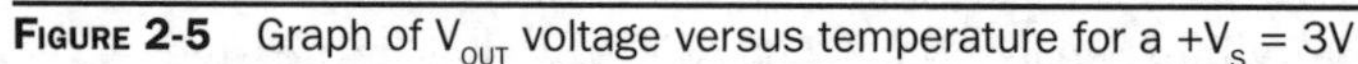

FIGURE 2-5 Graph of V_{OUT} voltage versus temperature for a $+V_S = 3V$

FIGURE 2-6 MCP3008 package form and pin-out

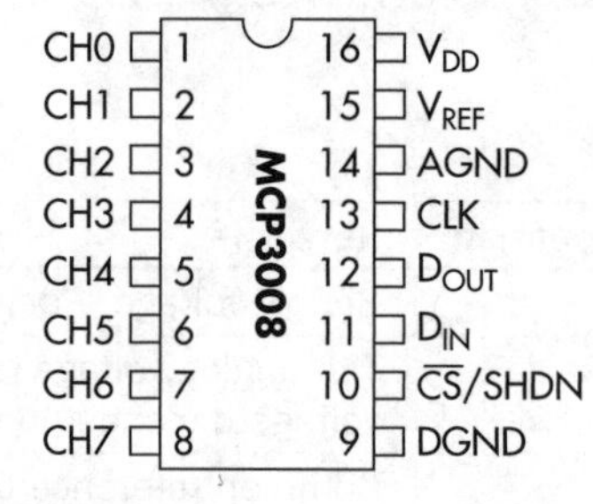

Inner Workings of the Microchip MCP3008 ADC

I will refer to the MCP3008 functional block diagram shown in Figure 2-7 throughout the following discussion.

The analog signal is first selected from one of eight channels that may be connected to the Input Channel Multiplexer. Using one channel at a time is called *operating in a single-ended mode*. The MCP3008 channels can be paired to operate in a differential mode if desired. A single configuration bit named SGL/DIFF selects single-ended or differential operating modes. Single-ended is the mode used in this project.

The selected channel is then routed to a Sample and Hold circuit, which is one input to a Comparator. The other input to the Comparator is from a Digital to Analog Converter (DAC) that receives its input from a 10-bit Successive Approximation Register (SAR).

Basically, the SAR starts at 0 and rapidly increments to a maximum of 1023, which is the largest number that can be represented with 10 bits. Now each increment increases the voltage appearing at the DAC's comparator input. The Comparator will trigger when the DAC voltage precisely equals the sampled voltage, and this will stop the SAR from incrementing. The digital number that exists on the SAR at the moment the Comparator "trips" is the ADC value. This number is then outputted, one bit at a time through the SPI circuit discussed in Serial Peripheral Interface. All this takes place between sample intervals. The actual voltage represented by the ADC value is a function of the reference voltage, V_{REF}, connected to the MCP3008. In this case, V_{REF} is 3.3V; thus, each bit represents 3.3÷1024 or approximately 3.223 millivolts. For example, an ADC value of 500 would represent an actual voltage of 1.612V, which was computed by multiplying .003223 by 500.

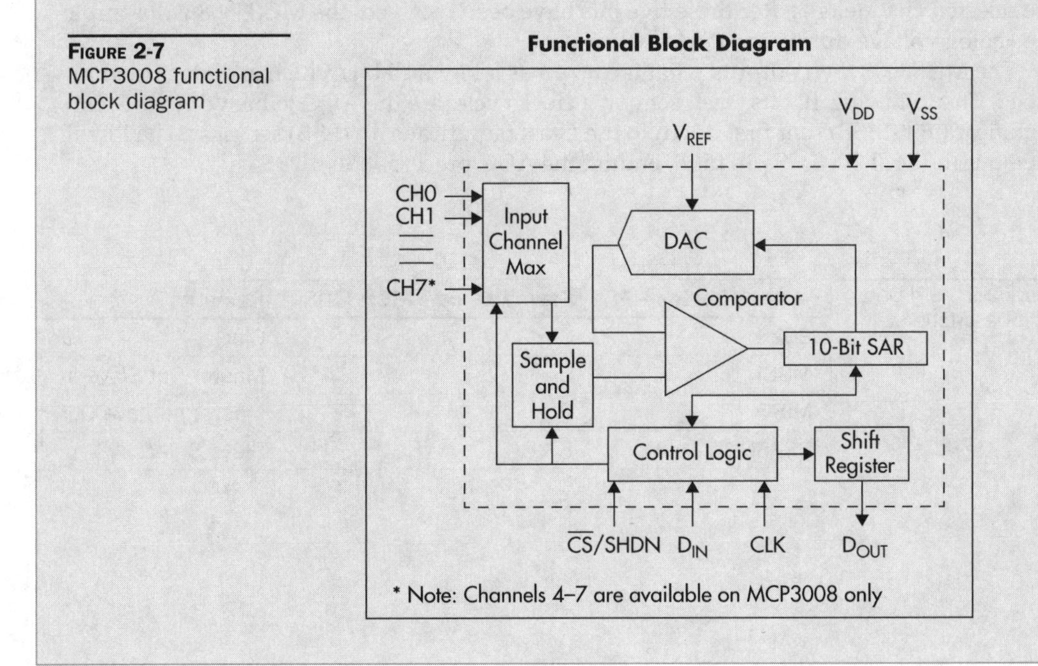

FIGURE 2-7 MCP3008 functional block diagram

Serial Peripheral Interface

The Serial Peripheral Interface (SPI) is one of several data communication channels that the Pi supports. It is a synchronous serial data link that uses one master device and one or more slave devices. There is a minimum of four data lines used with SPI; Table 2-2 shows the names associated with the master (Pi) and the slave (MCP3008) devices.

There are usually two shift registers involved in the data link, as shown in Figure 2-8.

These registers may be hardware or software depending upon the devices involved. The Pi implements its shift register in software while the MCP3008 has a hardware shift register. In either case, the two shift registers form what is known as an inter-chip circular buffer arrangement that is the heart of the SPI.

Figure 2-8 is a simplified block diagram showing the principal components used in an SPI data link.

Data communications is initiated by the master by first selecting the required slave. The Pi selects the MCP3008 by bringing the SS line to a low state or 0 VDC. During each clock cycle, the master sends a bit to the slave, which reads it from the MOSI line. Concurrently, the slave sends a bit to the master, which reads it from the MISO line. This operation is known as *full duplex communication*, i.e. simultaneous reading and writing between master and slave.

The clock frequency used is dependent primarily upon the slave's response speed. The MCP3008 can easily handle bit rates up to 3.6 MHz if powered at 5V. Because we are using 3.3V, the maximum rate is slighty less at approximately 2 MHz. This is still very quick and will process the Pi input without losing any data.

The first clock pulse received by the MCP3008, with its CS held low and D_{in} high, constitutes the start bit. The SGL/DIFF bit follows next and then three bits that represent the selected channel(s). After these five bits have been received, the MCP3008 will sample the analog voltage during the next clock cycle.

The MCP3008 then outputs what is known as a low null bit, which is disregarded by the Pi. The following 10 bits, each sent on a clock cycle, are the ADC value with the Most Significant Bit (MSB) sent first down to the Least Significant Bit (LSB) sent last. The Pi will then put the MCP3008 CS pin high, ending the ADC process.

TABLE 2-2 SPI Data Line Descriptions

Master Device—Pi	Slave Device—MCP3008	Remarks
SCLK	CLK	Clock
MOSI	D_{in}	Master Out Slave In
MISO	D_{out}	Master In Slave Out
CS/SHDN	SS	Slave Select

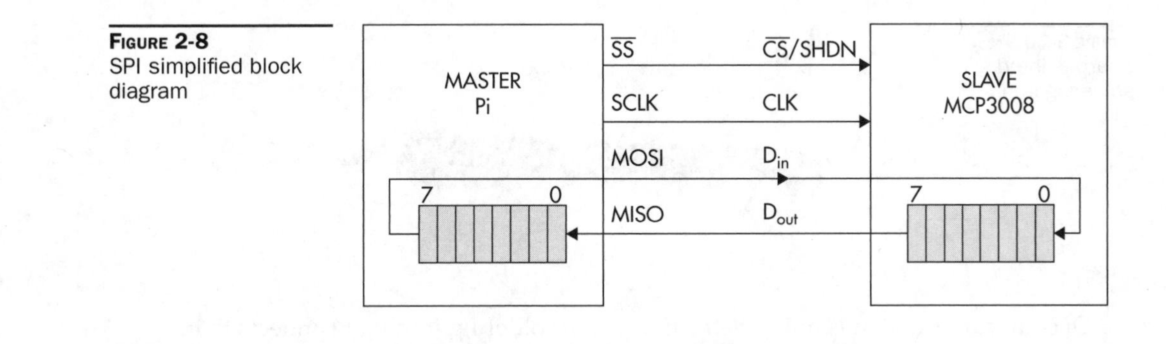

FIGURE 2-8 SPI simplified block diagram

Connecting and Testing the MCP3008 with the Raspberry Pi

The MCP3008 is connected to the Pi using the Pi Cobbler prototype tool along with a solderless breadboard. The Pi Cobbler is available from a variety of sources, but it must be assembled, which will require some soldering. There are instructions available on the Adafruit website that show you, step-by-step, how to assemble the Pi Cobbler. Soldering is a fun activity provided you have the right equipment and skill. I recently acquired a comparatively inexpensive digitally controlled soldering workstation, which is shown in Figure 2-9. It may be set to precise temperatures that enable very nice solder joints to be made with ease and repeatability.

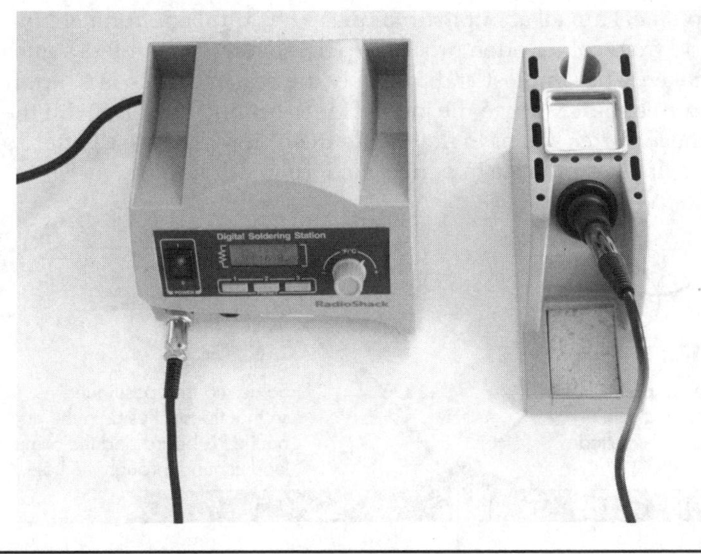

FIGURE 2-9 Digital soldering station

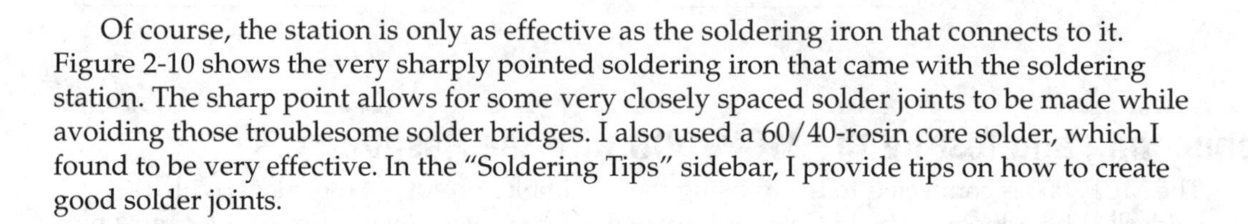

Figure 2-10 Sharp pointed soldering iron

Of course, the station is only as effective as the soldering iron that connects to it. Figure 2-10 shows the very sharply pointed soldering iron that came with the soldering station. The sharp point allows for some very closely spaced solder joints to be made while avoiding those troublesome solder bridges. I also used a 60/40-rosin core solder, which I found to be very effective. In the "Soldering Tips" sidebar, I provide tips on how to create good solder joints.

Soldering Tips

The key to quality soldering work is to have good soldering technique, keep the soldering iron tip clean, and use the highest quality solder available. Figure 2-11 shows the essence of good soldering technique. It is vital that the solder joint be hot enough for solder to flow easily. It takes practice to apply just the right amount of solder; too little may result in a cold solder joint and too much could lead to a short between closely spaced components.

Another issue regarding a good solder joint is the use of lead-free solder. Now, don't get down on me; I am all about maintaining a healthful environment but the elimination of lead from solder often produces poor solder joints unless some extra precautions are taken. The simplest and probably the best approach is to apply a high-quality, acid-free solder flux to the joint prior to heating the joint with the iron. This will allow the lead-free solder to flow more freely and produce a better soldered connection. Again, it takes practice to perfect soldering techniques.

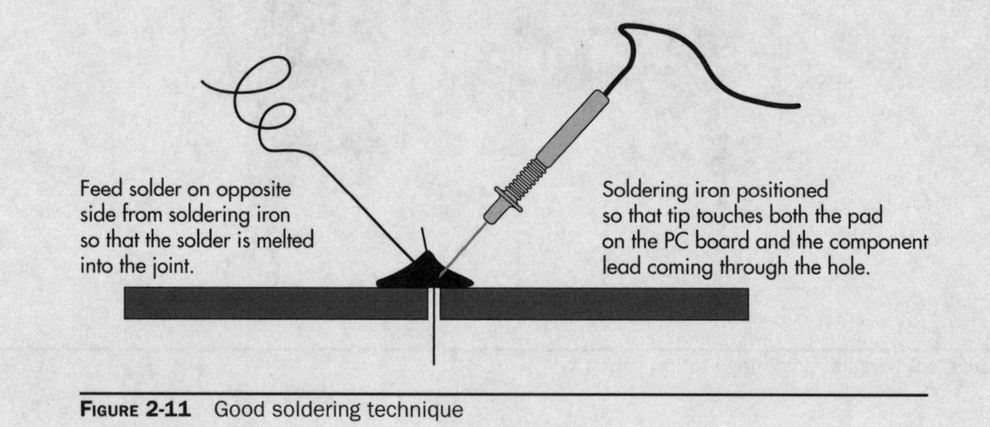

Feed solder on opposite side from soldering iron so that the solder is melted into the joint.
Soldering iron positioned so that tip touches both the pad on the PC board and the component lead coming through the hole.

Figure 2-11 Good soldering technique

One final thought that relates to solder joints as well as other types of electrical connections: There is a long running anecdotal observation that 90 percent of all electrical/electronic malfunctions are related to connection malfunctions. This makes a lot of sense when you think about it. We live in an oxygen rich atmosphere and oxygen is a great reduction agent; it wants to oxidize every element it can possibly chemically combine with. Metal oxides are reasonably good insulators as some of their free electrons have been "taken" up by oxygen molecules. This leads to higher and higher resistance being built up in a connection, which eventually will cause a failure. Of course, current flowing through a resistance produces heat, which in turn can cause a fire if the currents are sufficiently high. So what is the solution? One expensive solution is to gold plate electrical contact surfaces. Gold doesn't oxidize and is not subject to this type of failure. It is, of course, very expensive and not practical for large scale connectors. For the type of projects that I work on, I can only ensure that solder joints are sound from both a mechanical and electrical perspective. I also inspect electrical connections for oxidation and foreign matter and take appropriate action to repair or replace the component.

Initial Test

Initial testing involves both creating a hardware circuit and establishing the proper Python software environment. The circuit and software setups discussed in this chapter are based in large part on the excellent tutorial available from Matt Hawkins's blog, www.raspberrypi-spy.co.uk/tag/tmp36/, in which he discusses both the MCP3008 and the TMP36 sensors as well as the Python software.

Hardware Setup

I will first discuss the hardware circuit as that is relatively straightforward. Figure 2-12 shows the test schematic for the Pi Cobbler, MCP3008, and TMP36. I connected the TMP36 V_{out} lead to the MCP3008 Channel 0 input, which is pin 1. The actual physical setup is shown in Figure 2-13.

On the left side of the breadboard, you can see the TMP36 sensor connected with three jumper wires to the breadboard. Incidentally, I find using commercial jumper wires very useful and more reliable than using homemade jumpers constructed from hookup wire. There is almost nothing more frustrating than finding that a poor wiring connection due to a broken jumper wire was responsible for a non-functioning circuit. Besides, a set of jumper wires is quite inexpensive and lends a professional look to your project.

The hardware setup should proceed very quickly and the next portion of the test concerns the software.

Software Setup

The SPI hardware circuits that are part of the Pi must be enabled before executing any code that relies on those circuits. Initially you should check to determine if the native SPI device is available. Enter the following command at a terminal window command prompt and check to see if there is an "spi_bcm2708" in the list that is displayed.

```
lsmod ↵
```

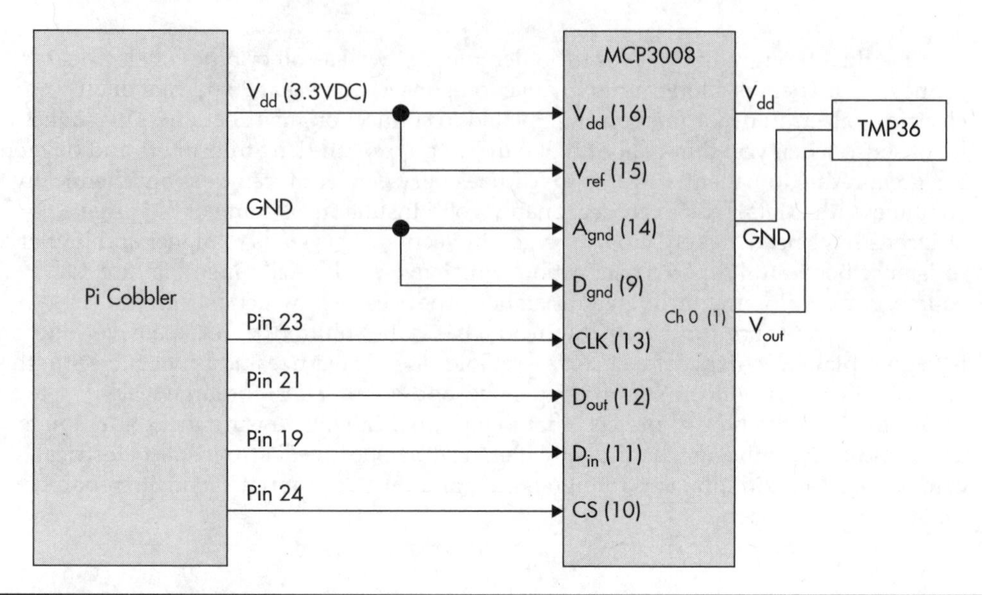

FIGURE 2-12 Test schematic

If there is, skip the next procedure or else edit the raspi-blacklist.conf as follows:

1. Enter **`sudo nano/etc/modprobe.d/raspi-blacklist.conf`**.
2. Add the # symbol in front of the line `spi-bcm2708`.
3. Use CTRL-O to save and CTRL-X to exit the nano editor.
4. Reboot the Pi by entering **`sudo reboot`**.

FIGURE 2-13 Physical test setup

Try the `lsmod` command again and you should see the spi-bcm2708 device listed. You now need to load the Python libraries that will allow programs to be run by the SPI circuits you just enabled using the following steps:

1. Install the Python development libraries by entering the following:

   ```
   sudo apt-get install python-dev ↵
   ```

2. After the install finishes, you need to create a special directory in which to create and run the SPI Python programs. From the Home directory, which should be at /home/pi, enter the following:

   ```
   mkdir py-spidev ↵
   ```

3. Change into the newly created directory:

   ```
   cd py-spidev ↵
   ```

4. Now download a Python script, which will automatically create the necessary SPI development environment:

   ```
   sudo wget https://raw.github.com/doceme/py-spidev/master/setup.py ↵
   ```

5. Download an additional file, which is required before the setup can begin:

   ```
   sudo wget https://raw.github.com/doceme/py-spidev/master/spidev_
   module.c ↵
   ```

6. Run the script and create the SPI development environment by entering the following:

   ```
   sudo python setup.py install ↵
   ```

The following test program displays a continuous stream of temperature values generated by the TMP36 sensor. The program is named SingleSensorTest.py and is available for download on this book's companion website. The code follows the MCP3008 ADC configuration guidelines and SPI protocols as discussed previously.

```
#!/usr/bin/python

import spidev
import time
import os

#open the SPI bus
spi = spidev.SpiDev()
spi.open(0,0)

#define a function to read the MCP3008 ADC value
#channel must be an integer between 0 and 7
def ReadChannel(chan):
    if((chan < 0) or (chan > 7)):
        return -1
    adc =spi.xfer2([1, (8 + chan) << 4, 0])
    data = ((adc[1]&3) << 8) + adc[2]
    return data
```

```
#define a function that converts data to voltage levels
#round to a specified number of decimal places
def ConvertVolts(data, places):
    volts = (data * 3.3) / 1023
    volts = round(volts, places)
    return volts

#define a function to calc temperature from TMP36 data
#round to a specified number of decimal places
def ConvertTemp(data, places):

    #ADC Value       Temp (°C)        Volts
    #        0             -50         0.00
    #       78             -25         0.25
    #      155               0         0.50
    #      233              25         0.75
    #      310              50         1.00
    #      465             100         1.50
    #      775             200         2.50
    #     1023             280         3.30

    #NOTE: The original constant value of 330 was changed to 358.


See text for reason.
    temp = ((data * 358)/1023) - 50
    temp = round(temp, places)
    return temp

#define temp channels
temp_chan0 = 0

#define time between readings
delay = 5

#print column headers
print "temp_level        temp_volts       temp"

#main loop
while True:

    #read the temp channel
    temp_level = ReadChannel(temp_chan0)
    temp_volts = ConvertVolts(temp_level, 2)
    temp = ConvertTemp(temp_level, 2)

    #display results
    print "------------------------------------"
    print temp_level, "         ", temp_volts, "      ", temp

    #delay before taking next measurement
    time.sleep(delay)
```

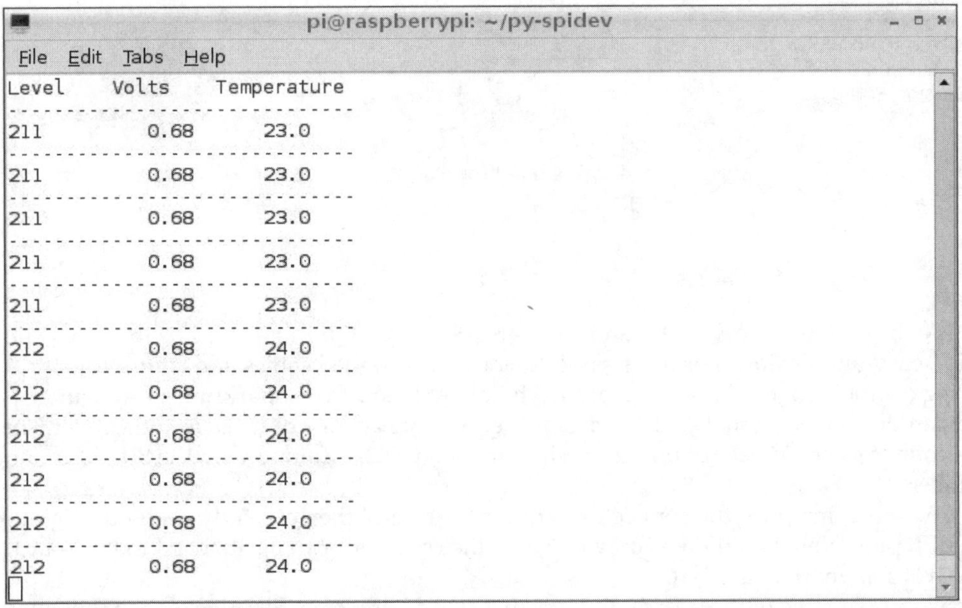

```
pi@raspberrypi: ~/py-spidev
File Edit Tabs Help
Level     Volts     Temperature
--------------------------------
211        0.68       23.0
--------------------------------
211        0.68       23.0
--------------------------------
211        0.68       23.0
--------------------------------
211        0.68       23.0
--------------------------------
211        0.68       23.0
--------------------------------
212        0.68       24.0
--------------------------------
212        0.68       24.0
--------------------------------
212        0.68       24.0
--------------------------------
212        0.68       24.0
--------------------------------
212        0.68       24.0
--------------------------------
212        0.68       24.0
```

FIGURE 2-14 Initial test results

Run the preceding program by entering the following:

```
sudo python SingleSensorTest,py ↵
```

Figure 2-14 is a screenshot of a portion of the program output with the TMP36 sensor measuring ambient room temperature.

I put a comment in the code listing explaining that I needed to adjust a constant in the temperature conversion function from the original 330 value to a 358 value. I did this because the MCP3008 chip was reporting a voltage of 0.68 while the real voltage as measured with a digital voltmeter was approximately 0.74V. This caused the temperature to be underreported by approximately 5°C. I also used a non-contact, precision infrared temperature meter to measure the real temperature. By adjusting the constant, I forced the function to calculate the correct temperature. I do not know why the MCP3008 chip was not accurately converting the TMP36 V_{out} but I did confirm that the error was linear and constant, thus easily corrected in the conversion formula. I also tried another MCP3008 chip and observed the same behavior so I concluded the error must be related to the sampling function. Mr. Hawkins discusses it in his blog, so I'll refer you there for further investigation.

Multiple Sensor System

It is time to build a three-channel system now that the single-channel test has proven that the ADC and sensor and supporting software function as expected. I will still use a wired system, as mentioned earlier in the chapter, because it simplifies the design and allows the

FIGURE 2-15 Sparkfun RJ45 breakout board and connector

focus to be on the sensors, ADC, and software. However, I did that to make the connections quick, easy, and flexible. For these same reasons, I chose RJ45 cables and connectors for the wiring component. I did find that an RJ45 breakout board from Sparkfun, along with a companion connector, makes the cable and sensor connections quite easy and convenient. The connector and breakout board are shown in Figure 2-15, along with the Sparkfun model numbers.

You must first push the connector onto the board and then carefully solder all eight of the PCB pins. Note that there is only one way the connector can be attached to the board, which is shown in Figure 2-16.

Next, take three of the boards and attach a row of single header pins, which will allow the boards to be directly plugged into a solderless breadboard. Figure 2-17 shows one of these boards with the pins attached. The other three boards have a TMP36 sensor directly attached to them, as shown in Figure 2-18. Ensure you solder the sensor to the left side breakout board pins with the sensor's flat side pointing up, as shown in the figure.

FIGURE 2-16 RJ45 connector attached to a breakout board

Figure 2-17 RJ45 board with attached header pins

Figure 2-18 RJ45 board with an attached TMP36 sensor

You will now need the interconnecting RJ45 cables, which may be bought or made. I would suggest that you purchase them if you do not have any experience in making up this cable type. It does require a special tool along with the Cat 5 or Cat 6 cable and ready-to-assemble snap-on connectors. The cable lengths depend upon the spacing between sensor locations and the Pi's location. I used three six-foot cables for my setup as it was a temporary demonstration system and not a permanent one. However, I did find that there is a tremendous difference in cable quality where one cable would reduce the output voltage by 16 mV while another would hardly have any effect. A 16 mV drop would cause the temperature to be measured 2°C less than the real temperature. The drop in voltage is likely due to the very limited output current capacity of the TMP36 sensor along with the variability on how cables are manufactured—some with greater capacitance and inductive loading than others. I did check the TMP36's manufacturer's datasheet where the limited current capacity was acknowledged. There is a circuit in the datasheet shown in Figure 2-19 that will boost the current level to a full-scale maximum of 2 mA, which should be more than sufficient to drive any cable attached to the sensor.

Instead of building the boost circuit, I found it simpler just to sort through all my spare RJ45 patch cables until I found three that did not significantly affect the sensor. However, the current boost circuit is likely essential if you plan on setting up a long cable run of more than 10 feet. The complete system schematic is shown in Figure 2-20. It is the same as the Figure 2-12 schematic with two additional sensors connected to the MCP3008 channels 1 and 2. The physical setup with the Pi is shown in Figure 2-21. Notice how I arranged the RJ45 connectors on the breadboard for easy hookup and cable attachment. This completes the system hardware configuration and it is time to focus on expanding the software to accommodate two additional sensors.

Figure 2-19 TMP36 current boost circuit

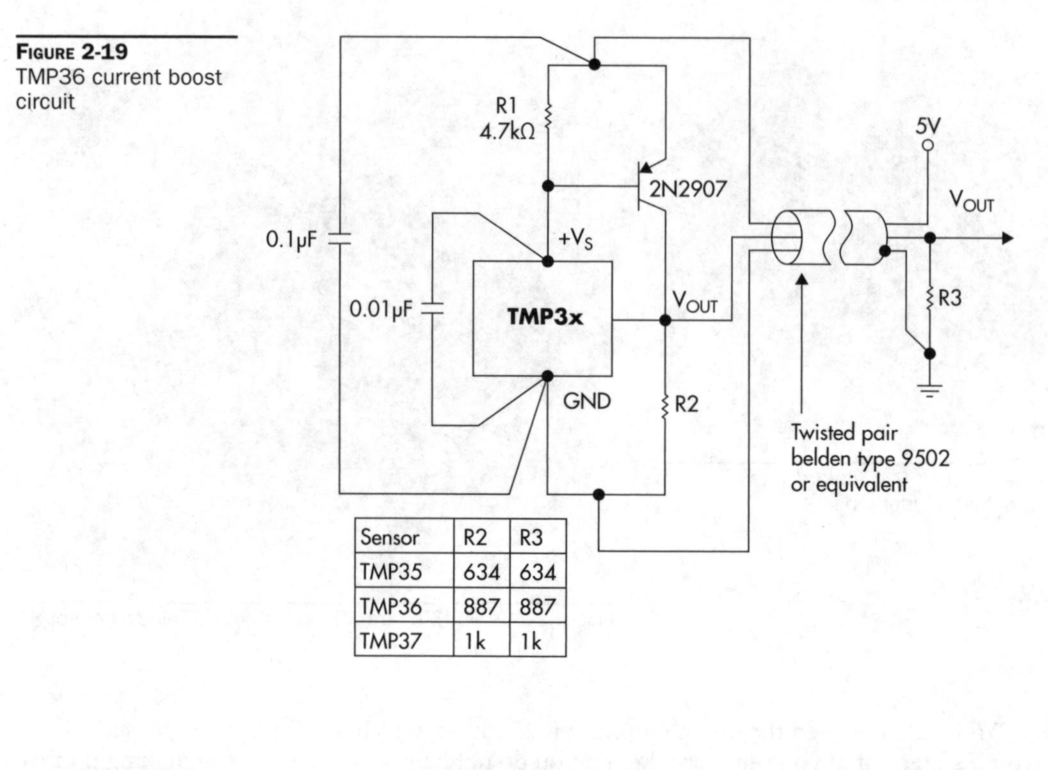

Sensor	R2	R3
TMP35	634	634
TMP36	887	887
TMP37	1k	1k

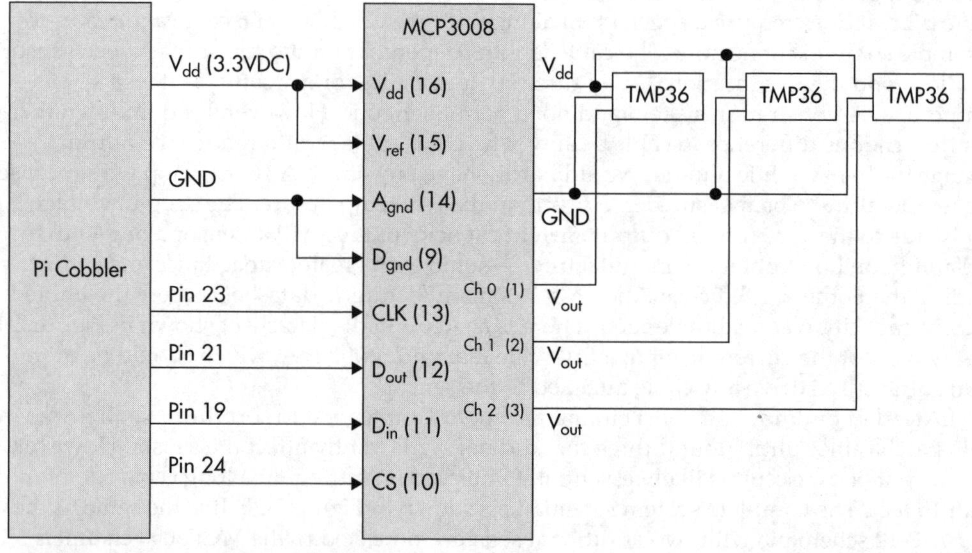

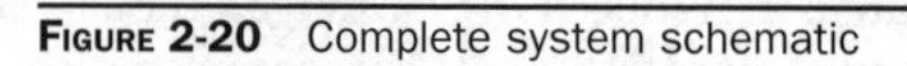
Figure 2-20 Complete system schematic

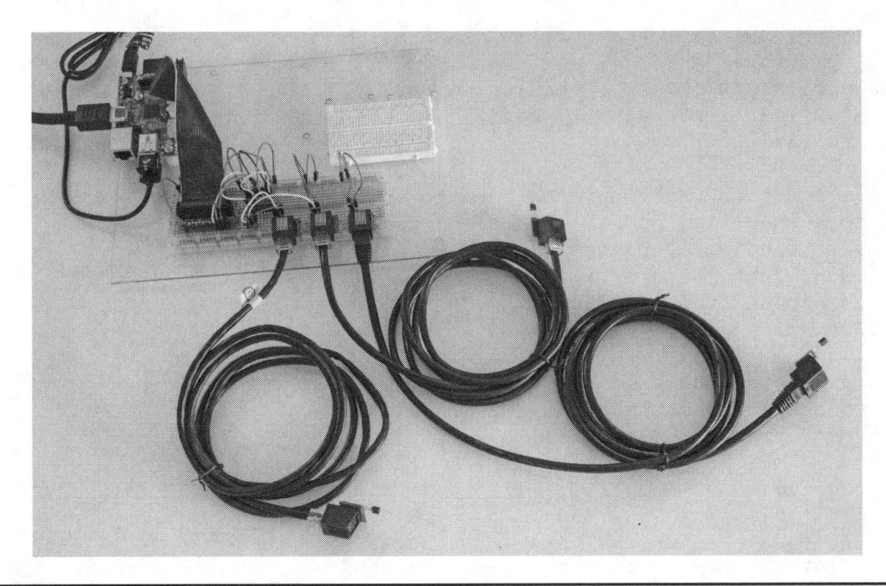

Figure 2-21 Physical system arrangement

Multiple Sensor Software

The software controlling the multiple sensor system is essentially the same as the single sensor version except for the two additional sensors. However, the program will be revised to display the date, time, and channel number as well as the temperature. I also do not display the `temp_level` or `temp_volts` variables in this version. All the new data will eventually be required for the database version that follows later in the chapter. In addition, I still "hard code" the temperature channels in to the program, i.e., channel 0 is sensor 1, channel 1 is sensor 2, and channel 2 is sensor 3. I recognize that this limits the program's flexibility but in the interests of simplicity I don't believe it is too much of a compromise. I also doubled the delay time to reduce the data flow a bit.

The new program is named MultipleSensorTest.py and is available on the book's companion website. A sample program display is shown in Figure 2-22.

```
#!/usr/bin/python

import spidev
import time
import os

#open the SPI bus
spi = spidev.SpiDev()
spi.open(0,0)

#define a function to read the MCP3008 ADC value
#channel must be an integer between 0 and 7
def ReadChannel(chan):
```

```
    if((chan < 0) or (chan > 7)):
        return -1
    adc =spi.xfer2([1, (8 + chan) << 4, 0])
    data = ((adc[1]&3) << 8) + adc[2]
    return data

#define a function that converts data to voltage levels
#round to a specified number of decimal places
def ConvertVolts(data, places):
    volts = (data * 3.3) / 1023
    volts = round(volts, places)
    return volts

#define a function to calc temperature from TMP36 data
#round to a specified number of decimal places
def ConvertTemp(data, places):

    #ADC Value       Temp (°C)        Volts
    #        0             -50          0.00
    #       78             -25          0.25
    #      155               0          0.50
    #      233              25          0.75
    #      310              50          1.00
    #      465             100          1.50
    #      775             200          2.50
    #     1023             280          3.30

    temp = ((data * 330)/1023) - 50
    temp = round(temp, places)
    return temp

#define temp channels
temp_chan0 = 0
temp_chan1 = 1
temp_chan2 = 2

#define time between readings
delay = 10

#print column headers
print "Date       Time      Channel      Temp"

#main loop
while True:

    #get the current system date in the format dd/mm/yyyy
    sample_date = time.strftime("%d/%m/%Y")

    #get the current system time in the 24 hour format hh:mm:ss
    sample_time = time.strftime("%H:%M:%S")
    #you change the format to 12 hour by substituting %I for %H

    #read all three temp channels
```

```
    temp_level1 = ReadChannel(temp_chan0)
    temp_level2 = ReadChannel(temp_chan1)
    temp_level3 = ReadChannel(temp_chan2)

    temp1 = ConvertTemp(temp_level1, 2)
    temp2 = ConvertTemp(temp_level2, 2)
    temp3 = ConvertTemp(temp_level3, 2)

    #display results
    print "------------------------------------------------------------"
    print sample_date, "      ", sample_time, "      ", temp_chan0 +1,

"     ", temp1
    print sample_date, "      ", sample_time, "      ", temp_chan1 +1,

"     ", temp2
    print sample_date, "      ", sample_time, "      ", temp_chan2 +1,

"     ", temp3

    #delay before taking next measurement
    time.sleep(delay)
```

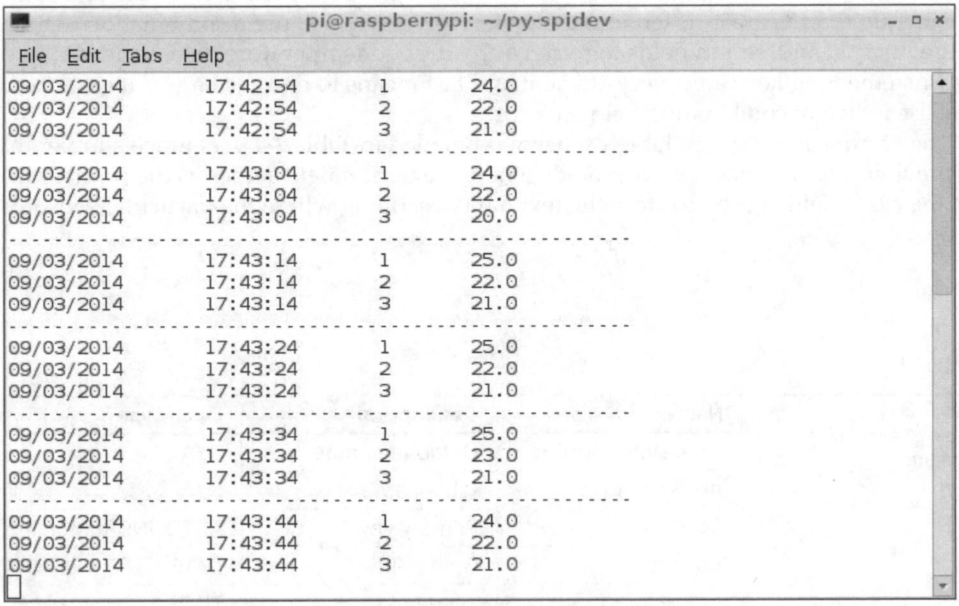

```
pi@raspberrypi: ~/py-spidev
File  Edit  Tabs  Help
09/03/2014       17:42:54      1       24.0
09/03/2014       17:42:54      2       22.0
09/03/2014       17:42:54      3       21.0
------------------------------------------------------------
09/03/2014       17:43:04      1       24.0
09/03/2014       17:43:04      2       22.0
09/03/2014       17:43:04      3       20.0
------------------------------------------------------------
09/03/2014       17:43:14      1       25.0
09/03/2014       17:43:14      2       22.0
09/03/2014       17:43:14      3       21.0
------------------------------------------------------------
09/03/2014       17:43:24      1       25.0
09/03/2014       17:43:24      2       22.0
09/03/2014       17:43:24      3       21.0
------------------------------------------------------------
09/03/2014       17:43:34      1       25.0
09/03/2014       17:43:34      2       23.0
09/03/2014       17:43:34      3       21.0
------------------------------------------------------------
09/03/2014       17:43:44      1       24.0
09/03/2014       17:43:44      2       22.0
09/03/2014       17:43:44      3       21.0
```

FIGURE 2-22 Sample display from the MultipleSensorTest.py program

I also wanted to point out that I deliberately chose not to use iteration to sample and display all the sensors as there were only three sensors and the memory saved was not as important to me as the program efficiency gained by "unrolling" the loops. I would definitely use loops if five or more sensors were utilized as the program would otherwise become quite large and tedious to enter.

I will now change focus slightly and discuss how to create a database to store the temperature data for eventual retrieval using a web browser.

Temperature Database

I will be using the concepts and procedures I previously discussed in Chapter 1 to set up a new database to store the temperature data generated by the three-sensor system. The *test* database structure will mostly be used as a template for the new database with some modifications. I will also set up a new user for this database as it would be a serious security issue to allow root access to a database that is also accessible via a web browser.

The new database is named HomeTempSystem and will have two tables in it named `sensorTemp` and `channelLocation`. The `channelLocation` table will enable a convenient method of describing a sensor's location and will enable you to change the sensor as needed. The schema or structure for the `sensorTemp` table is shown in Table 2-3.

The "id" is a new addition in the field listing as compared to the original *test* database and it is described as a Primary key. This is an important designation as there cannot be any duplicate records contained in any relational database table. The `id` field is simply an integer that is automatically incremented every time a new record is added. I will not explicitly use the `id` field in this project but it is a good security feature that ensures that only unique records are inserted into the table. However, I will use a combination of the `tdate`, `ttime`, and `tchan` fields to retrieve any desired temperature data. Incidentally, auto-incrementing keys are a very efficient and fast means to retrieve large data sets from a database in lieu of combination field lookups.

The `channelLocation` table's schema is detailed in Table 2-4. It is much simpler than the previous one as it has only two fields where the `tchan` field serves as the primary key and the `tloc` field is used to store the text data describing where the particular sensor is set up.

TABLE 2-3 `sensorTemp` Table Structure

Name	Description	Data Type
HomeTempSystem	Database name	N/A
`sensorTemp`	table name	N/A
`id`	Primary key	AUTO_INCREMENT
`tdate`	Date field	DATE
`ttime`	Time field	TIME
`tchan`	Channel number	NUMERIC
`ttemp`	Temperature field	NUMERIC

TABLE 2-4 channelLocation Table Structure

Name	Description	Data Type
HomeTempSystem	Database name	N/A
channelLocation	Table name	N/A
tchan	Primary key	NUMERIC
tloc	Sensor location	TEXT

To create the database and tables, you must follow the following steps:

1. Start the MySQL program by entering the following:

```
mysql -u root -p ↵
```

2. Enter the password you created when you first installed MySQL, as discussed in the previous chapter.
3. Create an empty database named HomeTempSystem by entering the following:

```
CREATE DATABASE HomeTempSystem; ↵
```

4. Switch over to the new database:

```
USE HomeTempSystem; ↵
```

5. Create the `sensorTemp` table containing all of the fields, as detailed in Table 2-3:

```
CREATE TABLE sensorTemp (id MEDIUMINT  AUTO_INCREMENT,
tdate DATE NOT NULL, ttime TIME NOT NULL, tchan NUMERIC NOT NULL,
ttemp NUMERIC NOT NULL, PRIMARY KEY(id)) ENGINE=MyISAM; ↵
```

   All the fields have been designated as `NOT NULL`, which means that a proper value has to be present or else the record will not be entered into the table. It is another means to ensure the table is not populated with garbage data.

6. Create the `channelLocation` table containing the two fields, as detailed in Table 2-4

```
CREATE TABLE channelLocation(tchan NUMERIC NOT NULL, tloc TEXT,

PRIMARY KEY(tchan)); ↵
```

Note that the `tchan` field must match the field description as specified in the `sensorTemp` table. This is necessary because this field is the logical link between the two tables. It is the primary key in the `channelLocation` table and is also known as a foreign key in the `sensorTemp` table. This arrangement is very useful given that the channel numbers are repeated many times in the `sensorTemp` table but only one instance of a text channel description is needed as the `tchan` field links the two. This table linkage is one of the most valuable features of relational databases. I provide a concrete example of this feature later in this chapter.

The `channelLocation` table should have data manually inserted into it as data will not be programmatically inserted. This type of data is considered static and unchanging and is normally provided for descriptive purposes, in this case, just the text locations where the sensors are located. I used three INSERT statements to manually populate the table, one of which is

```
INSERT INTO channelLocation (tchan, tloc) VALUES ('0', 'kitchen');
```

You will readily see the other two locations from the SELECT figures shown when I discuss the ViewRecords program later in the chapter.

Next, a new user must be created for the security reasons mentioned earlier. I named the new user "TempUser1," but it could be any name that suits your purposes or needs. This new user will have complete read and write privileges to only the HomeTempSystem database and to no others. Enter the following to first set up this user:

```
CREATE USER 'TempUser1' IDENTIFIED BY 'Px158qqr'; ↵
```

Note that a new password must be added for this user, as shown in the preceding command. You now need to associate this new user to the designated database by entering the following:

```
GRANT ALL PRIVILEGES ON HomeTempSystem.*  TO  'TempUser1'; ↵
```

Finally, you need to execute the following command to actually set up the user's privileges, as discussed in Chapter 1:

```
FLUSH PRIVILEGES; ↵
```

The last command completes the new database setup along with the new tables and user. You should next close the MySQL program by entering:

```
EXIT; ↵
```

You can now optionally test the new database by following the procedures detailed in the previous chapter. However, I will now proceed to demonstrate how to populate the table programmatically using a Python connection and the data generated by the MultipleSensorTest program.

Python Database Connection

An open-source Python package named python-mysqldb contains all the libraries necessary to establish connectivity between a Python program (script) and a MySQL database. You will need to install the package by entering this command:

```
sudo apt-get install python-mysqldb ↵
```

Next, you should create a test program named mysqlTest.py to confirm that the Python-to-MySQL connection works. Enter the following to start the nano editor and then enter the code, which follows the command. This code is also available on the book's companion website.

NOTE *I used the test database created at the root level in Chapter 1 for this test. It was quick and convenient, and it already had some sample data in the* `tempData` *table. I also logged in as the tester1 user, which I had previously added to the database.*

```
sudo nano mysqlTest.py ↵

#!/usr/bin/python
import MySQLdb
#this creates the basic connection object
db = MySQLdb.connect( host='localhost', user='tester1',passwd='password',db
='test')
#a cursor object is required to execute SQL commands on the database
cur = db.cursor()
#this SQL command retrieves all the records from the tempData table
cur.execute('SELECT * FROM tempData')
#print the field headers
print "DATE           TIME          LOCATION     TEMP (C)"
#display all the data record after record
for row in cur.fetchall():
    print row[0], "    ", row[1], "    ", row[2], "    ", row[3]
```

Execute the mysqlTest.py program by entering the following at the command line:

```
sudo python mysqlTest.py ↵
```

Figure 2-23 shows the resulting display after the program is run. All three records that were previously manually entered are displayed, one record at a time.

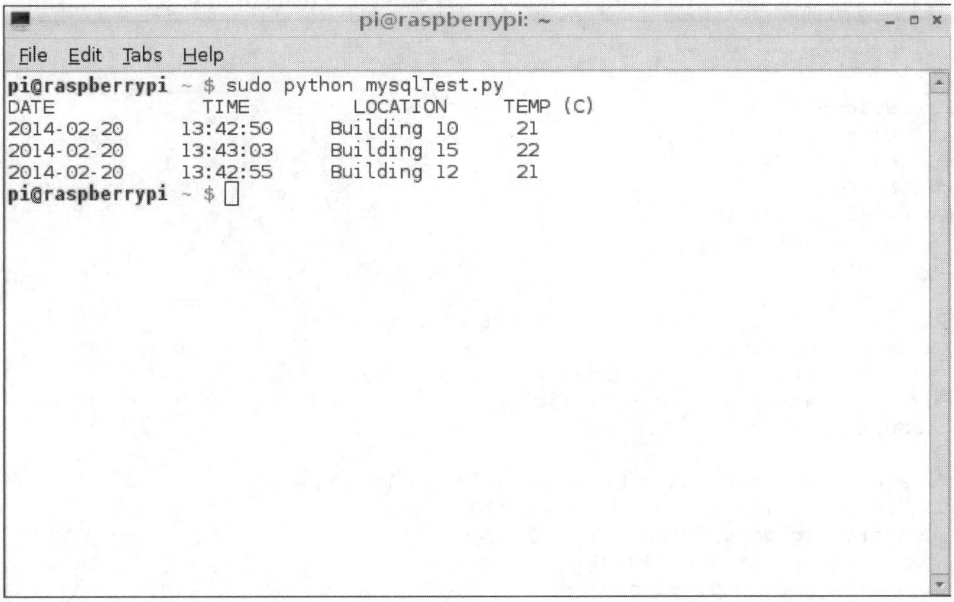

```
pi@raspberrypi: ~
File  Edit  Tabs  Help
pi@raspberrypi ~ $ sudo python mysqlTest.py
DATE            TIME         LOCATION      TEMP (C)
2014-02-20     13:42:50     Building 10     21
2014-02-20     13:43:03     Building 15     22
2014-02-20     13:42:55     Building 12     21
pi@raspberrypi ~ $
```

FIGURE 2-23 mysqlTest program results

Running this program has demonstrated how relatively easy it is to instantiate a MySQL database connection to a Python application and retrieve the data records. The next step in developing the web-based multiple sensor application is to show how to programmatically insert values into a MySQL database.

Inserting Data into a MySQL Database Using a Program

I will demonstrate how to insert data into the HomeTempData database by modifying the MultipleSensorTest.py program. You must have already created the database and table as well as added the new TempUser1 with the associated password, as shown in the previous section.

The program modifications consist of:

- Establishing a database connection
- Adding SQL statements to INSERT temperature data into the HomeTempSystem database
- Removing the console display statements
- Removing the channel number check
- Removing the data-to-voltage-level function
- Removing the comments ADC/Temp/Volts table
- Extending the delay between measurements to 60 seconds

I also renamed the modified program SensorDatabase.py. It is available on the book's companion website.

This program inserts the three sensors data directly into the HomeTempData database. It also inserts a date and time stamp that indicates when the measurements were taken.

```
#!/usr/bin/python

import spidev
import time
import os
import MySQLdb
#open the SPI bus
spi = spidev.SpiDev()
spi.open(0,0)

#define a function to read the MCP3008 ADC value
def ReadChannel(chan):
    adc =spi.xfer2([1, (8 + chan) << 4, 0])
    data = ((adc[1]&3) << 8) + adc[2]
    return data

#define a function to calc temperature from TMP36 data
#round to a specified number of decimal places
def ConvertTemp(data, places):
    temp = ((data * 346)/1023) - 50
    temp = round(temp, places)
    return temp
```

```
#define temp channels
temp_chan0 = 0
temp_chan1 = 1
temp_chan2 = 2

#define time interval for the countdown timer
delay = 1

#create the basic connection object
db = MySQLdb.connect(host='localhost', user='TempUser1', passwd='Px158qqr',
db='HomeTempSystem')
#a cursor object is required to execute SQL commands on the database
cur = db.cursor()
logTemp = "INSERT INTO sensorTemp (tdate, ttime, tchan, ttemp)
 VALUES (%s, %s, %s, %s)"
#main loop
while True:

    #get the current system date in the format yyyy/mm/dd
    sample_date = time.strftime("%Y/%m/%d")

    #get the current system time in the 24 hour format hh:mm:ss
    sample_time = time.strftime("%H:%M:%S")

    #read all three temp channels
    temp_level1 = ReadChannel(temp_chan0)
    temp_level2 = ReadChannel(temp_chan1)
    temp_level3 = ReadChannel(temp_chan2)

    temp1 = ConvertTemp(temp_level1, 2)
    temp2 = ConvertTemp(temp_level2, 2)
    temp3 = ConvertTemp(temp_level3, 2)

    #these statements insert all of the channel temp data into the
sensorTemp table
    cur.execute(logTemp, (sample_date, sample_time, temp_chan0, temp1))
    cur.execute(logTemp, (sample_date, sample_time, temp_chan1, temp2))
    cur.execute(logTemp, (sample_date, sample_time, temp_chan2, temp3))

    #countdown timer
    #delay before taking next measurements
    i = 60
    while (i != 0):
        time.sleep(delay)
        print i
        i -= 1
```

There is a hidden problem with the preceding program in that it will continue to run without any programmed way of stopping it until there is no more room for database records to be added. Such a situation would likely crash the Pi's OS. However, if each record is about 100 bytes in length, I estimate it would take over six years to fill-up 1GB of memory. I therefore have no problem in doing a CTRL-C to manually interrupt the program given this long time before calamity strikes. The worst that would happen is to corrupt one data record, which is quite acceptable during the development phase.

To run the program, first ensure that the multiple sensor system is attached and all the sensors are deployed as you want them to be placed. Next, enter the following command to start the logging of temperatures to the MySQL database:

```
sudo python SensorDatabase.py ↵
```

You should now see a countdown timer near the command prompt indicating that the program is running. This is really just a trivial add-on that is non-functional other than to indicate the program is running. I wanted to avoid the situation where the prompt simply disappeared and there was no indication of any activity. Let the program run for at least 30 minutes to build up a fair number of records before stopping it with the CTRL-C combination key press.

You can now start the MySQL program and look at the records generated by entering the following SQL command at the `mysql` prompt:

```
SELECT * FROM sensorTemp; ↵
```

Figure 2-24 shows a portion of the `sensorTemp` table results after I entered the preceding command.

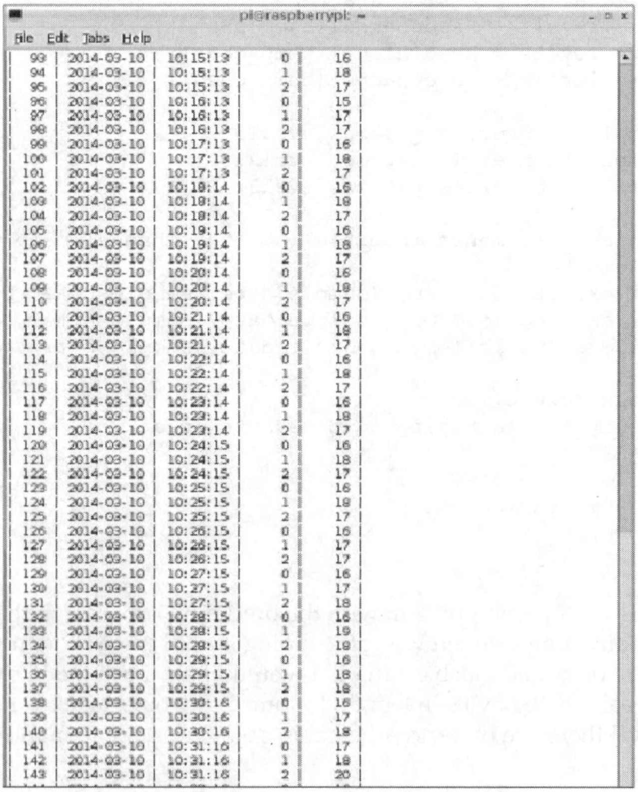

pi@raspberrypi: ~

File Edit Tabs Help

```
|  93 | 2014-03-10 | 10:15:13 |  0 |  16 |
|  94 | 2014-03-10 | 10:15:13 |  1 |  18 |
|  95 | 2014-03-10 | 10:15:13 |  2 |  17 |
|  96 | 2014-03-10 | 10:16:13 |  0 |  15 |
|  97 | 2014-03-10 | 10:16:13 |  1 |  17 |
|  98 | 2014-03-10 | 10:16:13 |  2 |  17 |
|  99 | 2014-03-10 | 10:17:13 |  0 |  16 |
| 100 | 2014-03-10 | 10:17:13 |  1 |  18 |
| 101 | 2014-03-10 | 10:17:13 |  2 |  17 |
| 102 | 2014-03-10 | 10:18:14 |  0 |  16 |
| 103 | 2014-03-10 | 10:18:14 |  1 |  18 |
| 104 | 2014-03-10 | 10:18:14 |  2 |  17 |
| 105 | 2014-03-10 | 10:19:14 |  0 |  16 |
| 106 | 2014-03-10 | 10:19:14 |  1 |  18 |
| 107 | 2014-03-10 | 10:19:14 |  2 |  17 |
| 108 | 2014-03-10 | 10:20:14 |  0 |  16 |
| 109 | 2014-03-10 | 10:20:14 |  1 |  18 |
| 110 | 2014-03-10 | 10:20:14 |  2 |  17 |
| 111 | 2014-03-10 | 10:21:14 |  0 |  16 |
| 112 | 2014-03-10 | 10:21:14 |  1 |  18 |
| 113 | 2014-03-10 | 10:21:14 |  2 |  17 |
| 114 | 2014-03-10 | 10:22:14 |  0 |  16 |
| 115 | 2014-03-10 | 10:22:14 |  1 |  18 |
| 116 | 2014-03-10 | 10:22:14 |  2 |  17 |
| 117 | 2014-03-10 | 10:23:14 |  0 |  16 |
| 118 | 2014-03-10 | 10:23:14 |  1 |  18 |
| 119 | 2014-03-10 | 10:23:14 |  2 |  17 |
| 120 | 2014-03-10 | 10:24:15 |  0 |  16 |
| 121 | 2014-03-10 | 10:24:15 |  1 |  18 |
| 122 | 2014-03-10 | 10:24:15 |  2 |  17 |
| 123 | 2014-03-10 | 10:25:15 |  0 |  16 |
| 124 | 2014-03-10 | 10:25:15 |  1 |  18 |
| 125 | 2014-03-10 | 10:25:15 |  2 |  17 |
| 126 | 2014-03-10 | 10:26:15 |  0 |  16 |
| 127 | 2014-03-10 | 10:26:15 |  1 |  17 |
| 128 | 2014-03-10 | 10:26:15 |  2 |  17 |
| 129 | 2014-03-10 | 10:27:15 |  0 |  16 |
| 130 | 2014-03-10 | 10:27:15 |  1 |  17 |
| 131 | 2014-03-10 | 10:27:15 |  2 |  18 |
| 132 | 2014-03-10 | 10:28:15 |  0 |  16 |
| 133 | 2014-03-10 | 10:28:15 |  1 |  19 |
| 134 | 2014-03-10 | 10:28:15 |  2 |  18 |
| 135 | 2014-03-10 | 10:29:15 |  0 |  16 |
| 136 | 2014-03-10 | 10:29:15 |  1 |  18 |
| 137 | 2014-03-10 | 10:29:15 |  2 |  18 |
| 138 | 2014-03-10 | 10:30:16 |  0 |  16 |
| 139 | 2014-03-10 | 10:30:16 |  1 |  17 |
| 140 | 2014-03-10 | 10:30:16 |  2 |  18 |
| 141 | 2014-03-10 | 10:31:16 |  0 |  17 |
| 142 | 2014-03-10 | 10:31:16 |  1 |  18 |
| 143 | 2014-03-10 | 10:31:16 |  2 |  20 |
```

FIGURE 2-24 Portion of `sensorTemp` records created by the SensorDatabase program

You can also view the records by running the following program, ViewRecords.py. It is also available on the companion website for this book.

```
#!/usr/bin/python
import MySQLdb
#this creates the basic connection object
db = MySQLdb.connect(host='localhost', user='TempUser1', passwd='Px158qqr',
db='HomeTempSystem')
#a cursor object is required to execute SQL commands on the database
cur = db.cursor()
#this SQL command retrieves all the records from the sensorTemp table
cur.execute('SELECT * FROM sensorTemp')
#display all the data record after record
for row in cur.fetchall():
    print row[0], "    ", row[1], "    ", row[2], "    ", row[3], "    ",
row[4]
```

Execute the ViewRecords.py program by entering the following at the command line:

```
sudo python ViewRecords.py ↵
```

Figure 2-25 shows the resulting display after the program is run. A portion of the records that were generated by SensorDatabase.py are displayed.

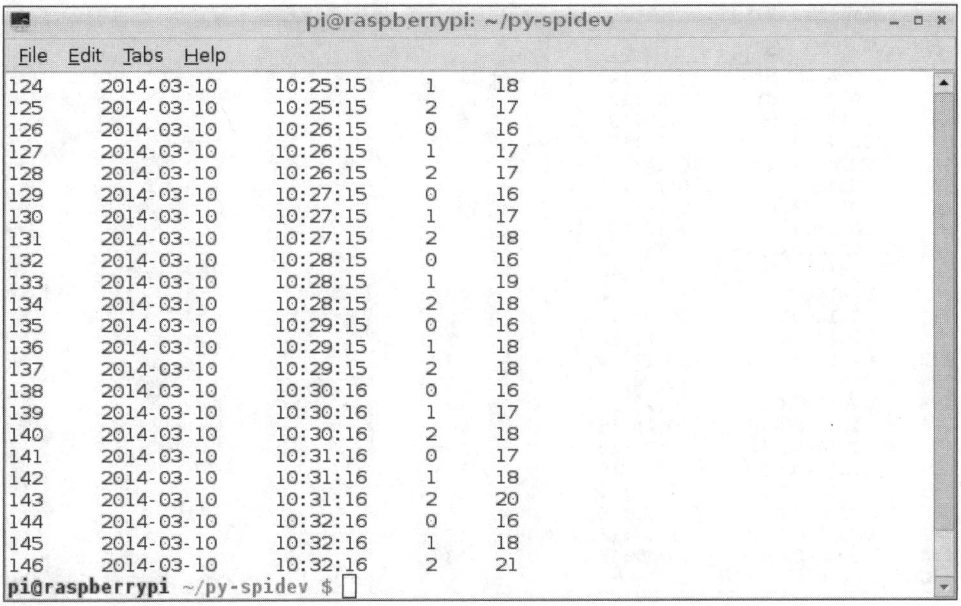

```
pi@raspberrypi: ~/py-spidev
File  Edit  Tabs  Help
124     2014-03-10     10:25:15     1     18
125     2014-03-10     10:25:15     2     17
126     2014-03-10     10:26:15     0     16
127     2014-03-10     10:26:15     1     17
128     2014-03-10     10:26:15     2     17
129     2014-03-10     10:27:15     0     16
130     2014-03-10     10:27:15     1     17
131     2014-03-10     10:27:15     2     18
132     2014-03-10     10:28:15     0     16
133     2014-03-10     10:28:15     1     19
134     2014-03-10     10:28:15     2     18
135     2014-03-10     10:29:15     0     16
136     2014-03-10     10:29:15     1     18
137     2014-03-10     10:29:15     2     18
138     2014-03-10     10:30:16     0     16
139     2014-03-10     10:30:16     1     17
140     2014-03-10     10:30:16     2     18
141     2014-03-10     10:31:16     0     17
142     2014-03-10     10:31:16     1     18
143     2014-03-10     10:31:16     2     20
144     2014-03-10     10:32:16     0     16
145     2014-03-10     10:32:16     1     18
146     2014-03-10     10:32:16     2     21
pi@raspberrypi ~/py-spidev $
```

FIGURE 2-25 SensorDatabase records displayed by the ViewRecords program

I next modified the ViewRecords program to display only the records from channel 0. This modification required inserting a conditional phrase in the SQL:

```
SELECT statement:
```

```
cur.execute('SELECT * FROM sensorTemp WHERE tchan = 0')
```

I renamed ViewRecords.py to Chan0ViewRecords.py. I did not provide a program listing as it requires only that one slight change to the `SELECT` statement. To run it, simply enter the following:

```
sudo python Chan0ViewRecords.py ↵
```

Figure 2-26 shows the output from the program where only the channel 0 records are displayed.

To view the sensor location text descriptions along with the temperature data, change the `SELECT` statement in the ViewRecords program to the following:

```
cur.execute('SELECT * FROM sensorTemp, channelLocation')
```

Figure 2-27 shows a portion of the output from the program where the location description is shown along with the temperature data.

```
pi@raspberrypi: ~/py-spidev
File Edit Tabs Help
78      2014-03-10      10:10:12      0      15
81      2014-03-10      10:11:12      0      16
84      2014-03-10      10:12:13      0      16
87      2014-03-10      10:13:13      0      15
90      2014-03-10      10:14:13      0      15
93      2014-03-10      10:15:13      0      16
96      2014-03-10      10:16:13      0      15
99      2014-03-10      10:17:13      0      16
102      2014-03-10      10:18:14      0      16
105      2014-03-10      10:19:14      0      16
108      2014-03-10      10:20:14      0      16
111      2014-03-10      10:21:14      0      16
114      2014-03-10      10:22:14      0      16
117      2014-03-10      10:23:14      0      16
120      2014-03-10      10:24:15      0      16
123      2014-03-10      10:25:15      0      16
126      2014-03-10      10:26:15      0      16
129      2014-03-10      10:27:15      0      16
132      2014-03-10      10:28:15      0      16
135      2014-03-10      10:29:15      0      16
138      2014-03-10      10:30:16      0      16
141      2014-03-10      10:31:16      0      17
144      2014-03-10      10:32:16      0      16
pi@raspberrypi ~/py-spidev $
```

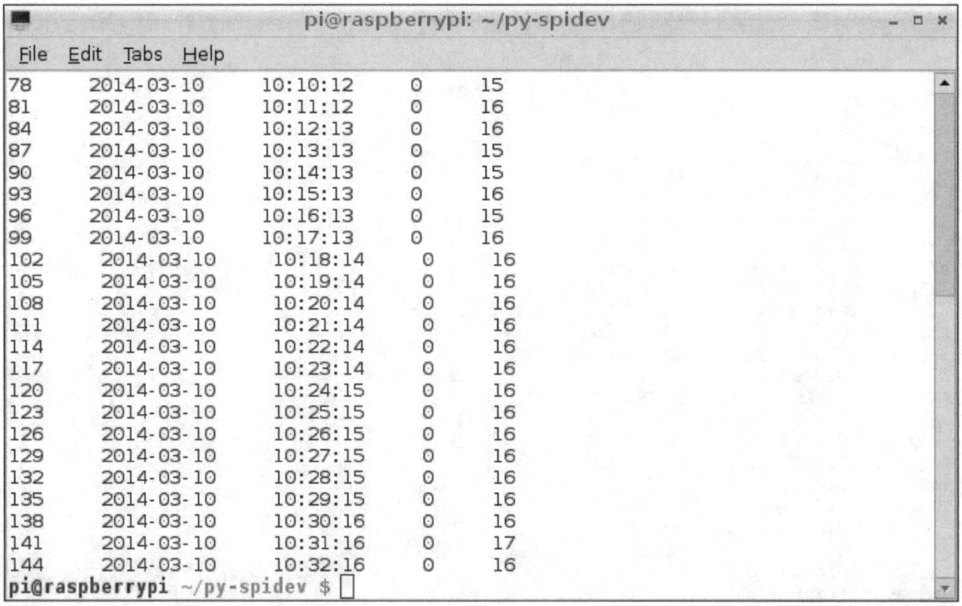

FIGURE 2-26 SensorDatabase records displayed by the Chan0ViewRecords program

```
| 133 | 2014-03-10 | 10:28:15 |  1 |  19 |  2 | bed room    |
| 134 | 2014-03-10 | 10:28:15 |  2 |  18 |  0 | kitchen     |
| 134 | 2014-03-10 | 10:28:15 |  2 |  18 |  1 | living room |
| 134 | 2014-03-10 | 10:28:15 |  2 |  18 |  2 | bed room    |
| 135 | 2014-03-10 | 10:29:15 |  0 |  16 |  0 | kitchen     |
| 135 | 2014-03-10 | 10:29:15 |  0 |  16 |  1 | living room |
| 135 | 2014-03-10 | 10:29:15 |  0 |  16 |  2 | bed room    |
| 136 | 2014-03-10 | 10:29:15 |  1 |  18 |  0 | kitchen     |
| 136 | 2014-03-10 | 10:29:15 |  1 |  18 |  1 | living room |
| 136 | 2014-03-10 | 10:29:15 |  1 |  18 |  2 | bed room    |
| 137 | 2014-03-10 | 10:29:15 |  2 |  18 |  0 | kitchen     |
| 137 | 2014-03-10 | 10:29:15 |  2 |  18 |  1 | living room |
| 137 | 2014-03-10 | 10:29:15 |  2 |  18 |  2 | bed room    |
| 138 | 2014-03-10 | 10:30:16 |  0 |  16 |  0 | kitchen     |
| 138 | 2014-03-10 | 10:30:16 |  0 |  16 |  1 | living room |
| 138 | 2014-03-10 | 10:30:16 |  0 |  16 |  2 | bed room    |
| 139 | 2014-03-10 | 10:30:16 |  1 |  17 |  0 | kitchen     |
| 139 | 2014-03-10 | 10:30:16 |  1 |  17 |  1 | living room |
| 139 | 2014-03-10 | 10:30:16 |  1 |  17 |  2 | bed room    |
| 140 | 2014-03-10 | 10:30:16 |  2 |  18 |  0 | kitchen     |
| 140 | 2014-03-10 | 10:30:16 |  2 |  18 |  1 | living room |
| 140 | 2014-03-10 | 10:30:16 |  2 |  18 |  2 | bed room    |
| 141 | 2014-03-10 | 10:31:16 |  0 |  17 |  0 | kitchen     |
| 141 | 2014-03-10 | 10:31:16 |  0 |  17 |  1 | living room |
| 141 | 2014-03-10 | 10:31:16 |  0 |  17 |  2 | bed room    |
| 142 | 2014-03-10 | 10:31:16 |  1 |  18 |  0 | kitchen     |
| 142 | 2014-03-10 | 10:31:16 |  1 |  18 |  1 | living room |
| 142 | 2014-03-10 | 10:31:16 |  1 |  18 |  2 | bed room    |
| 143 | 2014-03-10 | 10:31:16 |  2 |  20 |  0 | kitchen     |
| 143 | 2014-03-10 | 10:31:16 |  2 |  20 |  1 | living room |
| 143 | 2014-03-10 | 10:31:16 |  2 |  20 |  2 | bed room    |
| 144 | 2014-03-10 | 10:32:16 |  0 |  16 |  0 | kitchen     |
| 144 | 2014-03-10 | 10:32:16 |  0 |  16 |  1 | living room |
| 144 | 2014-03-10 | 10:32:16 |  0 |  16 |  2 | bed room    |
| 145 | 2014-03-10 | 10:32:16 |  1 |  18 |  0 | kitchen     |
| 145 | 2014-03-10 | 10:32:16 |  1 |  18 |  1 | living room |
| 145 | 2014-03-10 | 10:32:16 |  1 |  18 |  2 | bed room    |
| 146 | 2014-03-10 | 10:32:16 |  2 |  21 |  0 | kitchen     |
| 146 | 2014-03-10 | 10:32:16 |  2 |  21 |  1 | living room |
| 146 | 2014-03-10 | 10:32:16 |  2 |  21 |  2 | bed room    |
+-----+------------+----------+----+-----+----+-------------+
333 rows in set (0.04 sec)

mysql>
```

FIGURE 2-27 SensorDatabase records displayed by the ViewRecords program with the `channelLocation` table included

If you just want to check that the location information is being properly displayed, use the following `SELECT` statement and you will see only one set of records.

```
cur.execute('SELECT * FROM sensorTemp, channelLocation GROUP BY tchan')
```

Figure 2-28 shows the output from the program where the location description is shown along with the temperature data but grouped by location.

The previous demonstration finishes the discussion of creating database records from a sensor-based acquisition program. The next phase involves showing how to access and display selected records using a web browser.

Database Access Using a Web Browser

This is probably one of the easier portions of the project as it concerns the well-documented process of creating a dynamic website (HTML) that uses a web server language (PHP) to

```
pi@raspberrypi: ~
File Edit Tabs Help
mysql> INSERT INTO channelLocation (tchan, tloc) VALUES ('2', 'bed room');
Query OK, 1 row affected (0.02 sec)

mysql> SELECT * FROM channelLocation;
+-------+-------------+
| tchan | tloc        |
+-------+-------------+
|     0 | kitchen     |
|     1 | living room |
|     2 | bed room    |
+-------+-------------+
3 rows in set (0.00 sec)

mysql> SELECT * FROM sensorTemp,channelLocation GROUP BY tloc;
+----+------------+----------+-------+-------+-------+-------------+
| id | tdate      | ttime    | tchan | ttemp | tchan | tloc        |
+----+------------+----------+-------+-------+-------+-------------+
| 36 | 2014-03-10 | 09:56:10 |     0 |    15 |     2 | bed room    |
| 36 | 2014-03-10 | 09:56:10 |     0 |    15 |     0 | kitchen     |
| 36 | 2014-03-10 | 09:56:10 |     0 |    15 |     1 | living room |
+----+------------+----------+-------+-------+-------+-------------+
3 rows in set (0.08 sec)

mysql> 
```

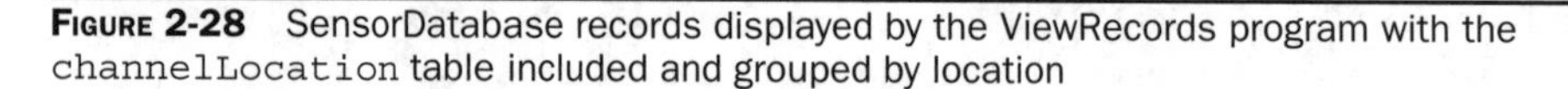

FIGURE 2-28 SensorDatabase records displayed by the ViewRecords program with the `channelLocation` table included and grouped by location

supply data from a relational database (MySQL) to a client (remote browser) upon demand. I am not including record locking discussion for now, which is a solution to the problem of two applications trying to access the same database record at the same time. I will just make the reasonable assumption that the database records are all accessible whenever the web server application needs them.

I will initially present a direct approach to creating a website, but many books and online tutorials are available on different approaches to creating a dynamic website. This website will not be fancy or flashy but simply serve up the desired temperature records in a tabular format. At the end of the chapter, I do provide a brief introduction to a Python-based microdevelopment framework named Flask, which will allow you to create fancier and more dynamic websites than are possible when using the simple PHP file that I discuss next.

The following is some straightforward PHP5 code that will display all the records in the `sensorTemp` table, which is part of the HomeTempSystem database. Notice that I logged in as TempUser1, which was added after creating the database for security reasons. To run it as shown, you will need to store this script as TempSensorTest.php in the /var/www directory. This code script is available on the book's companion website.

```
<?php
$username = "TempUser1";
$password = "Px158qqr";
$hostname = "localhost";
```

```
$database = 'homeTempSystem";

//connection to localhost
$con = mysqli_connect($hostname, $username, $password, $database);

//execute the SQL query and return records
$result = mysqli_query($con, "SELECT * FROM sensorTemp");

//fetch the data from the result recordset
while ($row = mysqli_fetch_array($result))
 {
  echo $row['id']."  ".$row['tdate']." ".$row['ttime']." ".$row['tchan']."
".$row['ttemp'];
  echo "<br>";
  }

//close the connection
mysql_close($con);
?>
```

To run the script on the Pi, you will need to open a web browser on the Pi. Simply type in the following in the browser's URL textbox: **http://localhost/TempSensorTest.php**.

If everything works as expected, you should see the database records that you earlier created displayed, as shown in Figure 2-29.

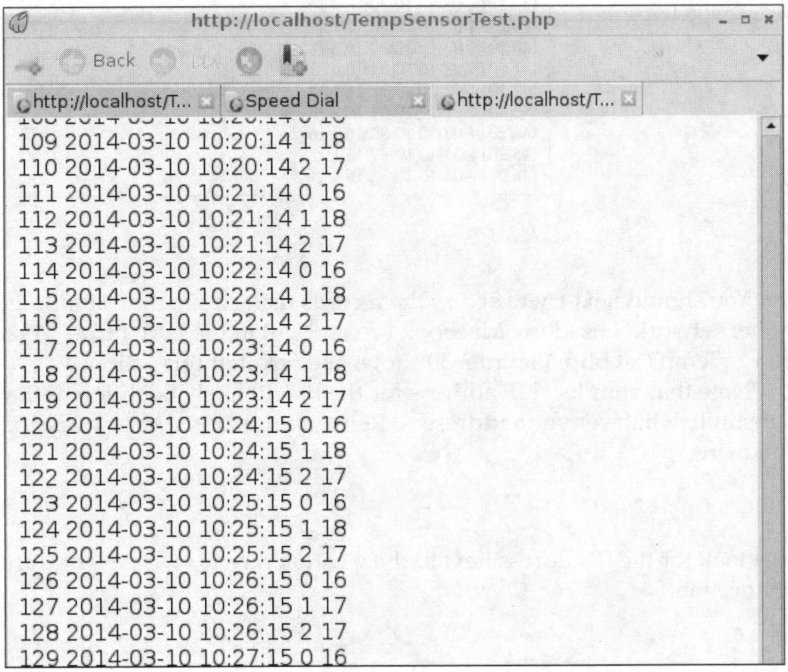

http://localhost/TempSensorTest.php

Back

http://localhost/T... | Speed Dial | http://localhost/T...

```
109 2014-03-10 10:20:14 1 18
110 2014-03-10 10:20:14 2 17
111 2014-03-10 10:21:14 0 16
112 2014-03-10 10:21:14 1 18
113 2014-03-10 10:21:14 2 17
114 2014-03-10 10:22:14 0 16
115 2014-03-10 10:22:14 1 18
116 2014-03-10 10:22:14 2 17
117 2014-03-10 10:23:14 0 16
118 2014-03-10 10:23:14 1 18
119 2014-03-10 10:23:14 2 17
120 2014-03-10 10:24:15 0 16
121 2014-03-10 10:24:15 1 18
122 2014-03-10 10:24:15 2 17
123 2014-03-10 10:25:15 0 16
124 2014-03-10 10:25:15 1 18
125 2014-03-10 10:25:15 2 17
126 2014-03-10 10:26:15 0 16
127 2014-03-10 10:26:15 1 17
128 2014-03-10 10:26:15 2 17
129 2014-03-10 10:27:15 0 16
```

FIGURE 2-29 Portion of the sensorTemp records displayed on the Pi web browser

FIGURE 2-30 Portion of the sensorTemp records displayed on a separate networked computer

192.168.0.13/TempSensorTest.php

192.168.0.13/TempSensorTest.php Reader

Apple iCloud Facebook Twitter Wikipedia Yahoo! News Popular

```
107 2014-03-10 10:19:14 2 17
108 2014-03-10 10:20:14 0 16
109 2014-03-10 10:20:14 1 18
110 2014-03-10 10:20:14 2 17
111 2014-03-10 10:21:14 0 16
112 2014-03-10 10:21:14 1 18
113 2014-03-10 10:21:14 2 17
114 2014-03-10 10:22:14 0 16
115 2014-03-10 10:22:14 1 18
116 2014-03-10 10:22:14 2 17
117 2014-03-10 10:23:14 0 16
118 2014-03-10 10:23:14 1 18
119 2014-03-10 10:23:14 2 17
120 2014-03-10 10:24:15 0 16
121 2014-03-10 10:24:15 1 18
122 2014-03-10 10:24:15 2 17
123 2014-03-10 10:25:15 0 16
124 2014-03-10 10:25:15 1 18
125 2014-03-10 10:25:15 2 17
126 2014-03-10 10:26:15 0 16
127 2014-03-10 10:26:15 1 17
128 2014-03-10 10:26:15 2 17
129 2014-03-10 10:27:15 0 16
130 2014-03-10 10:27:15 1 17
131 2014-03-10 10:27:15 2 18
132 2014-03-10 10:28:15 0 16
133 2014-03-10 10:28:15 1 19
134 2014-03-10 10:28:15 2 18
135 2014-03-10 10:29:15 0 16
136 2014-03-10 10:29:15 1 18
137 2014-03-10 10:29:15 2 18
138 2014-03-10 10:30:16 0 16
139 2014-03-10 10:30:16 1 17
140 2014-03-10 10:30:16 2 18
141 2014-03-10 10:31:16 0 17
142 2014-03-10 10:31:16 1 18
143 2014-03-10 10:31:16 2 20
144 2014-03-10 10:32:16 0 16
145 2014-03-10 10:32:16 1 18
146 2014-03-10 10:32:16 2 21
```

You should next try to access the records using a separate computer attached to your home network. I used my Macbook Pro and put in the URL http://192:168.0.13/SensorTempTest.php. Figure 2-30 shows the result of this action.

Note that your local IP address for the Pi will likely be different than my address. Just substitute whatever your address is. Remember that you can always find it by entering the following:

```
sudo ifconfig ↵
```

Just look for the IP address next to the `wlan0` entry for wireless or `eth0` for a wired connection.

Narrowing the Database Reports

It is not hard to imagine that the database size will grow rapidly as you accumulate more measurements over time. It would be a waste of time and bandwidth to have to sift through all the database records to find the specific data that you need to examine. I will show how to apply constraints to the database search by using the `WHERE` phrase that I had introduced earlier. The major complication is that you will be using a web browser to access the database and therefore will not have direct means to insert the `WHERE` phrase in the SQL query as was done in the earlier demonstration. Fortunately, this situation is well covered in the HTML and PHP area where a form will be created asking you for specific information to be sent to the web server for action. In this example, I am asking only for records of a specific channel number to be displayed, but the concepts can be readily extended to all the other database fields. There are two primary means by which data gets sent by a user to a web server. These are the GET and `POST` methods. Each has its advantages and disadvantages, but I have found that most developers prefer using the `POST` method so that is what I will implement. The code for a simple HTML form requesting a channel number is shown here:

```
<html>
<body>
<form action = http://192.168.0.13/TempSensorTestChan.php method = "POST">
Channel number: <input type = "text" name = "chan_no"><br>
<input type = "submit">
</form>
</body>
</html>
```

I named this code ChannelSelector.html and saved it on my laptop in the Documents folder. This program is really more of a script that the laptop client browser will use to send and receive data from the Pi web server. You should also notice that I hardcoded the Pi's IP address into the form along with a reference to a slightly modified version of the three-channel TempSensorTest program, which I discuss next. The key point regarding this HTML script is that it has a variable named `chan_no` that stores the channel number, and this variable is made available on the server side by the `POST` method. Figure 2-31 shows the form on a client-side browser.

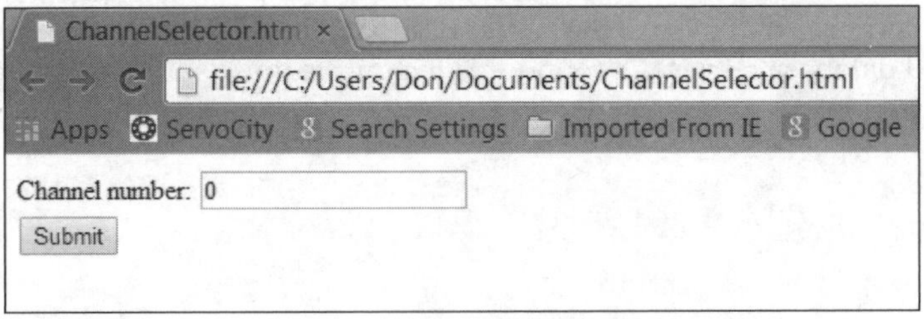

FIGURE 2-31 ChannelSelector form

Nothing will happen regarding database access until the Submit button is clicked. I will show the results of clicking on the button after I finish with the server program.

The server-side program is named TempSensorTestChan.py where the suffix "Chan" was added to indicate that the program constrains its output to the user-selected channel. The code shown here is identical to the TempSensorTest program with some modifications to accommodate the user-selected channel:

```
<?php
$username = "TempUser1";
$password = "Px158qqr";
$hostname = "localhost";
$database = 'homeTempSystem";

//connection to localhost
$con = mysqli_connect($hostname, $username, $password, $database);

//channel request
$channel = $_POST['chan_no'];

//execute the SQL query and return selected records
$result = mysqli_query($con, "SELECT * FROM sensorTemp WHERE tchan =
$channel");

//fetch tha data from the result recordset
while ($row = mysqli_fetch_array($result))
 {
  echo $row['id']."  ".$row['tdate']." ".$row['ttime']." ".$row['tchan']."
".$row['ttemp'];
  echo "<br>";
  }

//close the connection
mysql_close($con);
?>
```

Figure 2-32 shows the browser display after channel 0 was selected via the input form.

As I mentioned earlier, the constraints to be placed in the WHERE phrase can easily be extended to any or all of the other database fields using the same techniques I used for the channel selection. It is also possible to construct a complete query statement on the client side and pass that over to the server but that's best left to a more advanced study of web-based database retrieval. I found this technique quite suitable for these types of embedded applications.

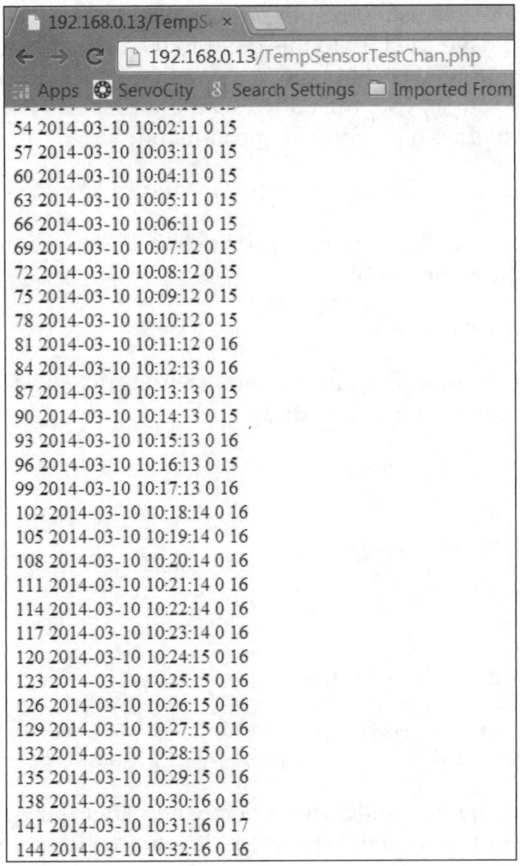
192.168.0.13/TempS
192.168.0.13/TempSensorTestChan.php
Apps
ServoCity
Search Settings
Imported From
54 2014-03-10 10:02:11 0 15
57 2014-03-10 10:03:11 0 15
60 2014-03-10 10:04:11 0 15
63 2014-03-10 10:05:11 0 15
66 2014-03-10 10:06:11 0 15
69 2014-03-10 10:07:12 0 15
72 2014-03-10 10:08:12 0 15
75 2014-03-10 10:09:12 0 15
78 2014-03-10 10:10:12 0 15
81 2014-03-10 10:11:12 0 16
84 2014-03-10 10:12:13 0 16
87 2014-03-10 10:13:13 0 15
90 2014-03-10 10:14:13 0 15
93 2014-03-10 10:15:13 0 16
96 2014-03-10 10:16:13 0 15
99 2014-03-10 10:17:13 0 16
102 2014-03-10 10:18:14 0 16
105 2014-03-10 10:19:14 0 16
108 2014-03-10 10:20:14 0 16
111 2014-03-10 10:21:14 0 16
114 2014-03-10 10:22:14 0 16
117 2014-03-10 10:23:14 0 16
120 2014-03-10 10:24:15 0 16
123 2014-03-10 10:25:15 0 16
126 2014-03-10 10:26:15 0 16
129 2014-03-10 10:27:15 0 16
132 2014-03-10 10:28:15 0 16
135 2014-03-10 10:29:15 0 16
138 2014-03-10 10:30:16 0 16
141 2014-03-10 10:31:16 0 17
144 2014-03-10 10:32:16 0 16

FIGURE 2-32 Portion of the channel 0 sensorTemp records displayed on a separate networked computer

Flask

Flask has been identified as one of a series of microdevelopment frameworks. I have used Flask and find it very useful as well as interesting. Strangely, it started out as an April Fool's joke that was placed before the embedded development community as a serious tool. And, in fact, it turned out to be quite a serious tool that quickly captured developers' imaginations and subsequently grew to have a large and energetic following. Flask can be thought of as a "lightweight" web server with limited but reasonable functionality as compared to the much larger and capable Apache web server. I think the best way to explain Flask is to present two simple examples that are based on material from Matt Richardson's excellent book *Getting Started with Raspberry Pi*. But first the Flask software must be loaded into the Pi. It is based on Python and has a good fit with the Pi Raspian OS.

Flask is available using the pip package manager. It is another package manager similar to apt, which I have used up to this point in the book. Pip functions with the Python Package Index (PyPI) repository, where the Flask package is stored and available for download. Of course, you must download pip first into the Pi, which is oddly enough done using the apt tool with the following command:

```
sudo apt-get install python-pip ↵
```

After pip is installed, you can use it to install Flask with all its dependencies. Enter the following at the command line:

```
sudo pip install flask ↵
```

Naturally, the first program that you should run is a Hello World type. Enter the following code using the nano editor:

```
sudo nano hello-flask.py ↵


from flask import Flask
app = Flask(__name__)
@app.route("/")

def hello():
    return "Hello World!"

if __name__ == "__main__":
    app.run(host='0.0.0.0', port = 81, debug=True)
```

Note that there are two underscores before and after the words "name" and "main."

Flask is set up in a client/server architecture with the server portion started on the Pi as a Python program:

```
sudo python hello-flask.py ↵
```

Next, start a browser on another computer on your home network and enter the Pi's address followed by the port number, which in my case is

```
192.168.0.13:81
```

If everything is all set, you should see the Hello World! greeting, as shown in Figure 2-33.

FIGURE 2-33 Flask's Hello World browser greeting

Some readers may have noticed that I used port 81 in lieu of port 80, which is the default port number for HTTP. This was needed because port 80 was already in use by the Apache web server and not available for the Flask web server.

There are also runtime messages being displayed on the Pi console screen as the Flask web server is running. Figure 2-34 shows these messages for the Hello World browser request being sent from the separate networked computer. Notice this networked computer is at address 192.168.0.2, as you can see in the message lines.

Flask also supports templates, which is a highly useful feature that allows you to quickly create web pages based on a number of different designs. The example shown here is named hello-template.py and simply displays the server's current date and time when queried from the remote browser:

```
sudo nano hello-template.py ↵
```

```
from flask import Flask, render_template
import datetime
app = Flask(__name__)
@app.route("/")

def hello():
    now = datetime.datetime.now()
    timeStr = now.strftime("%Y-%m-%d %H:%M")
    templateData =
    {
        'title' : "HELLO",
        'time' : timeStr
    }


    return render_template('main.html', **templateData)

if __name__ == "__main__":
    app.run(host='0.0.0.0', port = 81, debug=True)
```

You likely noticed that there is a file named "main.html" appearing in the `hello()` method return statement. This is the actual HTML template file and it must be placed in a subdirectory named "templates" from where the existing-template.py file is located, otherwise Flask cannot find it. Assuming you are in the Pi's home directory, enter the following to create the new subdirectory:

```
sudo mkdir templates ↵
```

Then change directories by entering cd templates.

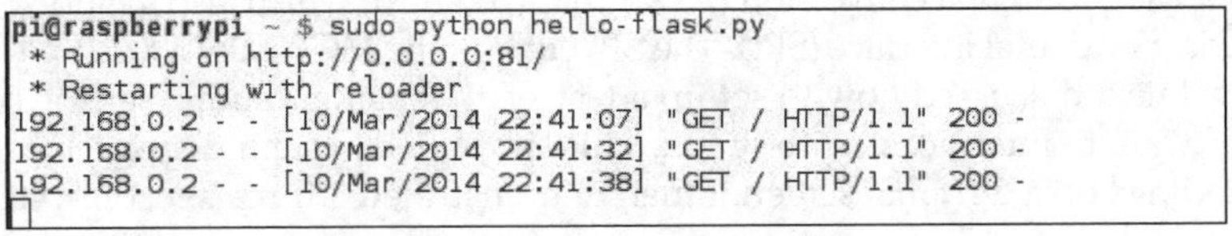

```
pi@raspberrypi ~ $ sudo python hello-flask.py
 * Running on http://0.0.0.0:81/
 * Restarting with reloader
192.168.0.2 - - [10/Mar/2014 22:41:07] "GET / HTTP/1.1" 200 -
192.168.0.2 - - [10/Mar/2014 22:41:32] "GET / HTTP/1.1" 200 -
192.168.0.2 - - [10/Mar/2014 22:41:38] "GET / HTTP/1.1" 200 -
```

Figure 2-34 Console messages from the Flask web server

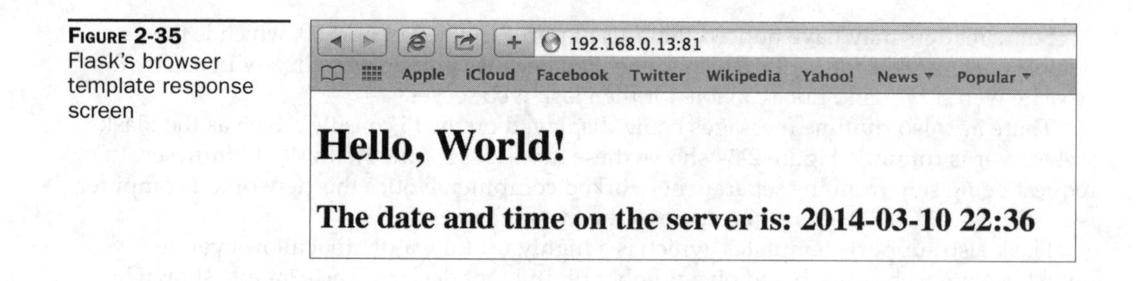

Figure 2-35 Flask's browser template response screen

Now that you are in the proper place, create the main.html file:

```
sudo nano main.html ↵

<!DOCTYPE html>
    <head>
        <title>{{ title }}</title>
    </head>

    <body>
        <h1>Hello World</h1>
        <h2>The date and time on the server is: {{ time }}<h2>
    </body>
</html>
```

Go back to the home directory after creating the file and run the hello-template program by entering

sudo python hello-template.py

The Pi will now be listening on port 81 for any browser client requests. As you did before, open a browser on another networked computer and enter the Pi's IP address with port 81 as a suffix. You should see the web browser display that is shown in Figure 2-35.

The previous two examples provide a brief introduction to Flask but should be sufficient to whet your appetite if you choose to use this clever developmental tool suite. There are many more available templates created by the open-source community, which will allow you to create very versatile and capable web pages.

Summary

The chapter began with a discussion of a three-sensor home temperature monitoring system. I showed how the basic sensor, a TMP36 from Analog Devices, functioned to provide an analog voltage proportional to the ambient temperature surrounding the small sensor. A discussion of analog-to-digital conversion (ADC) followed next, which also included an introduction to the serial peripheral interface (SPI). The SPI enables the MCP3008ADC chip to send data to the Pi.

I next discussed how to set up a test system using a single sensor with the necessary software to interface it with the Pi. This simplified system displayed a series of temperature readings on a terminal screen. I then demonstrated a three-sensor system that also displayed all the readings on a screen. I also showed you how to use RJ45 cables and breakout boards to mount and interconnect the sensors with the ADC chip.

The next portion of the chapter concerned how to set up a MySQL database to store all the temperature measurement data. This database is an essential element that must be in place before creating a web-based application. One of the key concepts in this discussion was how to logically connect the temperature measuring software with the database software. I showed you how to instantiate a Python database connector to meet this requirement.

I next showed you how to programmatically insert data into the database at predetermined intervals. I also included a program so that you could view these newly created records.

I then demonstrated how to access the database using a web browser. I used a PHP script to display data on a browser both on the Pi and also from a remote computer attached to the same home network as the Pi.

I next showed you how to "narrow" the data results using search criteria supplied via a client-based HTML form. This feature would be invaluable when searching a large database.

I concluded the chapter with a brief introduction to Flask, which is a "lightweight" web server based on Python. Flask was offered as an alternative to the much more comprehensive Apache web server, which also requires much more memory resources than Flask.

3 CHAPTER

Introduction to Object Orientation Programming (OOP) with Java

To begin your exploration into object orientation, pretend you have been transported to a virtual environment where objects are the primary life form. Let's call this environment Object Land.

Figure 3-1 shows a very abstract view of Object Land with two processes shown (small and big), each containing multiple objects that are in constant communication with one another to accomplish the overall process goals. The processes themselves are communicating with one another as needed to accomplish whatever needs to be done.

The key question arises: What is an object?

The textbook answer typically given is that an object is an instance of a class. Of course, this only further confuses the newcomer to Object Land where he or she doesn't know the definition of a class. Okay, what is the definition of a class, and more importantly, why should you care?

First, a quick quiz. What does Figure 3-2 represent?

- Bus
- Train
- Racecar
- Plane

The answer really lies with your life experience. Most people will know it is a racecar by its shape and the fact the driver is wearing a helmet. Others may recognize it by the process of elimination by recognizing that it is not a bus, train, or plane. We engage in this process continuously—that is, using models or abstractions to represent real-world things or objects.

FIGURE 3-1 Object Land

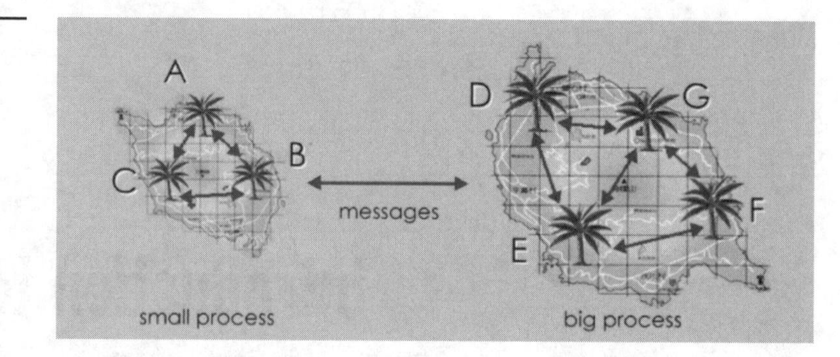

Similar activities are present in software design where you use abstractions to represent real-world things. This approach is much more relevant in developing software as compared to a much stricter procedural approach. Consider a situation in which you are at the train station exit having just arrived in New York City. You want to go to Radio City Music Hall and take in a show so you hail a cab. Once in the taxi, do you tell the driver, "Go to the end of the street, take a right, go through two sets of lights, take a left…" or do you simply say "Please take me to Radio City Music Hall"? The first approach is procedural while the latter is object oriented. In taking the OO approach, you are relying on that person object (the taxi driver) to be responsible to accept a message ("Please take me to Radio City Music Hall") and know how to accomplish the task. One very nice feature of this approach is that the object may have to change his implementation depending on traffic, street closure, and so on, but you as the message sender will not be aware of this change. You have to be aware of all the traffic conditions in New York City if you choose the procedural approach. Not a very appealing option!

Having established the fact that objects will, in fact, be useful to accomplish your goals in controlling sensors, it is time to examine some fundamental principles underlying all object-oriented programming paradigms. Table 3-1 lists the four bedrock principles that apply to all OO programming languages.

FIGURE 3-2 Unknown object

TABLE 3-1 Four Bedrock Object-Oriented Principles

OO Principle	Description
Abstraction	Modeling to represent real-world things
Polymorphism	Different object behavior generated by the same message
Inheritance	Shared common attributes and behaviors among objects
Encapsulation	Closing off the inner workings of objects from public view

A handy acronym to remember these principles is A PIE, taken from the beginning letter for each principle.

I decided to use a generic sensor as a model to demonstrate how to apply the OO approach. Determining basic sensor characteristics and behaviors is normally the first step in creating a class. The class is a data structure used to record these characteristics and behaviors. In formal OO terms, characteristics are known as attributes and behaviors are methods. Objects are created from classes. As mentioned earlier, an object is simply an instance of a class. How this is done depends upon the specific language being used. Many OO languages such as Java, C#, and C++ use the *new* operator to create a class instance. This process is known as instantiation.

It is often useful to refer back to basic definitions in developing class attributes and methods. A sensor has a fairly simple definition per the Merriam-Webster online dictionary: A device that responds to a physical stimulus (as heat, light, sound, pressure, magnetism, or a particular motion) and transmits a resulting impulse (as for measurement or operating a control).

The key is to try to encapsulate all the essential attributes and behaviors that are useful in describing a real-world object in a logical data structure such as a class. I also want to emphasize that there is really no single correct answer to creating a class. It turns out that some descriptions are better than others, and you will find that as you proceed with your design you will often turn back and revise your initial class definition. Experience in repeated OO design efforts will improve your initial efforts, and incorporating design patterns (DP), which I discuss shortly, will also help with the design. I have repeatedly told my beginning OO students that creating classes is probably the single hardest task to tackle in the whole OO approach.

An abstract class is often used to hold common attributes and behaviors that will be broadly applicable to a group of classes yet not hold enough specificity to allow a practical object instantiation. Abstract classes are useful only when used with the inheritance core principle.

Figure 3-3 shows a simple inheritance class diagram with an abstract parent class and four child classes. The parent class contains the general attributes and methods common to all the child sensor classes. Unified Modeling Language (UML) version 2 standards were followed in constructing Figure 3-3. UML is the software development industry's standard way of displaying graphical models. Knowing how to create useful UML diagrams promotes efficiency and effectiveness in communicating your design ideas to others in the development process.

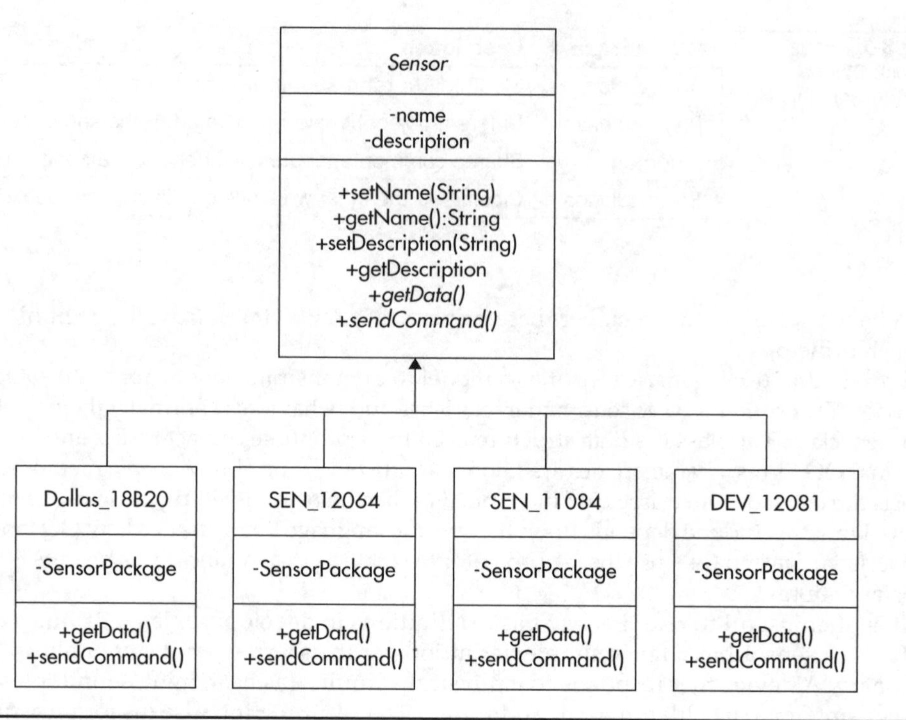

FIGURE 3-3 Sensor UML class diagram

The four child classes, namely `Dallas_18B20`, `SEN_12064`, `SEN_11084`, and `DEV_12081`, get their names either from the manufacturer's model number or the Sparkfun Breakout board number. These four `sensor` classes can be instantiated as they have specific implementations for the methods declared but not implemented in the parent abstract class. Inheritance is very useful in promoting software reuse but it does have its drawbacks. It must be used in a very considered approach to avoid situations where too many unique objects could be created by combining different parent case attributes and/or methods. Interfaces will be discussed as a more elegant way of creating objects without using inheritance.

The concept of *scope* is also important in OOP. Scope is the way OOP enforces encapsulation. Objects "know" things about themselves, i.e., their attributes, and they also know how to do things, their methods or behaviors. You don't want outside entities changing these properties without granting the entities permission. Scope enforces this constraint by setting attributes and methods as private, public, or protected. Private scope means exactly what its name states; attributes and methods are only available within the encompassing class and consequently all objects instantiated from that class. Outside entities cannot change, modify, or delete private attributes or methods. A – sign in front of a UML entry indicates private scope.

Attributes and methods marked as public are available to outside entities. However, attributes themselves are rarely made public as that typically destroys class encapsulation. Public methods, on the other hand, are the way classes allow messages to both be received and sent by class objects. This approach is termed the public interface and for the vast majority of situations, is the way classes are created. In some advanced OOP areas, there are inner classes that do not require a public interface to achieve the desired functionality. Inner classes will not be required for sensor programming. A plus (+) sign in front of a UML entry indicates public scope.

The final type of scope is protected. This is almost identical to private except child objects are permitted access to parent attributes and methods declared protected but no other outside entity is granted permission. Protected scope helps with inheritance class structure implementations, and child objects are always treated as the same type as the parent. From Figure 3-3, the statement can be made that "a `Dallas_18B20` or `SEN_12064` object is a `Sensor` object." Inheritance is always the "is a" relationship.

Another key concept to consider is composition, which is a situation where a class contains attributes that are objects instantiated from other classes. Composition allows you to build complex objects just as real-world things are made up of different components. Consider a car that naturally contains an engine. You can easily imagine a `Car` class, However, it would be a big mistake to create a child class named `Engine`. It would fail the commonsense inheritance test of stating "an Engine is a Car," which is required for true inheritance to exist. However, you can state with confidence, "a Car has an Engine." Composition is the "has a" relationship. Composition and the closely related Aggregation relationship concept are extremely helpful in creating useful and descriptive classes and are needed to successfully implement the interface concept mentioned earlier. I have included a composition class named `SensorPackage` in the child class definitions just to show how this works.

An interface is a specialized class that contains only methods but no attributes. Classes using interfaces can supplement the methods that are declared within the class or a parent class if inheritance is used. Interfaces also support inheritance to allow for specialized method implementations that fit specific subclass requirements. An example is really needed to clarify how interfaces are best used. But first, an explanation of how to set up an Integrated Development Environment (IDE) is in order as that will be the primary way Java code will be created for the Pi project.

Java Software Development Kit (SDK)

The Java SDK is the means by which you install Java on your laptop or desktop system. It is available from www.oracle.com/technetwork/java/javase/downloads/index.html and you should download and install the most current version. Having Java installed is a prerequisite to installing the Eclipse IDE. This IDE is available at www.eclipse.org. Go to the website and download it, just make sure that Java has already been installed on the computer.

Once you have the Eclipse installed and running, you will be able to create and run Java programs very easily. Eclipse also has many tools to help you debug your programs. All the following screenshots were taken from the Eclipse Kepler version 3.9 running on my MacBook Pro.

As is traditional, the first program will be a Hello World example. The Hello program code is shown here for your reference. It is very simple and short. Figure 3-4 shows this program after being run in the Eclipse IDE.

```
package hello;

public class Hello {

     public static void main(String[] args) {
          System.out.println("Hello World!");
     }
}
```

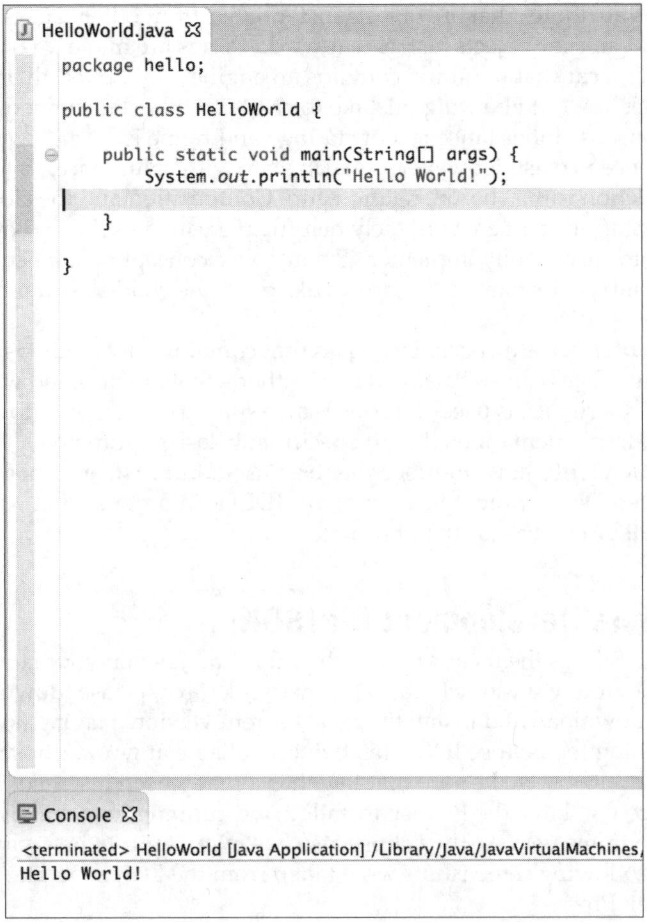

FIGURE 3-4 Eclipse screen for the Hello World program

```
HelloWorld.java

package hello;

public class HelloWorld {

    public static void main(String[] args) {
        System.out.println("Hello World!");

    }

}

Console

<terminated> HelloWorld [Java Application] /Library/Java/JavaVirtualMachines/
Hello World!
```

There are several things to discuss concerning this program:

- The program was created in a Java Project named Hello. All programs created in the Eclipse IDE require a project to be created to contain the program class.
- The class itself is named Hello. No spaces are allowed in a class name, and class names typically start with a capital letter. This class is also contained in a package named hello. Packages are useful for containing multiple classes. The hello package statement is "package hello;" and is always at the start of the coding.
- `public static void main(String() args)` is a method signature that every Java application requires. It is the starting point of any Java application.
- All the method code, referred to as the body, is contained between a pair of braces { … }.
- The only operational code (that does something) is `System.out.println ("Hello World");`. This statement uses the `out` object part of the System class that is part of Java's built-in libraries. The `println()` method displays the text within `println` parentheses to the Console window that is below the code editor window.

The Sensor Abstract Class

The following code is my first attempt at modeling a generic sensor using the guidelines I discussed earlier. There are only two key behaviors defined that will cause the Pi to send a command to the sensor and receive data in response. Both of these behaviors, or methods as I will call them now, are declared as abstract, meaning that any child classes must provide an implementation or the project will not compile and run. Abstract class and method names should be italicized in UML diagrams, as you saw in Figure 3-3.

NOTE *All the code is contained in the sensor package and is available on the book's companion website.*

```
package sensor;
public abstract class Sensor {
      String name, description;
      abstract public void getData();
      abstract public void sendCommand();

      public void setName(String newName) {
            name = newName;
      }
      public String getName() {
            return name;
      }
      public void setDescription(String newDescription) {
            description = newDescription;
      }
      public String getDescription() {
            return description;
      }
}
```

There is a problem inherent in this somewhat naive code, which I will point out later in the discussion. Meanwhile, let's assume everything will function as planned and next work on creating two of the four child classes for demonstration purposes.

Child Classes

I will next demonstrate how to create `Dallas_18B20` and `SEN_12064` child classes and a class to instantiate and test these classes. I'll use println statements as that is the simplest and most effective way of showing object behavior without using actual hardware.

I will also use four of the methods detailed in the `Sensor` class diagram as this will be sufficient to illustrate object interaction and behavior. These methods are

- `getData()`
- `sendCommand()`
- `setName()`
- `setDescription()`

The Java keyword to establish inheritance is "extends," as you can see in the top line of the following class definition for the `Dallas_18B20` sensor object.

```
package sensor;

public class Dallas_18B20 extends Sensor {

      public Dallas_18B20() {
            String desc = "I2C temperature sensor, 5V, range -25 to +125C";
            super.setDescription(desc);
            String name = "18B20 sensor located in the garage";
            super.setName(name);
      }

      SensorPackage pkge = new SensorPackage(0.1, 0.1, 0.1, 1.5);

      public double[] getPackageInfo() {
            return pkge.getPackage();
      }

      public void getData() {
            System.out.println("The 18B20 object would be sending data
to the Pi at this point");
      }

     public void sendCommand() {
      System.out.println("The 18B20 object would be receiving
command data from the Pi at this point");
     }
}
```

The method `Dallas_18B20()` is known as a constructor and is called any time a new object is created from the `Dallas_18B20` class. This constructor as defined takes no arguments. The parent class is referred to as `super` in the constructor.

Two concrete implementations are provided for all the abstract methods defined in the abstract `Sensor` parent class. Java would flag an error (called an exception) if you did not provide them.

All methods are public, which allows an external entity to send and receive messages from a `Dallas_18B20` class object. In this case, the external entity will be an object of the test class.

The `SEN_12064` class is set up in a very similar fashion to the `Dallas_18B20` class and is shown here:

```
package sensor;

public class SEN_12064 extends Sensor{

      public SEN_12064() {
            String desc = "I2C humidity sensor, 5V, range 0 to 100% RH";
            super.setDescription(desc);
            String name = "SEN_12064 sensor located in the garage";
            super.setName(name);
      }

      SensorPackage pkge = new SensorPackage(0.28, 1.2, 0.1, 2.4);

      public double[] getPackageInfo() {
            return pkge.getPackage();
      }

      public void getData() {
            System.out.println("The SEN_12064 object would be sending data
to the Pi at this point");
      }

    public void sendCommand() {
      System.out.println("The SEN_12064 object would be receiving command
data from the Pi at this point");
    }
}
```

The test class shown is named `TestSensors` and is the only class in the package that contains the `main()` method, which is the Java application's starting point.

```
package sensor;

public class TestSensors {

      public static void main(String[] args) {
            String desc;
            //instantiate a garage temperature sensor name gts
            Dallas_18B20 gts = new Dallas_18B20();
```

```
            //instantiate a garage humidity sensor name ghs
            SEN_12064 ghs = new SEN_12064();
            //get the description from the gts object
            desc = gts.getDescription();
            System.out.println(desc);
            //send a start command to the gts object
            gts.sendCommand();
            //have the gts object send its data
            gts.getData();
            //line space here to pretty up the display
            System.out.println();
            //get the description from the ghs object
            desc = ghs.getDescription();
            System.out.println(desc);
            //send a start command to the ghs object
            ghs.sendCommand();
            //have the ghs object send its data
            ghs.getData();
      }
}
```

The Console output for this test class is shown in Figure 3-5. It should be easy to see how messages are sent and received between the instantiated objects, `Dallas_18B20` and `SEN_12064` in this case.

The following is an example of sending a message:

```
gts.sendCommand();
```

Here, the `TestSensors` class is requesting `gts`, an object of the `Dallas_18B20` class, to accept the `sendCommand()`. A preset message will be displayed and no real sensor data will be taken. As mentioned earlier, objects know how to do things and, in this case, initiate a temperature measurement.

You may have noticed that sending the same message to `Dallas_18B20` and `SEN_12064` sensor objects produces different displays. This is an example of polymorphic behavior that is another of the fundamental OO principles. The behavior changes are hard coded into the `sendCommand()` and `getData()` methods for each child class. This way of

```
Console
<terminated> TestSensors [Java Application] /Library/Java/JavaVirtualMachines/jdk1.7.0_40.jdk/Contents/Home/bin/java (Mar 15, 2014, 3:35:23 PM)
I2C temperature sensor, 5V, range -25 to +125C
The 18B20 object would be receiving command data from the Pi at this point
The 18B20 object would be sending data to the Pi at this point

I2C humidity sensor, 5V, range 0 to 100% RH
The SEN_12064 object would be receiving command data from the Pi at this point
The SEN_12064 object would be sending data to the Pi at this point
```

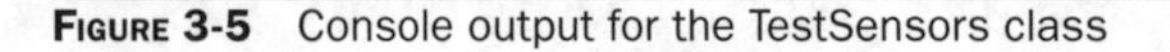

FIGURE 3-5 Console output for the TestSensors class

structuring your code leads to substantial difficulties if you decide to add additional sensor types with different ways of triggering measurement cycles (`sendCommand()`) and reporting the actual data (`getData()`) behaviors. Code designed this way is considered fragile and is easily broken in the sense that it is hard to make changes and is difficult to maintain. The following approach is offered as another way to add different class behaviors that is both easy to maintain and very adaptable to change.

Break out the trigger measurement and data reporting behaviors and put them into two abstract interface classes. Then create a series of implementation classes that extend the behavior interfaces with each implementation class supporting the type of behavior desired for a `Sensor` subclass. Figure 3-6 illustrates this interface design. It is the same basic class diagram as shown in Figure 3-3 with the interfaces added. Note the dashed line associations between the `Sensor` child classes and the interface child classes. This is how the `Dallas_18B20` and `SEN_12064` will get their implementations regarding the `start()` command, which is simply the `sendCommand()` relabeled as it makes more sense. Similarly, I renamed the `getData()` method to `transfer()`, which more precisely describes its behavior.

Don't be deterred by the apparent complexity shown in Figure 3-6. This design is very robust and may easily accommodate adding different sensor classes with varying `CommandBehavior` and `DataBehavior` interfaces. For instance, you can easily add an

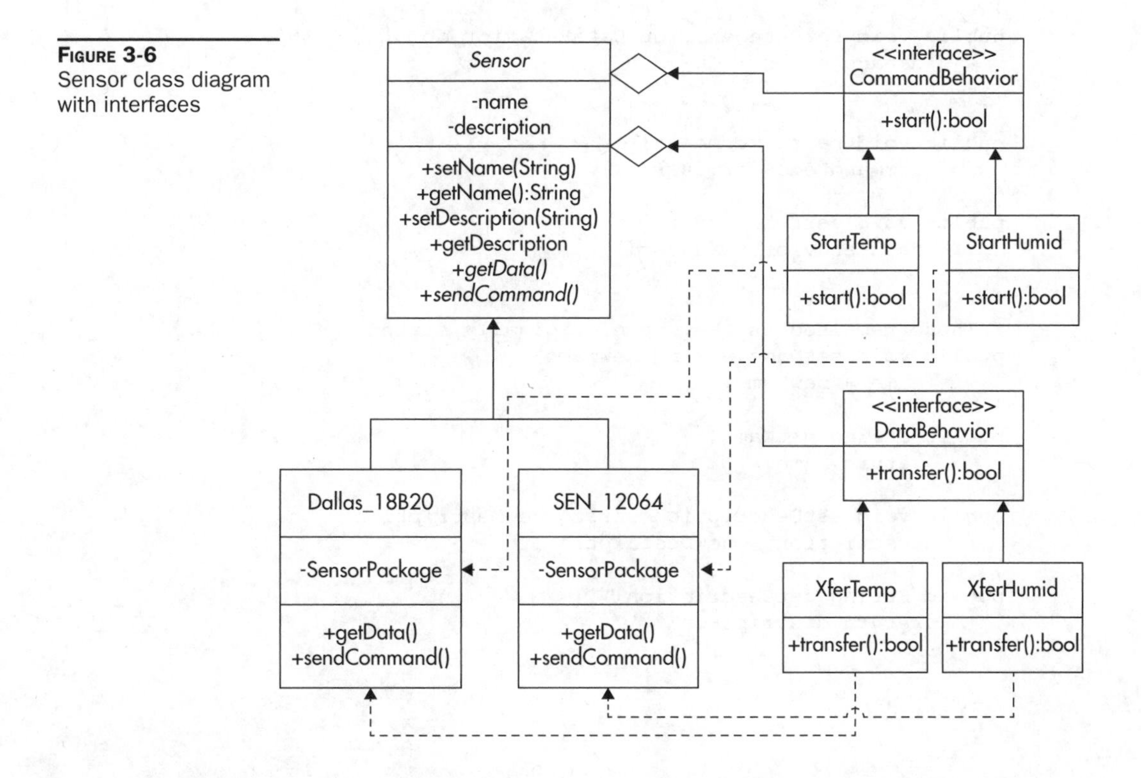

FIGURE 3-6 Sensor class diagram with interfaces

anemometer that measures wind speed, which certainly functions much differently than a Dallas_18B20 or SEN_12064 sensor. All you need to do is add additional implementation classes to the CommandBehavior and DataBehavior interfaces and a new subclass to the abstract Sensor parent class.

The next sequence of listings shows some of the code created to show how the interfaces are created and connected to the classes that require specific behaviors. The following code is the revised Sensor class with two new interface references, commandBehavior and dataBehavior, declared near the beginning of the class. The Sensor subclasses will use these references to associate to the correct command or data behaviors appropriate to their sensor type.

Incorporating these two references is an example of composition, which means that Sensor objects have both command and data behaviors whose actual implementations will be provided by other classes.

```
package sensor;

public abstract class Sensor {
      String name, description;

      CommandBehavior commandBehavior;
      DataBehavior dataBehavior;

      public void setCommandBehavior(CommandBehavior cb) {
            commandBehavior = cb;
      }
      public void setDataBehavior(DataBehavior db) {
            dataBehavior = db;
      }

      public void performCommand() {
            commandBehavior.start();
      }
      public void performData() {
            dataBehavior.transfer();
      }

      //These remained as they were originally stated
      public void setName(String newName) {
            name = newName;
      }
      public String getName() {
            return name;
      }
      public void setDescription(String newDescription) {
            description = newDescription;
      }
      public String getDescription() {
            return description;
      }
}
```

There are also two new methods in this class: `performCommand()` and `performData()` will be used in the test class to actually invoke the appropriate behavior via the `Sensor` superclass.

The following code is the revised `Dallas_18B20` class that is used to instantiate `Dallas_18B20` objects. Notice that the `sendCommand` and `getData` methods have been commented out as these behaviors are now delegated to interfaces. Also, in the constructor I have set the interface references `commandBehavior` and `dataBehavior` to TempStart and XferTemp child classes, respectively. This sets up the correct responses to command and data messages.

```
package sensor;

public class Dallas_18B20 extends Sensor {

      public Dallas_18B20() {
            String desc = "I2C temperature sensor, 5V, range -25 to +125C";
            super.setDescription(desc);
            String name = "18B20 sensor located in the garage";
            super.setName(name);
            //Select the appropriate starting behavior child class
            commandBehavior = new TempStart();
            //Select the appropriate data transfer behavior
            dataBehavior = new XferTemp();
      }

      /*no longer needed
      public void getData() {
            System.out.println("The SEN_12064 object would be sending data
to the Pi at this point");
      }
          public void sendCommand() {
                System.out.println("The SEN_12064 object would be
receiving command data from the Pi at this point");
         }*/

      //Remains the same
      SensorPackage pkge = new SensorPackage(0.1, 0.1, 0.1, 1.5);

      public double[] getPackageInfo() {
            return pkge.getPackage();
      }
}
```

One of the new high-level interfaces is shown in the following listing. It is very simple, consisting of only one method that will be implemented in a lower class:

```
package sensor;

public interface CommandBehavior {
      public Boolean start();
}
```

One of the new lower level interface implementation classes is shown in the following listing. It is associated with the `commandBehavior` interface, which can readily be seen by the `implements` keyword in the class definition header:

```
package sensor;

public class TempStart implements CommandBehavior {
      public Boolean start() {
            System.out.println("Starting the temperature sensor to acquire
data");
            return true;
      }
}
```

The test class is similar to the previous test class used without the interfaces. The next listing shows the complete class, which is used to demonstrate how the `Sensor` subclasses and the interfaces work together to produce the desired results.

```
package sensor;

public class TestSensorsWithInterfaces {

      public static void main(String[] args) {
            String desc;
            //instantiate a garage temperature sensor name gts
            Dallas_18B20 gts = new Dallas_18B20();
            //instantiate a garage humidity sensor name ghs
            SEN_12064 ghs = new SEN_12064();
            //get the description from the gts object
            desc = gts.getDescription();
            System.out.println(desc);
            //start the gts using an interface
            gts.performCommand();
            //have the gts object send its data
            gts.performData();
            //line space here to pretty up the display
            System.out.println();
            //get the description from the ghs object
            desc = ghs.getDescription();
            System.out.println(desc);
            //start the ghs using an interface
            ghs.performCommand();
            //have the ghs object send its data
            ghs.performData();

      }
}
```

The Console output generated by executing the preceding `TestSensorsWithInterfaces` program in the Eclipse IDE is shown in Figure 3-7.

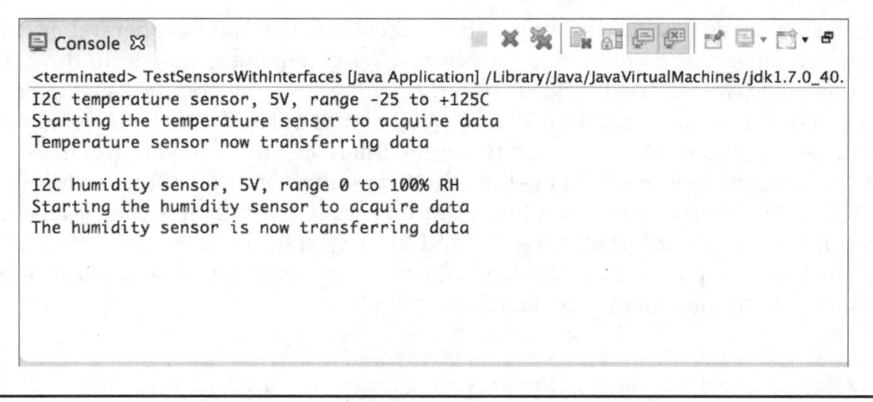

FIGURE 3-7 Console output for the `TestSensorsWithInterfaces` class

The previous discussion was lengthy but necessary to illustrate a good approach to creating code that is understandable and fairly easy to change to accommodate changing requirements. Another important principle underpinning this methodology is to:

Favor composition over inheritance.

Using the interfaces with the child implementation classes is a concrete example of applying this principle. The code was made robust by recognizing that command and data behaviors will be different for different sensor types. So instead of trying to hard-code specific types of behavior in each of the concrete `Sensor` subclasses, I choose to pull that all out and place it in the interfaces. This apparent recognition of object differences leads to one more principle:

Identify what varies and encapsulate it.

Command and data vary for each type of sensor so those behaviors were pulled out of the `Sensor` class and delegated to interfaces. Then appropriate interface implementation subclasses were created to invoke the specific behavior for that `Sensor` object with that type of behavior. Very simple and somewhat elegant!

I have saved the best for last in this discussion. The approach I have taken with creating this code is an example of the Strategy design pattern. Design patterns (DP) are often introduced into advanced computer science courses, but I believe that if you learn the patterns early on, you will develop good code design practices that will help you in all your future development efforts. DPs in software development have been around since 1995 when the *Design Patterns*[1] book was published. DPs are not specific solutions to specific problems but are a methodical approach for creating good solutions given a general type of problem domain. They are based on years of great software development by masters in this field. As the old saying goes, "standing on the shoulders of giants."

In the `Sensor` class definition, two additional methods were added that have not yet been discussed: `setCommandBehavior(CommandBehavior cb)` and `setDataBehavior(DataBehavior db)`. These methods allow `Sensor` objects to reference both command

[1] Gamma, Helm, Johnson and Vlissides, *Design Patterns*, Addison-Wesley Pub., 1995

and data behaviors as needed during program execution. This modus operandi is called *dynamic* or *late binding* as compared to establishing fixed references that occur during the compilation stage, which is called *static* or *early binding*. Dynamic binding is interesting from the perspective that sensors can change their command and data behaviors to suit real-time conditions. I have created two more interface implementation classes to illustrate dynamic binding. Don't get too excited as all they do is output a slightly different Console message compared to the original class. The two new classes created are `StartExtendedTempDataOnTheFly` and `StartExtendedHumidDataOnTheFly`. The demonstration is part of the `TestSensorsWithInterfacesDynamic` test class where the following statements do the dynamic binding:

```
gts.setCommandBehavior(new StartExtendedTempDataOnTheFly);
ghs.setCommandBehavior(new StartExtendedHumidDataOnTheFly);
```

The revised test class is shown here, and the corresponding Console output is shown in Figure 3-8.

```
package sensor;

public class TestSensorsWithInterfacesDynamic {

        public static void main(String[] args) {
                String desc;
                Sensor senseItem;
                //instantiate a garage temperature sensor name gts
                Dallas_18B20 gts = new Dallas_18B20();
                //instantiate a garage humidity sensor name ghs
                SEN_12064 ghs = new SEN_12064();
                //get the description from the gts object
                desc = gts.getDescription();
                System.out.println(desc);
                //start the gts using an interface
                gts.performCommand();
                //have the gts object send its data
```

```
Console
<terminated> TestSensorsWithInterfacesDynamic [Java Application] /Library/Java/JavaVirtualMachines/jdk1.7.0_40.jdk/Contents/Home/bin/java (Mar 16, 2014, 3:17:34 PM)
I2C temperature sensor, 5V, range -25 to +125C
Starting the temperature sensor to acquire data
Temperature sensor now transferring data

I2C humidity sensor, 5V, range 0 to 100% RH
Starting the humidity sensor to acquire data
The humidity sensor is now transferring data

Starting the temperature sensor to acquire extended data
This action initiated dynamically

Starting the humidity sensor to acquire extended data
This action initiated dynamically
```

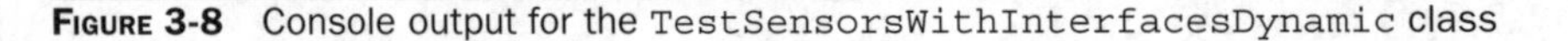

Figure 3-8 Console output for the `TestSensorsWithInterfacesDynamic` class

```
            gts.performData();
            //line space here to pretty up the display
            System.out.println();
            //get the description from the ghs object
            desc = ghs.getDescription();
            System.out.println(desc);
            //start the ghs using an interface
            ghs.performCommand();
            //have the ghs object send its data
            ghs.performData();
            System.out.println();
            //Now do some dynamic allocations to the existing objects gts
            //and ghs
            gts.setCommandBehavior(new StartExtendedTempDataOnTheFly());
            ghs.setCommandBehavior(new StartExtendedHumidDataOnTheFly());
            //Display the dynamically allocated behavior for the
            //temperature sensor
            senseItem = gts;
            senseItem.performCommand();
            System.out.println();
            //Display the dynamically allocated behavior for the humidity
            //sensor
            senseItem = ghs;
            senseItem.performCommand();

      }
}
```

Some additional text, "This action initiated dynamically," is displayed immediately after the line "Starting the temperature sensor to acquire extended temperature data." Those statements came from the `StartExtendedTempDataOnTheFly` class. A similar display is shown for the `StartExtendedHumidDataOnTheFly` case. Dynamic binding is also considered polymorphic behavior in that sending exactly the same message invokes a different response from the target objects. In this case, the message (or method call) is `senseItem.performCommand()` called on the `gts (Dallas_18B20)` and the `ghs (SEN_12064) Sensor` objects. Notice that I used a parent class reference named `senseItem` that I declared is the beginning of the class and set it first to `gts` and then to `ghs` prior to calling the `performCommand` method. You may use a parent reference to refer to any of its child classes but the opposite is not valid and will generate a compile error.

On-the-fly reconfiguration of sensor behavior provides a huge amount of flexibility in coping with real-time situations. Consider the case where a reconnaissance sensor has been deployed into a battlefield situation. Sometimes field conditions change, such as the addition of smoke or fog, making the normal sensors ineffective. Reconfiguring the sensors to accommodate the new environment would allow the recon sensor to continue its mission. Installing new behavior to match the mission is easily accomplished with dynamic binding.

Dynamic binding is simply not possible using a strict inheritance structure. The added flexibility that interfaces provide is yet another reason that most developers favor using interfaces over inheritance.

Real-World Controls

All the actions in the previous code examples have been in the form of print line statements, which are fine for explaining the concepts and principles but not so great in actually interfacing with real sensors. In the next chapter, I will be using code that will directly function with sensors to both initialize and receive data from them. I will be using a great library named Pi4J, which contains all the needed code to make it work as desired. I cover this library in greater depth in the following chapter.

Threads

I have included this discussion of Java threads because I use them in the following projects and because the feature is an important one as using threads promotes efficiency in program operations. Programs typically start with some initialization code, proceed through a sequence of instructions, and terminate with a stop sequence. This is called a program's *execution path* and is also known as a *process* or *task*. Having a program run as a single task or thread may be effective but not too efficient. Fortunately, Java supports multiple threads that enable very efficient program execution. Achieving efficiency would happen by decomposing a program into multiple tasks where each task can be assigned to a dedicated thread.

Threads can be created by either inheritance or implementing an interface. Extending the Java Thread class would be one way to create a threaded class but is not the recommended approach. Strict thread inheritance limits application flexibility for the same reasons mentioned earlier in the chapter. Using interfaces is the preferred way of creating threads. Java provides the Runnable interface, which has the thread behavior desired.

The following is a code snippet that shows how a class named `ThreadDemo` could extend `Thread`. `ThreadDemo` is also known as a target class in this development. The complete set of source code is available on the book's companion website.

```
import java.lang.*;

public class ThreadDemo extends Thread
{
        public void run()
        {
           // code for specific run behavior
        }
}
```

The `run()` method overrides the `Thread` class `run()` method with specific behavior associated with the `ThreadDemo` class. Implementing specific behavior in this manner creates the same situation that existed with the abstract `Sensor`, `Dallas_18B20`, and `SEN_12064` classes with the `sendCommand()` and `getData()` methods. The target class, having extended the thread class, cannot extend any other class, thus limiting the choices in developing the sensor application. The key to improving this situation is to use an interface that will encapsulate the `run()` behavior. The following code snippet shows how this may be accomplished:

```
import java.lang.*;

public class ThreadDemo implements Runnable
{
        Thread thread;

        public void run()
        {
            // code for specific run behavior
        }
}
```

There is now much greater flexibility using the Runnable interface as the target class may now extend another class and you can create a variety of Runnable implementation classes. The Runnable interface is very simple, as you can see from the Java language definition:

```
package java.lang;

public interface Runnable {
        public abstract void run();
}
```

Another interesting fact is that the `Thread` class itself also implements the Runnable interface. This proves that you gain nothing by extending the `Thread` class, yet you lose the flexibility of extending the target class. The following is one of two target classes created for a thread demo:

```
package threads;

public class JobA implements Runnable{
      public void run(){
            System.out.println("Now printing from A");
      }

}
```

Here's the second:

```
package threads;

public class JobB implements Runnable{
      public void run(){
            System.out.println("Now printing from B");
      }
}
```

Here is the ThreadDemo test class:

```
package threads;

public class ThreadDemo{

      Thread threadA, threadB;

      public static void main(String[] args) {
            System.out.println("Start of thread demo");
            Runnable threadJobA = new JobA();
            Runnable threadJobB = new JobB();

            Thread threadA = new Thread(threadJobA);
            Thread threadB = new Thread(threadJobB);

            threadA.start();
            threadB.start();

            System.out.println("Back in main");

      }

      public void run(){

      }

}
```

What follows now are transcribed snippets from the Console displays. I did this to minimize the number of screenshots in this section.

```
Start of thread demo
Back in main
Now printing from A
Now printing from B
```

Here's another:

```
Start of thread demo
Back in main
Now printing from B
Now printing from A
```

Here's another:

```
Start of thread demo
Now printing from A
Back in main
Now printing from B
```

Not quite what you expected! The output can vary depending on how the JVM controls the threads. A thread scheduler within the JVM controls when threads are run and when they are not. Threads have three states:

- new
- runnable
- running

A thread is in a *new* state after being instantiated but not yet "started." A thread is in the *runnable* state when the `start()` method has been called on it, but the JVM is not running it. Of course, a thread is in the *running* state when the JVM is running it.

You cannot force the thread scheduler to run a specific thread at a given time. This can lead to unpredictable program behavior, which can have problematic outcomes given a high-speed real-time operating system. But there is a solution to this dilemma that can bring a degree of uniformity to controlling threads. Although you cannot directly control the thread scheduler, you can put a thread to sleep such that the controller cannot put it into a running state until the sleep period has expired. Sleep times are integer numbers representing milliseconds that the thread is sleeping or suspended. Putting one thread to sleep in the ThreadDemo program ensures a consistent output.

Here are the modified target classes:

```
package threads;

public class JobA implements Runnable{
      public void run(){
            try{
                  Thread.sleep(50);
                  System.out.println("Now printing from A");
            }
            catch(InterruptedException e){
                  e.printStackTrace();
            }
      }

}
```

```
package threads;

public class JobB implements Runnable{
      public void run(){
            try{
                  Thread.sleep(100);
                  System.out.println("Now printing from B");
            }
            catch(InterruptedException e){
                  e.printStackTrace();
            }
      }
}
```

Notice that the JobA thread sleeps for 50 ms and the JobB thread sleeps for 100 ms. This yields the following consistent Console output:

```
Start of thread demo
Back in main
Now printing from A
Now printing from B
```

The output from main is always first because it doesn't sleep while JobA is sleeping for 50 ms and JobB sleeps for 100 ms. Now swapping the sleep times between the two target classes yields the following consistent Console output:

```
Start of thread demo
Back in main
Now printing from B
Now printing from A
```

Again, this makes sense if JobA is sleeping for 100 ms while JobB is sleeping for 50 ms.

Java provides yet another way of controlling thread behavior to some degree: *method synchronization*. A synchronized method cannot be interrupted, once started; it must run to completion. This constraint of running without interruption is what makes the method "atomic." This is not atomic in the classic nuclear sense, but in the idea that its operation is indivisible. Using synchronized methods in our example doesn't make much sense as the target classes contain only one method that does anything: `println`. Let's keep the synchronized modifier in mind as the sensor control methods or behaviors become more complex and you need to use threads for program efficiency.

Java Database Connector

In this section, you will create a Java database connection and execute queries on it, as demonstrated in the last chapter, which was done using a Python database connector. I will be using the laptop's OS version of the database connector to show you how this all works, but in the next chapter I will revert to the Raspberry Pi version to implement a database connection for the chapter project.

The Java database connector is aptly named Connector/J and is available for download at https://dev.mysql.com/downloads/connector/j/.

Ensure that you download the appropriate version for your OS, which in my case was mysql-connectorjava-5.1.29-bin.jar for the MacBook Pro. There are specific installation instructions at the download website on how to properly install the connector to ensure it works with the JDK. I did a bit of research and found that I only had to copy the Connector/J jar file into the following Mac directory to have it function perfectly with the JDK:

```
/Macintosh HD/Library/Java/Extensions
```

I performed the following procedure to test the new MySQL installation, which I highly recommend so you will be comfortable dealing with the database connectivity concepts. You need to first start the MySQL server, which you can do in a variety of ways depending upon the OS it is running on. Using the Terminal window in the MacBook Pro, I started the server by entering the following:

```
sudo /usr/local/mysql/support-files/mysql.server start ↵
```

Once the server is started, you need to log in to it. I logged in as root with no password as this was a "closed" development process and I had no intention of remotely accessing the database. Figure 3-9 shows the Terminal screen for starting the MySQL server and logging in as the "root" user.

I next selected a database that I previously created named `test` with the following command:

```
USE test ↵
```

I also created a table named `tempData` with several fields that match the same ones I used in the previous chapter. I then manually inserted 20 "dummy" records into the table using exactly the same procedures that I demonstrated in Chapter 2. Finally, I displayed all of the table records using the `SELECT *`, which is a type of SQL command. Figure 3-10 shows the Terminal screenshot for the preceding steps. It shows the results of all the actions discussed in this section including table creation, manual record insertion, and record recall and display.

Incidentally, you can log off MySQL by either entering `EXIT`, as I did in the figure, or by entering `quit`, as `EXIT` is just the MySQL alias for the `quit` command. If you wish to stop or restart the MySQL server, enter one of the following as appropriate:

```
sudo /usr/local/mysql/support-files/mysql.server stop ↵
sudo /usr/local/mysql/support-files/mysql.server restart ↵
```

```
donnorris — mysql — 96×24
Last login: Mon Mar 17 11:44:50 on ttys000
Dons-MacBook-Pro:~ donnorris$ sudo /usr/local/mysql/support-files/mysql.server start
Starting MySQL
. SUCCESS!
Dons-MacBook-Pro:~ donnorris$ sudo /usr/local/mysql/bin/mysql
Welcome to the MySQL monitor.  Commands end with ; or \g.
Your MySQL connection id is 1
Server version: 5.6.16 MySQL Community Server (GPL)

Copyright (c) 2000, 2014, Oracle and/or its affiliates. All rights reserved.

Oracle is a registered trademark of Oracle Corporation and/or its
affiliates. Other names may be trademarks of their respective
owners.

Type 'help;' or '\h' for help. Type '\c' to clear the current input statement.

mysql> 
```

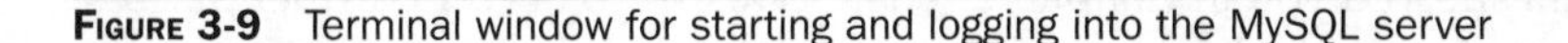

Figure 3-9 Terminal window for starting and logging into the MySQL server

FIGURE 3-10 Terminal window showing the database procedural steps

```
Copyright (c) 2000, 2014, Oracle and/or its affiliates. All rights reserved.

Oracle is a registered trademark of Oracle Corporation and/or its
affiliates. Other names may be trademarks of their respective
owners.

Type 'help;' or '\h' for help. Type '\c' to clear the current input statement.

mysql> USE test;
Reading table information for completion of table and column names
You can turn off this feature to get a quicker startup with -A

Database changed
mysql> SHOW TABLES;
+----------------+
| Tables_in_test |
+----------------+
| tempData       |
+----------------+
1 row in set (0.00 sec)

mysql> SHOW FIELDS FROM tempData;
+-------+--------------+------+-----+---------+----------------+
| Field | Type         | Null | Key | Default | Extra          |
+-------+--------------+------+-----+---------+----------------+
| id    | mediumint(9) | NO   | PRI | NULL    | auto_increment |
| tdate | date         | NO   |     | NULL    |                |
| ttime | time         | NO   |     | NULL    |                |
| tchan | text         | NO   |     | NULL    |                |
| ttemp | text         | NO   |     | NULL    |                |
+-------+--------------+------+-----+---------+----------------+
5 rows in set (0.01 sec)

mysql> SELECT * FROM tempData;
+----+------------+----------+-------+-------+
| id | tdate      | ttime    | tchan | ttemp |
+----+------------+----------+-------+-------+
|  1 | 2014-03-15 | 16:25:49 | 0     | 20.48 |
|  2 | 2014-03-15 | 16:25:59 | 0     | 20.46 |
|  3 | 2014-03-15 | 16:26:09 | 0     | 20.45 |
|  4 | 2014-03-15 | 16:26:19 | 0     | 20.45 |
|  5 | 2014-03-15 | 16:26:29 | 0     | 20.46 |
|  6 | 2014-03-15 | 16:26:39 | 0     | 20.47 |
|  7 | 2014-03-15 | 16:26:49 | 0     | 20.48 |
|  8 | 2014-03-15 | 16:26:59 | 0     | 20.47 |
|  9 | 2014-03-15 | 16:27:09 | 0     | 20.46 |
| 10 | 2014-03-15 | 16:27:19 | 0     | 20.45 |
| 11 | 2014-03-15 | 16:27:29 | 0     | 20.44 |
| 12 | 2014-03-15 | 16:27:39 | 0     | 20.43 |
| 13 | 2014-03-15 | 16:27:49 | 0     | 20.42 |
| 14 | 2014-03-15 | 16:27:59 | 0     | 20.41 |
| 15 | 2014-03-15 | 16:28:09 | 0     | 20.40 |
| 16 | 2014-03-15 | 16:28:19 | 0     | 20.40 |
| 17 | 2014-03-15 | 16:28:29 | 0     | 20.41 |
| 18 | 2014-03-15 | 16:28:39 | 0     | 20.42 |
| 19 | 2014-03-15 | 16:28:49 | 0     | 20.43 |
| 20 | 2014-03-15 | 16:28:59 | 0     | 20.43 |
+----+------------+----------+-------+-------+
20 rows in set (0.00 sec)

mysql> EXIT;
Bye
Dons-MacBook-Pro:~ donnorris$
```

You will find that the interactive session as demonstrated is almost identical to the Pi's interactive session that I showed you in Chapter 2. There is really no need to further discuss this type of session as you should be fairly comfortable in using it. It is time to move on to the more interesting topic of creating and using a Java program database connection.

Using the Java Connector in a Program

This section demonstrates how to create a Java program using Eclipse that connects with the test database and executes SQL statements to access and display all of the tempData records. The following program is named JDBCExample.java and is available on the book's companion website. I have included comments after the program output figure to help explain some of the key program statements.

```
package database;

import java.sql.*;

public class JDBCExample {
      // JDBC driver name and database URL
         static final String JDBC_DRIVER = "com.mysql.jdbc.Driver";
         static final String DB_URL = "jdbc:mysql://localhost/";

         //  Database credentials
         static final String USER = "root";
         static final String PASS = "";

         public static void main(String[] args) {

               Connection conn = null;
               Statement stmt = null;

         try{
            //Register JDBC driver
            Class.forName("com.mysql.jdbc.Driver");

            //Open a connection
            System.out.println("Creating a connection object");
            conn = DriverManager.getConnection(DB_URL, USER, PASS);

            //Execute a query
            System.out.println("Connecting to the test db");
            stmt = conn.createStatement();

            String sql = "USE test";
            stmt.executeUpdate(sql);

            sql = "SELECT id, tdate, ttime, tchan, ttemp FROM tempData";
            ResultSet rs = stmt.executeQuery(sql);
            //Extract data from result set
            while(rs.next()){
               //Retrieve by column name
               int id  = rs.getInt("id");
               String date = rs.getString("tdate");
               String time = rs.getString("ttime");
               String chan = rs.getString("tchan");
               String temp = rs.getString("ttemp");
```

```
               //Display values
               System.out.print(id);
               String out = "   " + date + "   " + time + "   " + chan + "   "
+ temp;
               System.out.println(out);
            }
            rs.close();
            System.out.println();
         }

         catch(SQLException se){
            //Handle errors for JDBC
            se.printStackTrace();
         }catch(Exception e){
            //Handle errors for Class.forName
            e.printStackTrace();
         }finally{
            //finally block used to close resources
            try{
               if(stmt!=null)
                  stmt.close();
            }catch(SQLException se2){
            }// nothing we can do
            try{
               if(conn!=null)
                  conn.close();
            }catch(SQLException se){
               se.printStackTrace();
            }//end finally try
         }//end try
         System.out.println("Goodbye!");
      }//end main
      }//end JDBCExample
```

Figure 3-11 shows the console output that results from executing the program. Notice that the record listing is almost exactly the same as the Figure 3-10 listing, which I manually initiated in an interactive session.

These two declarative statements set up references for `Connection` and `Statement` objects, which are required to establish a database connection:

```
Connection conn = null;
Statement stmt = null;
```

The JDBC driver must be registered before either one of these objects can be activated. The driver registration process consists of loading the Oracle driver's class file into memory so it can be utilized as a JDBC interface implementation.

Registration is only required once in the program. The common approach to register the driver is to use Java's `Class.forName()` method to dynamically load the driver's class file into memory. This action automatically registers it. This method is preferable because it allows the driver registration to be both configurable and portable.

```
Class.forName("com.mysql.jdbc.Driver");
```

FIGURE 3-11 Console output that results from executing the program.

```
Console
<terminated> JDBCExample [Java Application] /Library/Java/JavaVirtualMachines/
Creating a connection object
Connecting to the test db
1  2014-03-15  16:25:49  0  20.48
2  2014-03-15  16:25:59  0  20.46
3  2014-03-15  16:26:09  0  20.45
4  2014-03-15  16:26:19  0  20.45
5  2014-03-15  16:26:29  0  20.46
6  2014-03-15  16:26:39  0  20.47
7  2014-03-15  16:26:49  0  20.48
8  2014-03-15  16:26:59  0  20.47
9  2014-03-15  16:27:09  0  20.46
10  2014-03-15  16:27:19  0  20.45
11  2014-03-15  16:27:29  0  20.44
12  2014-03-15  16:27:39  0  20.43
13  2014-03-15  16:27:49  0  20.42
14  2014-03-15  16:27:59  0  20.41
15  2014-03-15  16:28:09  0  20.40
16  2014-03-15  16:28:19  0  20.40
17  2014-03-15  16:28:29  0  20.41
18  2014-03-15  16:28:39  0  20.42
19  2014-03-15  16:28:49  0  20.43
20  2014-03-15  16:28:59  0  20.43

Goodbye!
```

The `Connection` and `Statement` objects can now be assigned after the driver class file is successfully registered. Assignment is done by the following statements:

```
conn = DriverManager.getConnection(DB_URL, USER, PASS);
stmt = conn.createStatement();
```

SQL queries can now be executed against the "connected" database once these objects are assigned. The following statement performs the actual query:

```
stmt.executeUpdate(sql);
```

where `sql` represents a SQL string command such as `"USE test"` or `"SHOW TABLES"`.

The records actually extracted from a database table are in the form of a resultset, which is an array of records that match the conditions in the SQL query string. For this example program, the following statements extract all the records from the `tempData` table:

```
sql = "SELECT id, tdate, ttime, tchan, ttemp FROM tempData";
ResultSet rs = stmt.executeQuery(sql);
```

I could also have restated the `sql` string as:

```
sql = "SELECT * FROM tempData";
```

This would have returned the same resultset but I wanted to emphasize all the fields that were involved in the original query statement.

Using the record data is quite straightforward as you can see from a portion of the loop that extracts individual field data from a record:

```
while(rs.next()){
     //Retrieve by column name
     int id  = rs.getInt("id");
     String date = rs.getString("tdate");
     String time = rs.getString("ttime");
```

The `rs.next()` method returns a record from the resultset as long records are still available that have not been read. When all the records have been accessed, the `next()` method returns a null and the `while` loop will stop. Meanwhile, the field variables such as `id`, `tdate`, and `ttime` now hold the current values from the selected record. They have to be converted to normal java variables, which you do with the loop assignment expressions shown in the preceding code snippet. These new variables can now be handled as ordinary Java variables without regard to their database origin. The `print` and `println` statements use these new variables to display all the selected records.

I would also like to point out the try and catch blocks that surround most of the code in the example program. One of Java's great features is the ability to deal with problems, more technically known as *exceptions*, during a program's execution. The try block covers the code that could encounter an exception, such as not finding the desired database. I intentionally changed the database name from test to test1 to test the program's exception handling. Figure 3-12 shows the result when the program cannot find the deliberately misnamed database.

The collection of error messages is known as *stack traceback*, which is very handy for Java developers to identify the source of the original exception that caused the program to abnormally stop or cease running. In this case it is easy to see the problem's source, which is shown on the top line of the traceback with the statement `"Unknown database 'test1' "`.

This concludes my introduction to a Java database connector, which is all that you really need before putting it to use in an interesting project, which I describe in the next chapter.

```
Console
<terminated> JDBCExample [Java Application] /Library/Java/JavaVirtualMachines/jdk1.7.0_40.jdk/Contents/Home/bin/java (Mar 17, 2014, 7:19:23 PM)
Creating a connection object
Connecting to the test db
com.mysql.jdbc.exceptions.jdbc4.MySQLSyntaxErrorException: Unknown database 'test1'
        at sun.reflect.NativeConstructorAccessorImpl.newInstance0(Native Method)
        at sun.reflect.NativeConstructorAccessorImpl.newInstance(NativeConstructorAccessorImpl.java:57)
        at sun.reflect.DelegatingConstructorAccessorImpl.newInstance(DelegatingConstructorAccessorImpl.java:45)
        at java.lang.reflect.Constructor.newInstance(Constructor.java:526)
        at com.mysql.jdbc.Util.handleNewInstance(Util.java:411)
        at com.mysql.jdbc.Util.getInstance(Util.java:386)
        at com.mysql.jdbc.SQLError.createSQLException(SQLError.java:1054)
        at com.mysql.jdbc.MysqlIO.checkErrorPacket(MysqlIO.java:4237)
        at com.mysql.jdbc.MysqlIO.checkErrorPacket(MysqlIO.java:4169)
        at com.mysql.jdbc.MysqlIO.sendCommand(MysqlIO.java:2617)
        at com.mysql.jdbc.MysqlIO.sqlQueryDirect(MysqlIO.java:2778)
        at com.mysql.jdbc.ConnectionImpl.execSQL(ConnectionImpl.java:2828)
        at com.mysql.jdbc.StatementImpl.executeUpdate(StatementImpl.java:1842)
        at com.mysql.jdbc.StatementImpl.executeUpdate(StatementImpl.java:1764)
        at database.JDBCExample.main(JDBCExample.java:30)
Goodbye!
```

FIGURE 3-12 Exception handling output

Summary

At the beginning of the chapter, I answered the question, "What is an object?" I also introduced the four core object-oriented (OO) principles that form the basis of all OO languages.

I next used a unified modeling language (UML) diagram to illustrate the higher-level structure of my first attempt to develop a Java application that would initialize and then read data from sensors.

The chapter then covered class definitions along with various levels of scope descriptions. Important inheritance and composition principles were discussed, and I pointed out that I would attempt to always use composition over inheritance. Doing so promotes program flexibility and maintainability.

I next discussed the Eclipse integrated development environment (IDE), which was used on a laptop to create and run the Java programs demonstrated in the chapter. I took you through all the steps of creating and executing a traditional "Hello World" program to illustrate how easy it is to use Eclipse.

I then showed you how to create a class named `Sensor` with appropriate child classes that were based solely on an inheritance class structure. I later showed you how such a structure is considered "fragile" and easily broken if changes are made to the software.

Next, I demonstrated a much more robust approach, which was based on using interfaces. Although it appeared a bit complex at the outset, I showed that it was relatively straightforward and much more robust compared to the original version that was based on strict inheritance.

I let you in on a little secret that the approach I used was based on the Strategy design pattern (DP), which is one of many powerful approaches that are available to software developers. DPs have been formally documented since the mid '90s but likely have been in existence for much longer. Using a DP approach to software development is a professional approach that I urge all readers to embrace if at all possible. This section concluded with a demonstration of how to add additional behaviors dynamically, or "on-the-fly," without breaking or modifying any existing code.

The next section dealt with threads, which I wanted to alert you to because they are very useful to "speed up" a program and they go a long way toward promoting program efficiencies. Improving program efficiency is an important topic for Raspberry Pi operations as its CPU normally operates at a 700 MHz clock rate, which is well below the speed of typical laptop/desktop systems.

The final chapter section concerned the Java database connector, which is required to logically connect with a MySQL database as was done in the previous chapter using a Python connector. I demonstrated essentially the same database application as I showed in the previous chapter except I used the Connector/J database connector in lieu of the Python variety.

You are now well prepared to tackle the next chapter's Java-based project.

CHAPTER 4

Home Weather Station

This chapter's project will show you how to build a home weather station that will report the basic meteorological conditions in your area, including temperature and pressure. It will also have provisions for an activation contact where you can remotely activate systems such as a garage door opener or an air conditioning system from a web browser. This project builds on many of the Java concepts and techniques introduced in the last chapter. I have a very simple reason for using Java in this project—to acquaint you with a small portion of the many thousands of classes that are already written and available to be used in your own applications. Another important rationale to use Java is the "write once, run every" (WORE) philosophy that is a bedrock Java principle. This means that you only have to create the Java source code one time and it is available to run on a multitude of different platforms. I will use the Eclipse IDE introduced in the last chapter to develop most of the non-GPIO source code on the laptop. An IDE such as Eclipse could be loaded directly onto the Pi but I do not recommend it. It would run very slowly and likely not support some of the Eclipse functionality found on a full-scale laptop version. Besides, why not take advantage of the key Java WORE concept I mentioned earlier.

Of necessity, there will be some source code that must be developed directly on the Pi, which will normally involve the Pi4J classes I introduce later in the chapter. This approach is required because of the low-level nature of the Pi4J classes, which means that they're primarily hardware driver–type classes. This approach is the command line method, which is how I have presented commands so far in this book.

This project will also incorporate a MySQL database as I saw no reason to introduce further complexity by using a different database application. Most of the relational open source databases function in a similar manner and use almost identical SQL statements. As the old saying goes, *You know one, you know them all.* I am being a bit facetious here but it is largely true for most object-oriented languages. Python and Java are quite a bit different in implementation but both hold true to the core object-oriented principles.

Java and GPIO

Part of the weather station project will use some of the Pi's general-purpose input/output (GPIO) pins, which are available via the 26-pin header, named P1 and shown in Figure 4-1.

Figure 4-1 Raspberry Pi 26 pin GPIO header connector

A GPIO pin is a single-bit digital port, which normally may act either as an input or an output, depending upon how it has been configured. The voltage levels that can be applied as an input or sourced as an output are either 0 or 3.3VDC. Exceeding 3.3V with an input level will likely permanently damage the specific pin and may even permanently damage the whole Pi unit. There are level converter chips available that safely convert between 5V sensors and the 3.3V that the Pi GPIO pins accept.

GPIO Pin Labeling

There has been a bit of confusion surrounding how the GPIO pins (and others) are identified since the Pi was introduced. The confusion arises from the way the manufacturer, Broadcom, labeled the pins and how the Raspberry Foundation also labeled the same pins. The Broadcom labeling is also referred to as the chipset or native pin labeling and is directly related to the actual pin labels on the system on a chip (SOC). These labels are generally referred to as the BCM2835 set or, more simply, the BCM numbers, while the Raspberry Pi Foundation labels are referred to as the RasPi GPIO pin label designations.

The whole pin labeling issue became a bit more muddled when the rev 2 boards were introduced and several P1 pins were relabeled. The wiringPi project introduced the abstract pin label layer concept, which is what you will use in this Java project.

Also note that the physical pin numbers only relate to the P1 connector and are not related to any logical pin designation. They are simply consecutive numbers from 1 to 26 and representing the physical pin position on the P1 connector. The first two and last two

FIGURE 4-2 Ribbon cable attached to the Raspberry Pi's P1 connector

numbers are shown on Figure 4-1 and are quite important in the sense that you must align the matching ribbon connector or you will misconnect the pins to the external circuit. The P1 connector does not have an alignment slot so you should use the stripe on the ribbon to identify pin 1. Figure 4-2 shows the proper alignment of a 26-wire ribbon cable attached to P1.

I have provided Table 4-1 for your reference, which shows the correspondence between the BCM2835 and RasPi pin labeling.

RasPi Pin Number	RasPi Description	BCM Description	RasPi Pin Number	RasPi Description	BCM Description
1	3V3	3V3	2	5V0	5V0
3	SDA0 *	GPIO0	4	5V0	5V0
5	SCL0 *	GPIO1	6	GND	GND
7	GPIO_GCLK	GPIO4	8	TXD0 *	GPIO14
9	GND	GND	10	RXD0 *	GPIO15
11	GPIO_GEN0	GPIO17	12	GPIO_GEN1	GPIO18
13	GPIO_GEN2	GPIO21	14	GND	GND
15	GPIO_GEN3	GPIO22	16	GPIO_GEN4	GPIO23
17	3V3	3V3	18	GPIO_GEN5	GPIO24
19	SPI_MOSI *	GPIO10	20	GND	GND
21	SPI_MISO *	GPIO9	22	GPIO_GEN6	GPIO25
23	SPI_SCLK *	GPIO11	24	SPI_CE0_N *	GPIO8
25	GND	GND	26	SPI_CE1_N *	GPIO7

TABLE 4-1 GPIO Pin Descriptions

Neither the BCM2835 or the RasPi pin labels are used in this chapter's project as the Java program code uses only the wiringPi designations, which also have been accepted and extended by the Java Pi4J code library discussed later in this chapter. BCM pin labeling is often used in Python program development, as you saw in Chapter 2. Table 4-2 shows the P1 connector with wiringPi pin designations alongside BCM and header pin designations.

Please note the differences between board revisions 1 and 2, shown as R1 and R2 in Table 4-2.

The following changes where effective for the Raspberry Pi PCB Revision 2:

- GPIO_GEN2 [BCM2835/GPIO27] was routed to P1 pin 13.
- Changed what was SCL0/SDA0 to SCL1/SDA1.
- SCL1 [BCM2835/GPIO3] was routed to P1 pin 5.
- SDA1 [BCM2835/GPIO2] was routed to P1 pin 3.

wiringPi Pin	BCM GPIO	Name	Header	Name	BCM GPIO	wiringPi Pin
–	–	3V3	**1 | 2**	5V0	–	–
8	R1:0/R2:2	SDA1	**3 | 4**	5V0	–	–
9	R1:1/R2:3	SCL1	**5 | 6**	GND	–	–
7	4	GPIO07	**7 | 8**	TXD0	14	15
–	–	GND	**9 | 10**	RXD0	15	16
0	17	GPIO00	**11 | 12**	GPIO01	18	1
2	R1:21/R2:27	GPIO02	**13 | 14**	GND	–	–
3	22	GPIO03	**15 | 16**	GPIO04	23	4
–	–	3V3	**17 | 18**	GPIO05	24	5
12	10	MOSI	**19 | 20**	GND	–	–
13	9	MISO	**21 | 22**	GPIO06	25	6
14	11	SCLK	**23 | 24**	CE0	8	10
–	–	GND	**25 | 26**	CE1	7	11

Legend

3V3 3.3VDC	**SDA*x*** I2C Data Bus Number *x*
5V0 5.0VDC	**SCL*x*** I2C Clock Bus Number *x*
GND Ground	**MOSI** Serial Peripheral Interface Master Out Slave In
GPIO*xx* General Purpose Input Output pin *xx*	**MISO** Serial Peripheral Interface Master In Slave Out
TXD0 UART Serial Transmit	**SCLK** Serial Peripheral Interface Clock
RXD0 UART Serial Receive	**CE*x*** Chip Enable Number *x*

TABLE 4-2 P1 GPIO Connector

- The power and ground connections previously marked "Do Not Connect" on P1 remain as connected, specifically:
  - **P1-04:** +5V0
  - **P1-09:** GND
  - **P1-14:** GND
  - **P1-17:** +3V3
  - **P1-20:** GND
  - **P1-25:** GND

Figure 4-3 is an excellent reference diagram that was taken from the Pi4J website, which clearly shows all the P1 pins along with the wiringPi designations that are used in the Java interface software.

You should crosscheck your connections with Table 4-2 and Figure 4-3 any time that you are directly interfacing to the P1 connector.

Raspberry Pi P1 Header

Pin #	Name			Name	Pin #
	3.3 VDC Power	1	2	5.0 VDC Power	
8	SDA0 (12C)	3	4	DNC	
9	SCL0 (12C)	5	6	0V (Ground)	
7	GPIO 7	7	8	TxD	15
	DNC	9	10	RxD	16
0	GPIO0	11	12	GPIO1	1
2	GPIO2	13	14	DNC	
3	GPIO3	15	16	GPIO4	4
	DNC	17	18	GPIO5	5
12	MOSI	19	20	DNC	
13	MISO	21	22	GPIO6	6
14	SCLK	23	24	CE0	10
	DNC	25	26	CE1	11

http://www.pi4j.com

FIGURE 4-3 P1 GPIO pin connector

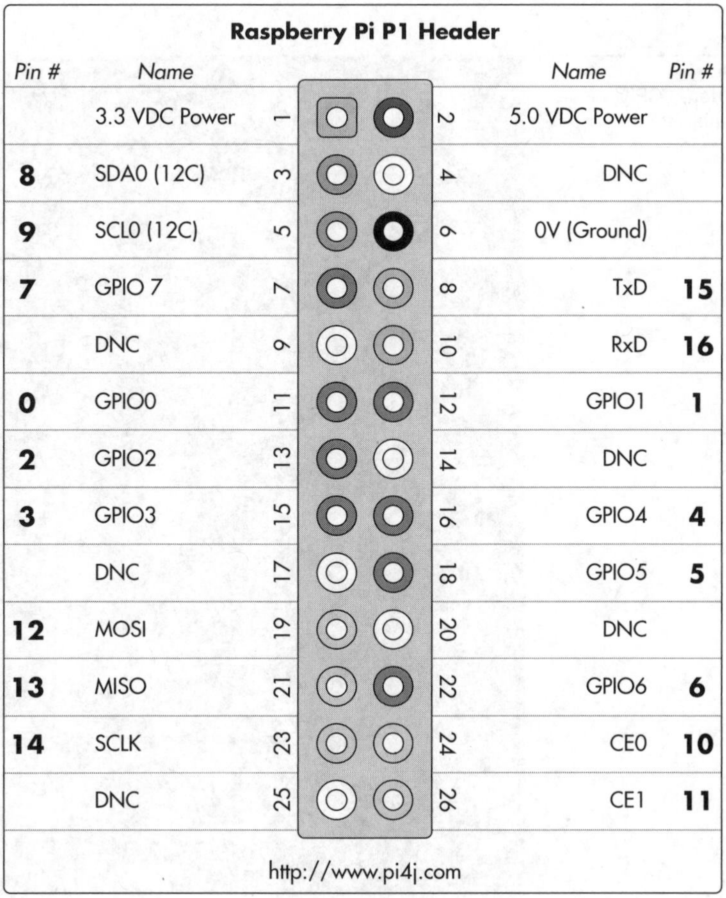

GPIO Pin Expansion

The Pi rev 2 boards also contain four additional GPIO pins that are available from unpopulated pin headers. There are eight plated PCB holes identified in the documentation as P5 and shown in Figure 4-4 located adjacent to the P1 GPIO connector.

Table 4-3 describes all these additional pins with their wiringPi and the BCM designations. Each side has three columns. The outermost column, wiringPi Pin, refers to the pin number in the wiringPi code. The middle one, BCM GPIO, refers to the pin number of the BCM2835 chip, which is the chipset address. The innermost column, Name, is the name of the function of the pin and is the same as the RasPi description.

Pin 1 is the square plated hole located in the upper-left corner of P5. You will need to install a 12-pin connector to access the pins. The connector is supposed to be installed on the board's underside per Note 3 on the rev 2.0 board schematic available at www.raspberrypi.org/wp-content/uploads/2012/10/Raspberry-Pi-R2.0-Schematics-Issue2.2_027.pdf. A suggested connector is shown in Figure 4-5. None of the projects in this book require these extra pins but it is nice to know that they are available if needed.

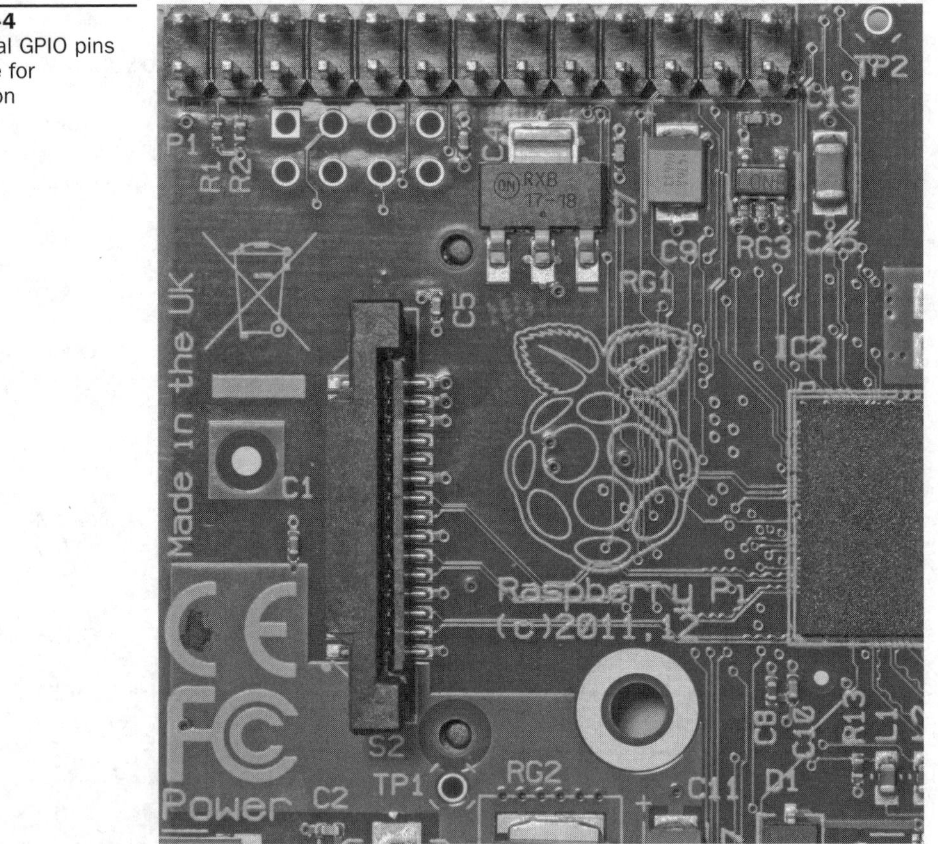

FIGURE 4-4 Additional GPIO pins available for expansion

wiringPi Pin	BCM GPIO	Name	Header	Name	BCM GPIO	wiringPi Pin
–	–	5V0	**1 | 2**	3V3	–	–
17	28	GPIO08	**3 | 4**	GPIO09	29	18
19	30	GPIO10	**5 | 6**	GPIO11	31	20
–	–	GND	**7 | 8**	GND	–	–
Legend						
3V3 3.3VDC						
5V0 5.0VDC						
GND Ground						
GPIO*xx* General Purpose Input Output pin *xx*						

TABLE 4-3 P5 Auxiliary GPIO Connector (Rev. 2 Boards Only)

Figure 4-6 is an excellent reference diagram that was taken from the Pi4J website, which clearly shows all the P5 pins. I do not use these pins but they are readily available for any project that requires these additional GPIO interfaces.

FIGURE 4-5 P5 GPIO pin expansion connector

FIGURE 4-6 P5 connector pins with labels

Raspberry Pi P5 Header

Pin #	Name			Name	Pin #
	3.3 VDC Power	2	1	5.0 VDC Power	
18	GPIO18	4	3	GPIO17	**17**
20	GPIO20	6	5	GPIO19	**19**
	0V (Ground)	8	7	0V (Ground)	

http://www.pi4j.com

Interrupts

Java programs created to interface with sensors and/or actuators are considered real-time programs as they must respond to ongoing conditions and events. This type of activity requires that the GPIO pins must be able to detect events that happen such as button or keystroke presses as well as preset changes in sensor values. There are two ways that the Pi can detect these changes:

- Polling
- Interrupts

Polling is a programmed function where the Pi is set in a loop and regularly tests a GPIO pin(s) to ascertain if the state (pin voltage) has changed. Naturally, this type of event programming is very CPU cycle intensive and probably occupies most of the Pi programmed activity. The program is designed such that once an event is detected, the program will branch to a designated logical location to handle the code that is associated with that particular event. Polling, while easy to implement, is quite wasteful in terms of program efficiency and somewhat limits the Pi in what else it can do besides performing a poll loop. Fortunately, there is an alternative to polling and that is the use of an interrupt. Every GPIO pin can also accommodate an interrupt. An interrupt is an event that stops or "interrupts" the normal programming flow and directs the microprocessor to execute some special handler program or code, depending upon the interrupt source. Pi interrupts may be triggered in several ways, as listed here:

- HIGH level detected
- LOW level detected
- HIGH to LOW transition detected
- LOW to HIGH transition detected

Using them will considerably improve performance but at the expense of adding a certain level of complexity to the software. In addition, there are some general guidelines to be considered when using interrupts:

- *Use an interrupt service routine table.* This is a list of logical locations that direct the program to branch to once a specific interrupt is detected. The Java Virtual Machine (JVM) must be configured to support this feature, but most are not. This feature will not be used in the example program.
- *Keep the handler code short and to the point.* The handler code should contain only the absolute minimum code necessary to service the event and nothing more.
- *Double check any initialization code.* Interrupt handlers are often created as threads with associated priorities. Ensure they are correct and check items such as proper GPIO pin assignments and data direction registers and initial state levels.
- *Minimize disabling interrupts.* Interrupts can be disabled, but this should not ordinarily be done except in the case of an actual interrupt. Interrupting another interrupt is typically not a good idea.

- *Use the key word "volatile" for a shared variable.* The volatile modifier tags a shared variable that might unexpectedly change value during a threaded operation or an interrupt sequence. It forces the JVM to reread the value before any access or assignment involving the shared variable.
- *Be aware of the number of clock cycles used in an interrupt sequence.* Programs may be created with certain expected delays that will not hold true if an interrupt occurs. Developers have to know the effect on preprogrammed timing that an interrupt could possibly cause.
- *Favor the use of interrupts in lieu of polling.* Using interrupts can allow a processor to enter a "sleep" mode to conserve energy and related battery life. Polling keeps the processor constantly "alive," defeating the sleep mode.

I will demonstrate a simple example program using a manually initiated interrupt after I discuss the key Pi4J library in the next section.

Pi4J Library

The Raspberry Pi development community is quite fortunate to have a talented developer named Robert Savage, who freely made available a fairly complete Java class library that implements GPIO functionality. This library includes both high-level application type classes as well as many low-level driver classes. The library is named Pi4J and is available for download at www.pi4j.com. The download and installation of this library on your Pi is crucial for this project to succeed. Please follow this procedure to set up your Pi to control the GPIO pins using Java:

1. Ensure that Oracle's Java JDK is already installed on the Pi. It should be if you are using a Wheezy distribution from September 2013 or later. Enter the following at a command line prompt:

   ```
   java -version ↵
   ```

   Figure 4-7 is the result of this command line query as to the Java version installed on the Pi. Your version may very well be different as upgraded Java versions are likely to be included in the Wheezy distribution in the future.
2. The next step is to download the Pi4J library. I found the simplest way to do this is to first download the SNAPSHOT release named pi4j-1.0-SNAPSHOT.deb onto my laptop from https://code.google.com/p/pi4j/downloads/list. I then copied it into the Pi's home directory using a thumb drive and the Pi's File Manager application.

```
pi@raspberrypi: /opt/pi4j/examples
File Edit Tabs Help
pi@raspberrypi /opt/pi4j/examples $ java -version
java version "1.7.0_40"
Java(TM) SE Runtime Environment (build 1.7.0_40-b43)
Java HotSpot(TM) Client VM (build 24.0-b56, mixed mode)
pi@raspberrypi /opt/pi4j/examples $
```

Figure 4-7 Java version query

3. Once in the home directory, enter the following command to install this SNAPSHOT into the appropriate locations in the Pi:

```
sudo dpkg -i pi4j-1.0-SNAPSHOT.deb ↵
```

NOTE *dpkg is a package manager application designed to unpack and install Debian formatted packages, i.e., those software packages with a .deb file extension.*

4. After the installation is completed, a new directory (pi4j) will be created with two new subdirectories (lib and examples) created within it as follows:

```
/opt/pi4j/lib
/opt/pi4j/examples
```

   The preceding step completed the Pi4J installation, but you should proceed with the next few steps to create all the needed class files and be ready to run the example program, which will in turn confirm that the library functions as expected and is usable for program development.

5. Change into the examples subdirectory by entering the following:

```
cd /opt/pi4j/examples ↵
```

6. Once in the directory, enter the following, which automatically builds all the class files from the existing downloaded source files:

```
./build ↵
```

There were 28 example source files in the download that I made. That number is subject to change as the developers who control the website add and subtract depending upon comments received from the active Pi4J community.

The `./build` command just shown causes a script to run that iterates through all the example source code files to produce corresponding class files. The actual compile command is shown next, which you must use to compile your own source file:

```
sudo javac -classpath .:classes:/opt/pi4j/lib/'*' -d . <sourcefilename>  ↵
```

It is very important that you pay attention to all the symbols and whitespace in the preceding command as leaving anything out or misaligning their placement will cause the compile to fail as I found out much to my frustration.

Enter the following to execute or run a class file:

```
sudo java -classpath .:/classes:/opt/pi4j/lib/'*' <classfilename> ↵
```

Note that you should not enter the `.class` extension in the class filename.

LED Blink Program

A blink program is the introductory embedded hardware equivalent of the "Hello World" type used in pure software development. The particular blink example I used is named BlinkGpioExample.java and written by Robert Savage. The program is part of the set of programs downloaded in the examples directory. This program is far from the simplistic code that ordinarily blinks an LED. It blinks two LEDs but also contains code to respond to a key press. The key press triggers an interrupt that is handled by a method that changes the blink rate of the LEDs. The code listing for this program is shown here, with further explanatory comments following it:

```
/*
* #%L
* **********************************************************************
* ORGANIZATION  :  Pi4J
* PROJECT       :  Pi4J :: Java Examples
* FILENAME      :  BlinkGpioExample.java
*
* This file is part of the Pi4J project. More information about
* this project can be found here:  http://www.pi4j.com/
* **********************************************************************
* %%
* Copyright (C) 2012 - 2013 Pi4J
 * %%
* Licensed under the Apache License, Version 2.0 (the "License");
* you may not use this file except in compliance with the License.
* You may obtain a copy of the License at
*
*      http://www.apache.org/licenses/LICENSE-2.0
*
* Unless required by applicable law or agreed to in writing, software
* distributed under the License is distributed on an "AS IS" BASIS,
* WITHOUT WARRANTIES OR CONDITIONS OF ANY KIND, either express or
*implied.
* See the License for the specific language governing permissions and
* limitations under the License.
* #L%
*/

import com.pi4j.io.gpio.GpioController;
import com.pi4j.io.gpio.GpioFactory;
import com.pi4j.io.gpio.GpioPinDigitalInput;
import com.pi4j.io.gpio.GpioPinDigitalOutput;
import com.pi4j.io.gpio.PinPullResistance;
import com.pi4j.io.gpio.RaspiPin;
import com.pi4j.io.gpio.event.GpioPinDigitalStateChangeEvent;
import com.pi4j.io.gpio.event.GpioPinListenerDigital;

/**
* This example code demonstrates how to perform simple
* blinking LED logic of a GPIO pin on the Raspberry Pi
* using the Pi4J library.
* @author Robert Savage
* some amplifying comments added by D. J. Norris
*/
```

```
public class BlinkGpioExample {

        public static void main(String[] args) throws InterruptedException {
           System.out.println("<--Pi4J--> GPIO Blink Example ... started.");

           // create gpio controller
           final GpioController gpio = GpioFactory.getInstance();

           // provision gpio pin #01 & #03 as an output pins and blink
           final GpioPinDigitalOutput led1 = gpio.provisionDigitalOutputPin
           (RaspiPin.GPIO_01);
           final GpioPinDigitalOutput led2 = gpio.provisionDigitalOutputPin
           (RaspiPin.GPIO_03);
           // provision gpio pin #02 as an input pin with its internal pull
           // down resistor enabled
           final GpioPinDigitalInput myButton = gpio.provisionDigitalInputP
           in(RaspiPin.GPIO_02, PinPullResistance.PULL_DOWN);

         // create and register gpio pin listener
         myButton.addListener(new GpioPinListenerDigital() {
               @Override
             public voidhandleGpioPinDigitalStateChangeEvent
            (GpioPinDigitalStateChangeEvent event)    {
                           if (event.getState().isHigh())
                               { led2.blink(200);
                                 System.out.println("Button pressed"); }
                           else
                               { led2.blink(1000); }
                      } // end of handleGpioPinDigitalStateChange Event method
                           // body
                 } // end of GpioPinListenerDigital implementation
             ); //end of myButton.addListener method signature

             // continuously blink the led1 every .5 seconds for 15 seconds
             led1.blink(500, 15000);
              // continuously blink the led2 every  second
              led2.blink(1000);
              System.out.println(" ... the LED will continue blinking until the
              program is terminated.");
              System.out.println(" ... PRESS <CTRL-C> TO STOP THE PROGRAM.");

              // keep program running until user aborts with a (CTRL-C)
              for (;;) {
                         Thread.sleep(500);
               } // end of the forever loop

              // stop all GPIO activity/threads
              // this method will forcefully shutdown all GPIO monitoring threads
              // and scheduled tasks
              // gpio.shutdown();   uncomment to terminate the Pi4J GPIO
              // controller
```

```
    } // end of main method
} // end of BlinkGpioExample class definition
```

The statement

```
final GpioController gpio = GpioFactory.getInstance();
```

is an excellent representation of a design pattern (DP), which was introduced, in the previous chapter. It represents the Abstract Factory DP, where an instance of a desired object is created on demand. For this case, the object instantiated is a `GpioController`, which is used to manage the GPIO pins. Incidentally, the "final" modifier used in this statement directs the JVM to prevent any modification of this object within the program.

The statement

```
final GpioPinDigitalOutput led1 = gpio.provisionDigitalOutputPin(RaspiPin
.GPIO_01);
```

instantiates the `led1` object, which is also GPIO01 set in an output mode. You need to refer to Figure 4-3 to actually determine which physical pin it is on the P1 header. Similar statements set up led2 as an output and the push button named myButton as an input.

The statement

```
myButton.addListener(new GpioPinListenerDigital() { @Overide
        public void handleGpioPinDigitalStateChangeEvent
        (GpioPinDigitalStateChangeEvent event)   {
        if (event.getState().isHigh())  { led2.blink(200);
        System.out.println("Button pressed"); }
                              else { led2.blink(1000); } }
        // end of handleGpioPinDigitalStateChange Event method
                   } // end of GpioPinListenerDigital implementation
            ); //end of myButton.addListener method signature
```

is a very complex one but can be easily understood if examined piece-by-piece. It is essentially an interrupt handler in the sense that it is the code that will process the event of pushing the external button. It is handy to think of user interactions with a program as events whether they be external button presses, mouse clicks, keyboard button presses, or something else. The `addListener` method is the overall handler but the external button press must be registered to it in order for the JVM to know where to go once this particular event happens or in developer's lingo, "fires off." An object that technically is described as an anonymous inner class because it is not assigned a name is instantiated by the new operator from the `GpioPinListenerDigital` class. This class contains only one method named `handleGpioPinDigitalStateChangeEvent`. It also comes right after the `@Overide` annotation, which is Java's way of saying that you best put something here on how you want to process the event. In the `handle...method` argument is an object named event, which refers to the `GpioPinDigitalStateChangeEvent` class. This object immediately becomes non-null after the button is pressed. The state of the event object may be tested to determine if it is a high or low level. If high, the LED blinking rate is set to 200 ms and if low remains at 1000 ms. I also put a `println` statement in the input pin level test statement code to display the text "Button pressed," just to confirm that it was working as planned.

The main operating code in this class is a forever loop as shown here:

```
// keep program running until user aborts with a (CTRL-C)
for (;;) {
      Thread.sleep(500);
} // end of the forever loop
```

This approach to embedded control programming is very common where the processor simply waits for user interaction with no other design responsibilities to undertake. It is perfectly suited for this test program but likely should be avoided for programs that have more complex requirements such as this chapter's weather station.

You will need to connect a few components to the Pi in order to test this program. Figure 4-8 is a schematic of how the components interconnect. I used a solderless breadboard with a Pi Cobbler to quickly hook up all the parts.

The schematic shows both the header pin numbers and the wiringPi logical names. I recommend that you refer back to Figure 4-3 as you wire this project just to avoid any mistakes in the connections. The actual physical setup is shown in Figure 4-9 with the Pi Cobbler ribbon cable removed for better breadboard visibility.

I ran the blink program for a short time and observed that the LEDs did blink as programmed and the `println` statement "Button pressed" appeared on a terminal window each time the button was pressed. Figure 4-10 is a terminal window screenshot showing a running blink program.

You should now have a properly operating Java development system running on the Pi if you have been successful in completing all the previous steps. The following sections demonstrate how to program the weather station sensors using Java and the Pi4J library.

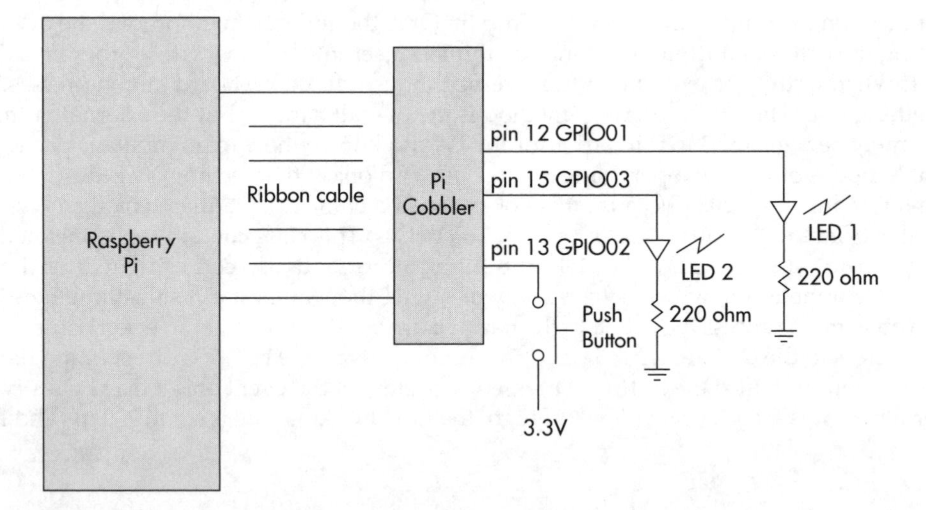

FIGURE 4-8 Blink program test schematic

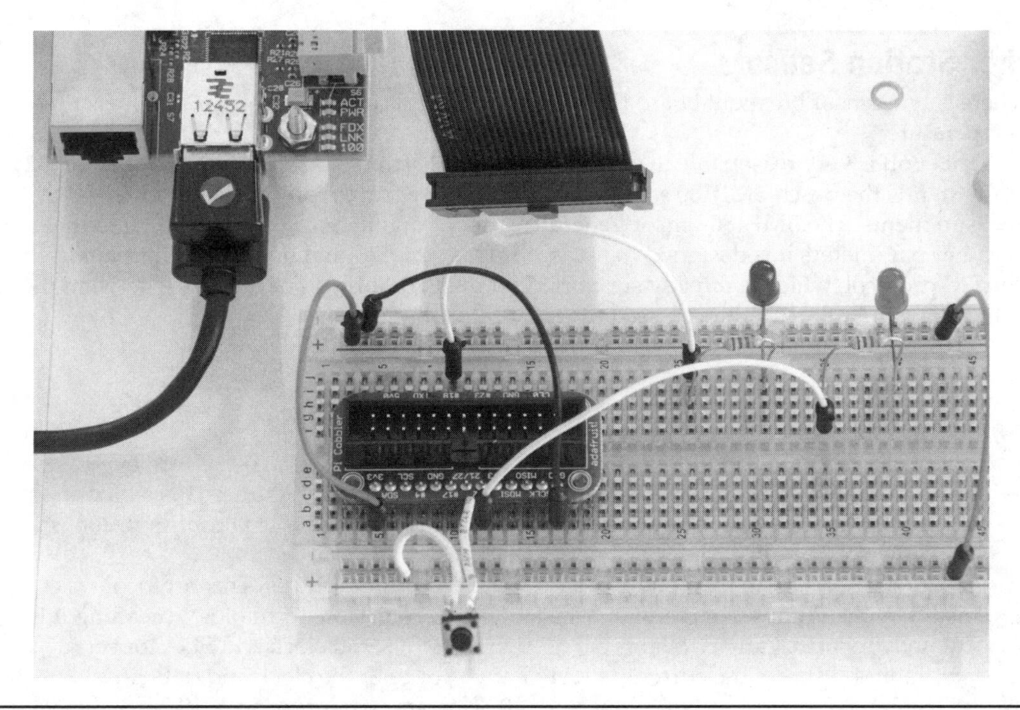

Figure 4-9 Physical setup for the Blink program

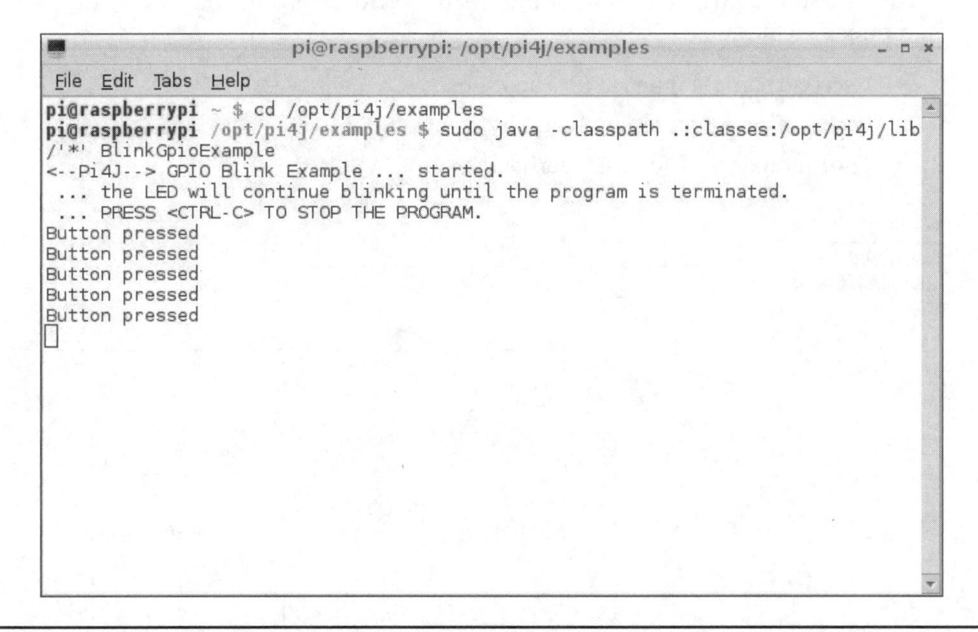

Figure 4-10 Terminal window screenshot for a running blink program

Weather Station Sensors

There is one sensor breakout board (BoB) included in this design: SEN-11824 – pressure & temperature.

This BoB is very reasonable in cost and is available from Sparkfun.com. The SEN-11824 BoB extends the Bosch BMP180 sensor chip for highly accurate pressure and temperature measurements. The BMP180 sensor generates a serial digital output proportional to the weather parameters it is designed to measure. This serial digital output data stream uses the I2C protocol, which is directly supported by the Pi4J library. For readers who want more information on the I2C protocol, see the "I2C Protocol" sidebar.

I2C Protocol

The Inter-Integrated Circuit interface or I2C (pronounced "eye-two-cee" or "eye-squared-cee) is also known as a synchronous serial data link. Figure 4-11 is a block diagram of the I^2C interface showing one master and one slave. This configuration is also known as a multi-drop or bus network.

I2C supports more than one master as well as multiple slaves. This protocol was created by the Philips Company in 1982 and is a very mature technology, meaning it is extremely reliable. Only two lines are used: SCL for a serial clock and SDA for serial data. Table 4-4 shows the current rev 2 Pi names for both the clock and data lines.

The I2C bandwidth is fairly low, with 400 Kbps maximum and 100 Kbps nominal specifications. However, the nominal is more than adequate for most systems as the typical sensor sample interval ranges from 1 to 10 seconds.

A quick bandwidth calculation for the weather station sensors follows. Assumptions include

- 1-second sample interval for each sensor.
- Each sensor generates an 8 data byte packet.
- 4 additional bytes for I2C overhead.

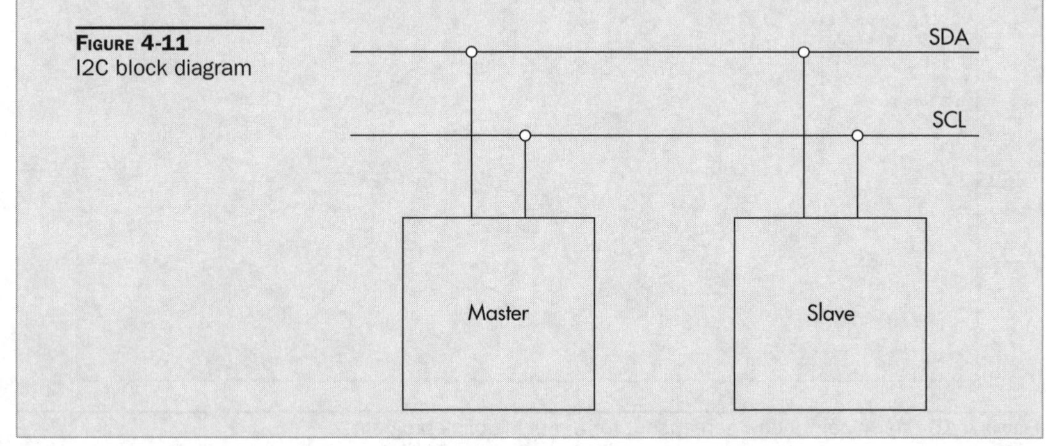

Figure 4-11 I2C block diagram

Signal Name	Description	Pi Name*
SCL	Clock	SCL1
SDA	Data	SDA1
* SCL0 and SDA0 for rev 1 boards		

TABLE 4-4 I2C Signal Lines

The calculation is as follows:
Total number of bits/sec (bps) = Total byte packets * 8 bits/byte * Number sensors * Samples/sec
= 12 * 8 * 3 * 1
= 216 bps or 0.216kps, which is far below the nominal 100 Kbps rate

Sensor Wiring Connections

The sensor BoB requires connections to both a power supply and I2C bus. Figure 4-12 shows the connection points on the SEN_11824. I also soldered header pins to the BoB to make it compatible for use in a solderless breadboard setup. Figure 4-13 shows the SEN_11824 with the header pins attached.

FIGURE 4-12 SEN_11824 connections

Figure 4-13 SEN_11824 with header pins attached

Figure 4-14 is a block diagram showing how the Pi is connected with the SEN_11824 BoB. The data and clock signal lines are connected in parallel. The board is also connected to the 3.3V and ground lines. The physical setup using a Pi Cobbler and a solderless breadboard is shown in Figure 4-15.

Everything from this point on depends upon the software, which is discussed in the following sections.

Figure 4-14 Weather system block diagram

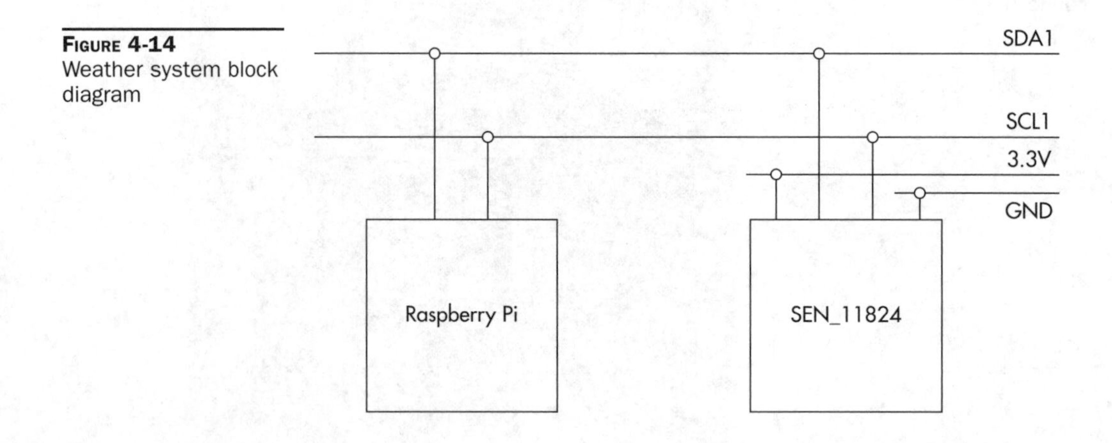

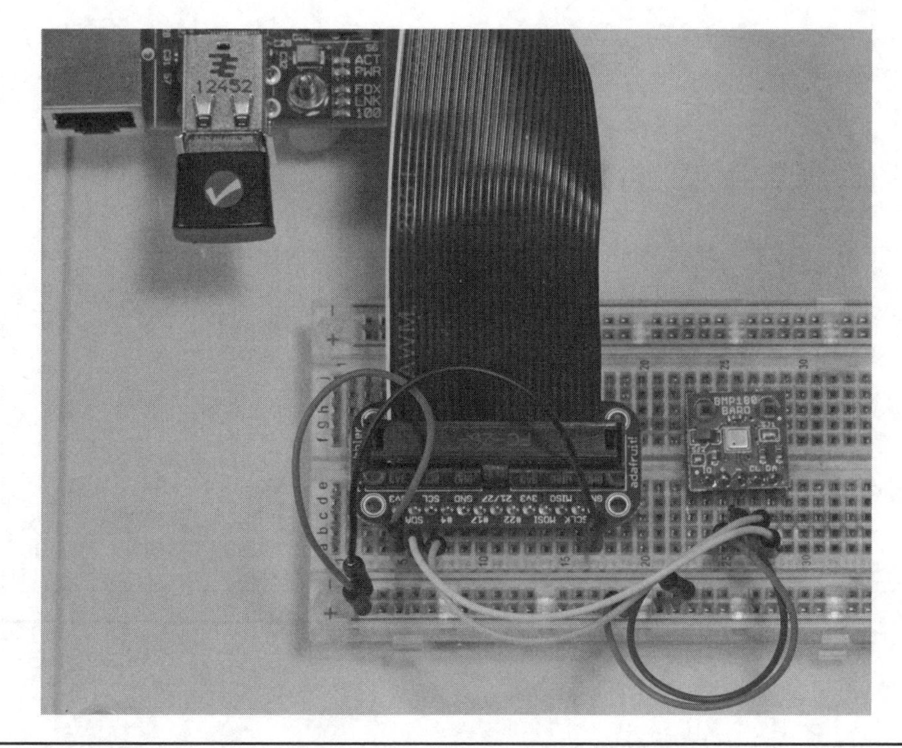

FIGURE 4-15 Physical test system setup

Weather Station Software

The first step in creating the weather station software is to enable the I2C protocol within the Wheezy distribution. It is a fairly simple process:

1. Append two lines to the module file located in the `etc` directory.

```
sudo nano /etc/module ↵
      i2c-bcm2708
      i2c-dev
```

2. Comment out the flowing line in the configuration file raspi-blacklist.conf. No action is required if this file does not already exist.

```
sudo nano /etc/modprobe.d/raspi-blaclklist-conf ↵
      #blacklist i2c-bcm2708
```

3. Install the i2c-tools package.

```
sudo apt-get install i2c-tools ↵
```

```
pi@raspberrypi: /opt/pi4j/examples
File  Edit  Tabs  Help
pi@raspberrypi /opt/pi4j/examples $ sudo i2cdetect -y 1
     0  1  2  3  4  5  6  7  8  9  a  b  c  d  e  f
00:          -- -- -- -- -- -- -- -- -- -- -- -- --
10: -- -- -- -- -- -- -- -- -- -- -- -- -- -- -- --
20: -- -- -- -- -- -- -- -- -- -- -- -- -- -- -- --
30: -- -- -- -- -- -- -- -- -- -- -- -- -- -- -- --
40: -- -- -- -- -- -- -- -- -- -- -- -- -- -- -- --
50: -- -- -- -- -- -- -- -- -- -- -- -- -- -- -- --
60: -- -- -- -- -- -- -- -- -- -- -- -- -- -- -- --
70: -- -- -- -- -- -- -- 77
pi@raspberrypi /opt/pi4j/examples $
```

FIGURE 4-16 Screenshot for the i2cdetect command

4. Test to see that i2c is enabled.

   `sudo i2cdetect -y 1 ↵   (replace the 1 with a 0 for rev 1 boards)`

5. If the test is successful you should see a terminal display similar to what is shown in Figure 4-16. This figure shows the I2C addresses for the SEN_11824 sensor at 0x77.

It is now time to create a Java program, which will read data from the sensors using the I2C protocol, assuming that the I2C address was detected, as shown previously.

Java Software

The Pi4J library makes reading data from the sensors fairly simple by hiding most of the low-level details and allowing the developer to focus on retrieving and processing the measurement data. The BMP180 pressure/temperature sensor that is mounted on the SEN_11824 BoB uses a series of registers to provide access to the sensor data. These register locations are specified in the sensor's technical datasheet, which must be used in order to access the correct data. Table 4-5 shows the registers I used to program the BMP180.

Sensor Classes

All the Java classes are included in a package named sensor and have been modeled after the class diagram shown in Figure 3-15. I have drastically modified the attributes and methods to match this project's minimal requirements, but I have still used an interface to encapsulate the data behavior. The temperature and pressure sensors require a different

TABLE 4-5 BMP180 Relevant Registers

Register Address	Name	Remarks
0xF8	out_xlsb	Read only, ADC bits 0–4
0xF7	out_lsb	Read only, ADC bits 5–12
0xF6	out_msb	Read only, ADC bits 12–20
0xF4	control	Read/write
0xE0	reset	Read/write
0xD0	id	Read only
0xAA to 0xBF	calibration data	Read only

method implementation to start data acquisition and transfer data for which this architecture is well suited. I have included most of the code listings in the text that follows to facilitate a brief discussion on the key points of the project. As always, the complete code sources are available from the book's companion website.

This is the abstract sensor class from which the concrete child sensor classes will inherit the common data behavior:

```
package sensor;

public abstract class Sensor {
      DataBehavior dataBehavior;

      public double performData() {
            double data;
            data = dataBehavior.read();
            return data;
      }
}
```

The following is the SEN_11824 class source code. Notice that the constructor has an integer argument that instantiates the appropriate object type, either temperature or pressure because each one has a different dataBehavior.

```
package sensor;

public class SEN_11824 extends Sensor {

      public SEN_11824(int type) {

            //Select the appropriate type data transfer behavior
            if(type == 0) {
dataBehavior = new XferTemp();
}
else {
      dataBehavior = new XferPress();
}
      }
}
```

The `DataBehavior` interface is shown here; it contains only one method, `read`, which returns a double if the call is successful.

```
package sensor;

public interface DataBehavior {
     public double read();
}
```

The `XferTemp` class is one of the two child classes of the DataBehavior interface that implement its `read` method. It is designed to read two temperature data bytes from the SEN_11824 BoB. This class imports all the relevant Pi4J library classes to communicate via I2C with the board. I provide some further discussion about the class methods after the code listing.

```
package sensor;

import com.pi4j.io.i2c.I2CBus;
import com.pi4j.io.i2c.I2CDevice;
import com.pi4j.io.i2c.I2CFactory;
import java.io.IOException;
import java.io.ByteArrayInputStream;
import java.io.DataInputStream;

public class XferTemp implements DataBehavior {

     public static I2CBus bus;
     public static I2CDevice device;
     final int I2CAddress = 0x77;
     final int errInt = -1;

     static final int startEEPROM = 0xAA;
     static final int endEEPROM = 0xBF;
     static final int numCalBytes = 0x16;
     static final int calOffset = 0;

     //EEPROM registers
     short AC1, AC2, AC3, B1, B2, MB, MC, MD;
     int AC4, AC5, AC6;

     //data variables
     int B5, UT;

     //control register address
     static final byte controlReg = (byte)0xF4;

     //command to get the temperature
     static final byte getTemp = 0x2E;

     //temperature data register address
     static final byte tempAddress = (byte)0xF6;
```

```
//temp data offset
static final int tempOffset = 0;

//error signal
static final double error = -99.0;

DataInputStream inputTempData, calData;

public XferTemp() {
      try {
            bus = I2CFactory.getInstance(I2CBus.BUS_1);
            device = bus.getDevice(I2CAddress);
            //load calibration data
            getCal();
      }
      catch(IOException e) {
            e.printStackTrace();
      }
}

public int getUT() {
      try {
            byte[] tempBytes = new byte[2];
            device.write(controlReg, getTemp);
            Thread.sleep(250);
            device.read(tempAddress, tempBytes, tempOffset, 2);
            inputTempData = new DataInputStream
           (newByteArrayInputStream(tempBytes));
            UT = inputTempData.readUnsignedShort();
            return UT;
      }
      catch(IOException e) {
            e.printStackTrace();
            return errInt;
      }
      catch(InterruptedException e) {
            e.printStackTrace();
            return errInt;
      }
} //end of getUT() method

public double read() {
      try {
            double temp;
            byte[] tempBytes = new byte[2];
            //write the getTemp command to the BMP180
            device.write(controlReg,  getTemp);
            //Give the chip some time to configure itself
            Thread.sleep(250);
            //read the uncompensated temperature and check if 2 bytes
            //were returned
```

```
            int rawDataCount = device.read(tempAddress, tempBytes,
            tempOffset, 2);

            if( rawDataCount != 2)
            {
                  System.out.println("Error: read " + rawDataCount +
                  " bytes");
                  //no sense in continuing
                  return error;
            }

            //get the actual temp data (UT) from the input byte stream
            inputTempData = new DataInputStream
            (newByteArrayInputStream(tempBytes));
            UT = inputTempData.readUnsignedShort();

            //calculate the compensated temperature
            int X1 = ((UT - AC6) * AC5) >> 15;
            int X2 = (MC << 11) / (X1 + MD);
            B5 = X1 + X2;
            temp = ((B5 + 8) >> 4) / 10;
              return temp;
      } // end try block
      catch(IOException e) {
            e.printStackTrace();
            return error;
      }
      catch(InterruptedException e) {
            e.printStackTrace();
            return error;
      }
}// end of read() method

      public void getCal() {
      try {
            byte[] calBytes = new byte[numCalBytes];
            //read cal bytes and check if the proper numbers were read
            int calTotal = device.read(startEEPROM, calBytes,
            calOffset, numCalBytes);

            if(calTotal != 0x16)
            {
                System.out.println("Read " + calTotal + " cal bytes");
            }

            //assign the calibration data variables to the data just
            //read
            calData = new DataInputStream(newByteArrayInputStream
            (calBytes));
            //first three byte pairs are signed int's
            AC1 = calData.readShort();
            AC2 = calData.readShort();
            AC3 = calData.readShort();
```

```
                //next three are unsigned int's
                AC4 = calData.readUnsignedShort();
                AC5 = calData.readUnsignedShort();
                AC6 = calData.readUnsignedShort();
                //last four byte pairs are signed int's
                B1 = calData.readShort();
                B2 = calData.readShort();
                MB = calData.readShort();
                MC = calData.readShort();
            } // end of try block
            catch(IOException e) {
                e.printStackTrace();
            }
        } // end of getCal() method
} // end of the class
```

This class is quite lengthy and I considered separating the calibration portion into a separate class but decided against that as it would overly complicate class design, and the calibration was a reasonably small snippet of code. This is one area where interested and motivated readers might attempt to refactor the code to "abstract out" the calibration routine and replace it with a separate interface in similar fashion to how the DataBehavior interface was structured. Making decisions as to what should or should not belong in a class definition is an issue that should always be considered as the software is developed. Just be mindful that a class should only have one or two responsibilities. Assigning many functions to a single class will enviably lead to overly complex code that is hard to debug and maintain.

The functions of the following three methods within this class are discussed next:

- `getCal()`
- `read()`
- `getUT()`

Method `getCal()` reads in 22 data bytes that are stored in the BMP180 chip EEPROM. A series of calibration constants are then calculated based on the 11 integer values derived from the 22 bytes. The rationale and procedure for the calibration factors are detailed in a PDF document, available from http://wmrx00.sourceforge.net/Arduino/BMP085-Calcs.pdf, that covers both the BMP180 temperature and pressure measurements.

Method `read()` is the override implementation of the DataBehavior interface method of the same name. It reads in 2 data bytes representing the BMP180 temperature, applies the calibration factors, and returns a double representing the compensated temperature.

Method `getUT()` returns the integer representing the raw, uncompensated temperature value. This value is required for the pressure calculation done in the `XferPres` class.

The other DataBehavior child class is named `XferPres`. The `XferPres` is similar to the `XferTemp` class as it also interfaces with same SEN_11824 BoB as does the `XferTemp` class. However, it transfers 3 data bytes to represent the 21-bit ADC pressure measurement as compared to the 2 bytes, or 16 bits, for the temperature measurement. The pressure calibration algorithm is considerably different than the temperature version, which lends some more support to the interface consideration mentioned previously. I have provided the complex code listing to show that it is involves many intricate calculations to provide the true BMP180 pressure measurement.

```
package sensor;

import com.pi4j.io.i2c.I2CBus;
import com.pi4j.io.i2c.I2CDevice;
import com.pi4j.io.i2c.I2CFactory;
import java.io.IOException;
import java.io.ByteArrayInputStream;
import java.io.DataInputStream;

public class XferPres implements DataBehavior {

      public static I2CBus bus;
      public static I2CDevice device;
      final int I2CAddress = 0x77;
      final int errInt = -1;

      static final int startEEPROM = 0xAA;
      static final int endEEPROM = 0xBF;
      static final int numCalBytes = 0x16;
      static final int calOffset = 0;

      //EEPROM registers
      short AC1, AC2, AC3, B1, B2, MB, MC, MD;
      int AC4, AC5, AC6, B5;

      double p, up, s, x, y, z, c3, c4, b1, c5, c6, mc, md, x0, x1, x2, y0,
      y1, y2, p0, p1, p2, T, a;

      //int to store the raw temp
      int UT;
      //control register address
      static final byte controlReg = (byte)0xF4;
      //command to get the pressure
      static final byte getPres = 0x34;
      //pressure data register address
      static final byte presAddress = (byte)0xF6;
      //pressure data offset
      static final int presOffset = 0;
      //error signal
      static final double error = -99.0;

      DataInputStream inputPresData, calData;

      XferTemp tempNow = new XferTemp();
      public XferPres() {
            try {
                  bus = I2CFactory.getInstance(I2CBus.BUS_1);
                  device = bus.getDevice(I2CAddress);
```

```
            //get the current uncompensated temp as an int
            UT = tempNow.getUT();
            //load calibration data
            getCal();
        }
        catch(IOException e) {
            e.printStackTrace();
        }
    }

    public double read() {
        try {
            byte[] presBytes = new byte[3];
            //write the getPres command to the BMP180
            device.write(controlReg, getPres);
            //Give the chip some time to configure itself
            Thread.sleep(250);
            //read the uncompensated pressure and check if 3 bytes
            //were returned
            int rawDataCount = device.read(presAddress,
            presBytes,presOffset, 3);

            if( rawDataCount != 3)
            {
                System.out.println("Error: read " + rawDataCount +
                " bytes");
                //no sense in continuing
                return error;
            }

            //get the actual pressure data (pu) from
            //the input byte stream
            inputPresData = new DataInputStream
            (new ByteArrayInputStream(presBytes));

            byte[] data = new byte[3];
            data[0] = inputPresData.readByte();
            int d0 = 0xFF & data[0];
            data[1] = inputPresData.readByte();
            int d1 = 0xFF & data[1];
            data[2] = inputPresData.readByte();
            int d2 = 0xFF & data[2];
            up = d0 * 256 + d1 + d2 / 256;
            a = c5 * (UT -c6);
            T = a + (mc / (a + md));
            s = T -25;
            x = (x2 * Math.pow(s, 2)) + (x1 * s) + x0;
            y = (y2 * Math.pow(s, 2)) + (y1 * s) + y0;
            z = (up  - x) / y;
            p = p2 * Math.pow(z, 2) + (p1 + z) + p0;
              return p;
```

```
        } // end try block
        catch(IOException e) {
             e.printStackTrace();
             return error;
        }
        catch(InterruptedException e) {
             e.printStackTrace();
             return error;
        }
    }// end of read() method

        public void getCal() {
        try {
             byte[] calBytes = new byte[numCalBytes];
             //read cal bytes and check if the proper numbers were read
             int calTotal = device.read(startEEPROM, calBytes,
             calOffset, numCalBytes);

             if(calTotal != 0x16)
             {
                 System.out.println("Read " + calTotal + " cal bytes");
             }

             //assign the calibration data variables to the data
             //just read
             calData = new DataInputStream(newByteArrayInputStream
             (calBytes));
             //first three byte pairs are signed int's
             AC1 = calData.readShort();
             AC2 = calData.readShort();
             AC3 = calData.readShort();
             //next three are unsigned int's
             AC4 = calData.readUnsignedShort();
             AC5 = calData.readUnsignedShort();
             AC6 = calData.readUnsignedShort();
             //last four byte pairs are signed int's
             B1 = calData.readShort();
             B2 = calData.readShort();
             MB = calData.readShort();
             MC = calData.readShort();
             MD = calData.readShort();

             //polynomial calculations
             c3 = 160.0 * Math.pow(2, -15) * AC3;
             c4 = Math.pow(10, -3) * Math.pow(2, -15) * AC4;
             b1 = Math.pow(160, 2) * Math.pow(2, -30) * B1;
             c5 = (Math.pow(2, -15) / 160) *AC5;
             c6 = AC6;
             mc = (Math.pow(2, 11) / Math.pow(160, 2)) *MC;
             md = MD /160.0;
```

```
                x0 = AC1;
                x1 = 160.0 * Math.pow(2, -13) * AC2;
                x2 = Math.pow(160, 2) * Math.pow(2, -25) * B2;
                y0 = c4 * Math.pow(2, 15);
                y1 = c4 * c3;
                y2 = c4 * b1;
                p0 = (3791.0 - 8.0) /1600.0;
                p1 = 1.0 - 7357.0 * Math.pow(2, -20);
                p2 = 3038.0 * 100.0 * Math.pow(2, -36);
            } // end of try block
            catch(IOException e) {
                e.printStackTrace();
            }
        } // end of getCal() method
    } // end of the class
```

The methods `getCal()` and `read()` perform similar functions for the `XferPress` class as the identical methods did in the `XferTemp` class. This pressure calibration method is considerably more complex compared to the temperature calibration, and the class also uses an uncompensated temperature value, as discussed previously.

The following is a simplified test class named `TestTemp1`, which returns both temperature and pressure readings at 10-second intervals. This test class checks if the I2C protocol was functioning as expected and sensor data was being generated. Enter the following at the command prompt in order to run this program:

```
sudo java -classpath .:/classes:/opt/pi4j/lib/'*' sensor/TestTemp1 ↵
```

This command presumes you have put the statement `package sensor` as the first line in each of the class source files. You will get the 'Java runtime class not found' error if you have missed including the package statement in any of the source files.

```
package sensor;

import java.text.*;

public class TestTemp1 {
    final static int TEMP_TYPE = 0;
    final static int PRES_TYPE = 1;
    final static double MB_TO_INHG = 0.0295299830714;
    static double tempData;
    static double presData;

    public static void main(String[] args) throws InterruptedException {

        SEN_11824 tempSensor = new SEN_11824(TEMP_TYPE);
        SEN_11824 presSensor = new SEN_11824(PRES_TYPE);
        DecimalFormat df = new DecimalFormat("#.##");
        while(true) {
            tempData = tempSensor.performData();
            presData = presSensor.performData();
            System.out.println("Temp = " + df.format(tempData) + "
            deg C");
```

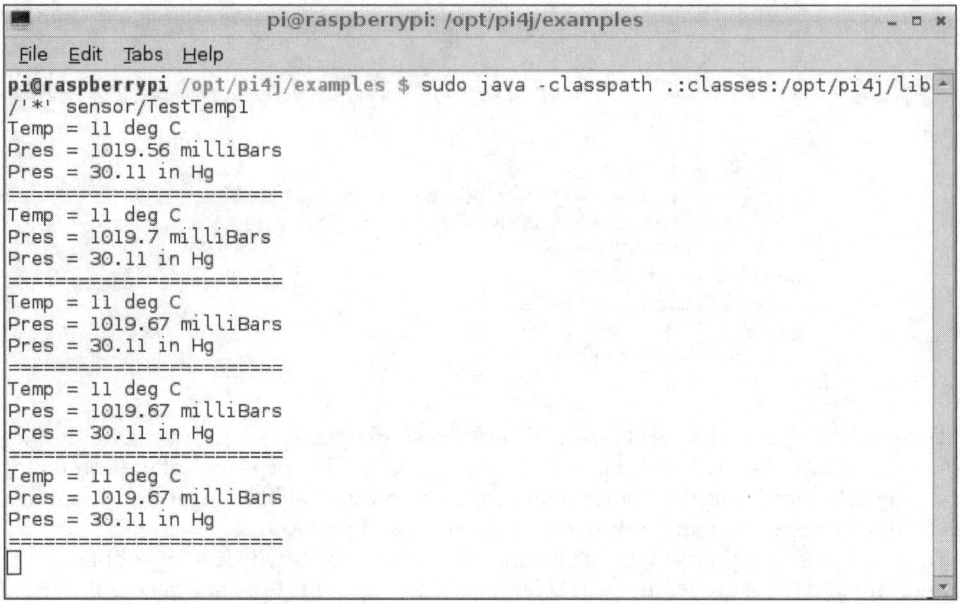

```
pi@raspberrypi: /opt/pi4j/examples
File Edit Tabs Help
pi@raspberrypi /opt/pi4j/examples $ sudo java -classpath .:classes:/opt/pi4j/lib
/'*' sensor/TestTemp1
Temp = 11 deg C
Pres = 1019.56 milliBars
Pres = 30.11 in Hg
========================
Temp = 11 deg C
Pres = 1019.7 milliBars
Pres = 30.11 in Hg
========================
Temp = 11 deg C
Pres = 1019.67 milliBars
Pres = 30.11 in Hg
========================
Temp = 11 deg C
Pres = 1019.67 milliBars
Pres = 30.11 in Hg
========================
Temp = 11 deg C
Pres = 1019.67 milliBars
Pres = 30.11 in Hg
========================
```

FIGURE 4-17 TestTemp1 program output

```
                System.out.println("Pres = " + df.format(presData) + "
                milliBars");
                System.out.println("Pres = " + df.format(MB_TO_INHG *
                presData) + " in Hg");
                System.out.println("========================");
                Thread.sleep(10000);
            }
    } // end of main
} // end of class
```

Figure 4-17 is a screenshot of the TestTemp1 program output after running for several minutes. And yes, it was a bit chilly in my development lab when I ran this program. To stop the program, you must press the CTRL-C key combination. I checked the pressure reading at a nearby airport's automated weather reporting system, which was 30.11 in Hg—exactly matching my system's measurement.

Thermostatic Application

In this section, I describe how to set up a simple thermostatic application. The temperature set point will be entered from the keyboard and the local temperature will be monitored by the BMP180 sensor. A GPIO pin will be set high if the measured temperature is below the set point. This is precisely how a normal thermostat functions.

The key to this simple application is to create a new client application that requests the user to enter a temperature value, which is then compared to the actual temperature. If the actual temperature is less than the requested temperature, the heating system should be turned on. Lacking a real heating system, I will simply turn on an LED and use a `println` statement to display the control action. The LED should be connected in exactly the same manner as LED 1 was, as shown in the Figure 4-8 schematic.

This new application uses the `XferTemp` class and is still part of the sensor package, which makes compiling and executing the code very easy. The `PiThermostat` class is shown here:

```
package sensor;

import com.pi4j.io.gpio.GpioController;
import com.pi4j.io.gpio.GpioFactory;
import com.pi4j.io.gpio.GpioPinDigitalOutput;
import com.pi4j.io.gpio.RaspiPin;
import java.util.Scanner;

public class PiThermostat {

      static double setPointT, currentT;
      static XferTemp xferTemp = new XferTemp();
      static Scanner console = new Scanner(System.in);
      final static GpioController gpio = GpioFactory.getInstance();
      final static GpioPinDigitalOutput led1 =
      gpio.provisionDigitalOutputPin(RaspiPin.GPIO_01);

      public PiThermostat() {
              }

          public static void main(String[] args) throws InterruptedException
{

            // get the set point temp from the user
            System.out.print("Enter set point temperature: ");
            setPointT = console.nextDouble();

                  // keep program running until user aborts (CTRL-C)
                  while(true) {
                  currentT = xferTemp.read();
                  System.out.println("Current temp = " + currentT);
                  if(currentT < (setPointT -0.5)) {
                        led1.high();
                        System.out.println("System on");
                  }
                  else {
                        led1.low();
                        System.out.println("System off");
                  }
                  Thread.sleep(15000);
            } // end while loop
      } // end main
} // end class definition
```

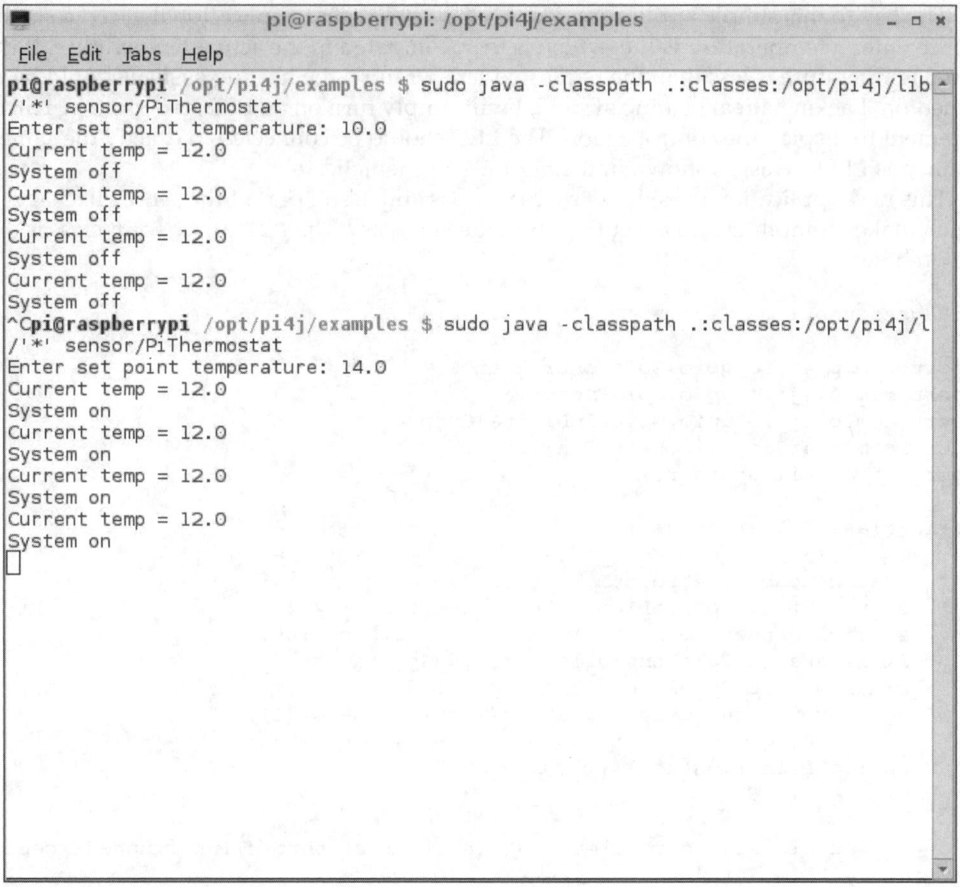

```
pi@raspberrypi: /opt/pi4j/examples
File Edit Tabs Help
pi@raspberrypi /opt/pi4j/examples $ sudo java -classpath .:classes:/opt/pi4j/lib
/'*' sensor/PiThermostat
Enter set point temperature: 10.0
Current temp = 12.0
System off
Current temp = 12.0
System off
Current temp = 12.0
System off
Current temp = 12.0
System off
^Cpi@raspberrypi /opt/pi4j/examples $ sudo java -classpath .:classes:/opt/pi4j/l
/'*' sensor/PiThermostat
Enter set point temperature: 14.0
Current temp = 12.0
System on
Current temp = 12.0
System on
Current temp = 12.0
System on
Current temp = 12.0
System on
```

FIGURE 4-18 PiThermostat program output

Figure 4-18 shows the screenshot from the PiThermostat program with two set points separately entered, one below and one above the current room temperature.

I also observed the LED turn on when the set point was set above the current room temperature, thus further confirming proper thermostat operation.

Setting the Thermostat Remotely

I will show you how to remotely enter the thermostat's set point as that is this book's central theme. I will use a simple PHP web page hosted on the Pi's Apache web server to accomplish this task. The generic file location and structure were thoroughly discussed in Chapter 1 so I will only show you the code and discuss how it functions.

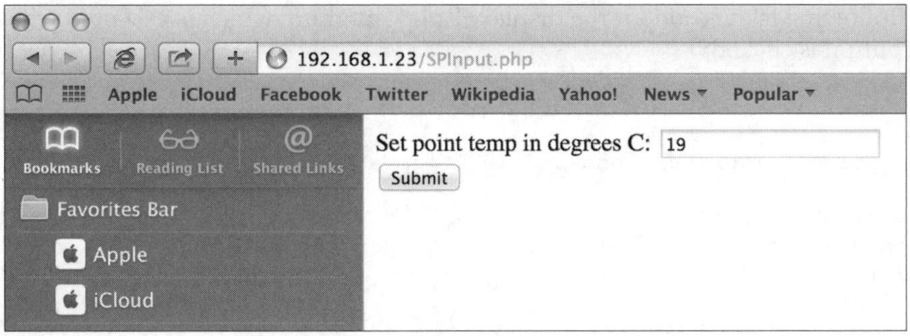

FIGURE 4-19 Browser screenshot for the initial SPInput PHP program

Two server files are required:

- SPInput.php
- displaySP.php

The following is the file listing for SPInput.php:

```
<html>
<body>
<form action="displaySP.php" method="post">
Set point temp in degrees C: <input type="text" name="temp"><br>
<input type="submit">
</body>
</html>
```

Note that this code calls the second PHP file, displaySP.php, when the user clicks the submit button. Figure 4-19 is a screenshot from a Mac Pro after I have put in my Pi's URL with the initial PHP program, which you can see in the URL box in the figure. I entered 19 as a set point temperature and clicked on the Submit button. Figure 4-20 shows the immediate result.

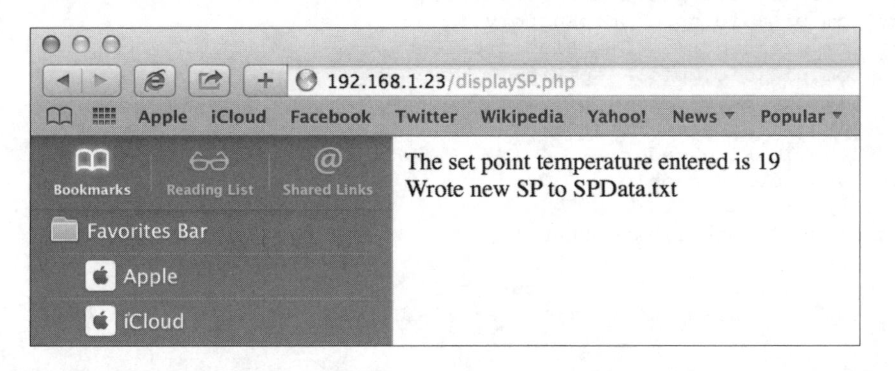

FIGURE 4-20 After clicking the Submit button

The following is the code for the displaySP.php program, which is executed when the Submit button is clicked:

```
<html>
<body>
The set point temperature entered is <?php echo $_POST["temp"]; ?><br>
<?php
$file = "/var/www/SPData.txt";
$fh = open($file, 'w') or die("could not open the file");
fwrite($fh, $_POST["temp"]);
fclose($fh);
?>
Wrote new SP to SPData.txt
</body>
</html>
```

This program not only echoes back the set point value the user entered but also saves that value in a small data file named SPData.txt. This is necessary in order to allow the thermostat application to access the desired set point at any time if it had been set days before the thermostat application was run. Note that a cookie could also have been set, but I felt this approach was simpler and less complex. You will have to create the file and modify file permissions before using the web server programs. The procedure is simple:

- Create the file by entering **sudo nano SPdata.txt** ↵ (I created it in the /var/www directory).
- Modify the permissions by entering **sudo chmod 777 SPdata.txt** ↵.

Failure to modify the permissions will cause the "die" message to appear as the displaySP.php program is not executing with root-level privileges.

The remaining task left in this process was to modify the PiThermostat program to use the stored set point in lieu of asking the user to enter one. The modified program was renamed RemotePiThermostat.java and is shown here:

```
package sensor;

import com.pi4j.io.gpio.GpioController;
import com.pi4j.io.gpio.GpioFactory;
import com.pi4j.io.gpio.GpioPinDigitalOutput;
import com.pi4j.io.gpio.RaspiPin;
import java.io.File;
import java.io.BufferedReader;
import java.io.FileReader;
import java.io.IOException;

public class RemotePiThermostat {

      static double setPointT, currentT;
      static XferTemp xferTemp = new XferTemp();
```

```
  final static GpioController gpio = GpioFactory.getInstance();
  final static GpioPinDigitalOutput led1 =
  gpio.provisionDigitalOutputPin(RaspiPin.GPIO_01);
  static BufferedReader br;

  public RemotePiThermostat() {
    }

public static void main(String[] args) throws IOException,
InterruptedException {

  try {
        // get the set point temp from the file /var/www/SPData.txt
        String fh = "/var/www/SPData.txt";
        String spText =  readFile(fh);
        setPointT = Double.parseDouble(spText);

        // keep program running until user aborts (CTRL-C)
        while(true) {
              currentT = xferTemp.read();
              System.out.println("Current temp = " + currentT);
              System.out.println("Stored set point temp = " +
              setPointT);
              if(currentT < (setPointT -0.5)) {
                    led1.high();
                    System.out.println("System on");
              }
              else {
                    led1.low();
                    System.out.println("System off");
              }
              Thread.sleep(15000);
         } // end of while loop
     } // end of try block
   catch(IOException ioe) {
        System.out.println("Threw an exception " + ioe);
   }
} // end of main

public static String readFile(String fileName) throws IOException {
  br = new BufferedReader(new  FileReader(fileName));
  try {
        StringBuilder sb = new StringBuilder();
        String line = br.readLine();

        while(line != null) {
              sb.append(line);
              sb.append("\n");
              line = br.readLine();
        }
```

```
            return sb.toString();
        } // end of try block
        catch(IOException ioe) {
            System.out.println("Threw an exception " + ioe);
            return "error";
        }
        finally {
            br.close();
        }
    } // end of readFile method

} // end of RemotePiThermostat class definition
```

Figure 4-21 shows the Pi's screen when the RemotePiThermostat program was run. Note that the system was "on" because the set point at 19°C was higher than the current measured temperature at 12°C.

Database Classes

It is now time to discuss the database classes that will store the measurements created by the sensors. I will use the MySQL classes shown in Chapter 3 as templates, including the Java database connector. The obvious first step is to define the MySQL schema or structure. I set up a similar database as was defined in Chapter 3 but added an additional field to store

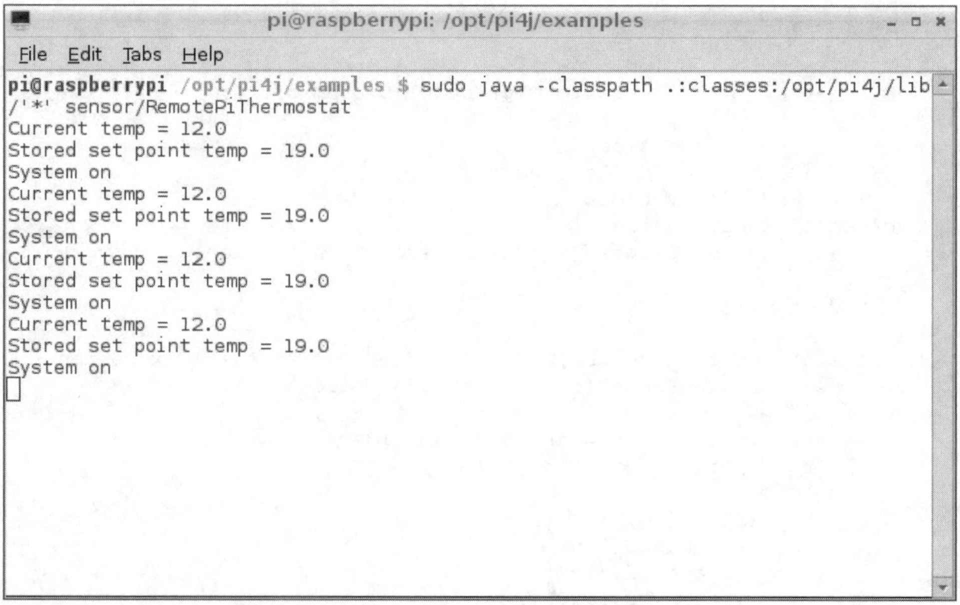

```
pi@raspberrypi: /opt/pi4j/examples
File Edit Tabs Help
pi@raspberrypi /opt/pi4j/examples $ sudo java -classpath .:classes:/opt/pi4j/lib
/'*' sensor/RemotePiThermostat
Current temp = 12.0
Stored set point temp = 19.0
System on
Current temp = 12.0
Stored set point temp = 19.0
System on
Current temp = 12.0
Stored set point temp = 19.0
System on
Current temp = 12.0
Stored set point temp = 19.0
System on
```

Figure 4-21 RemotePiThermostat program output display

Table 4-6 HomeWeatherStation Schema

Name	Description	Data Type
HomeWeatherStation	Database name	N/A
sensorMeasurements	table name	N/A
id	Primary key	MEDIUMINT AUTO_INCREMENT
tdate	Date	TEXT
ttime	Time	TEXT
ttemp	Temperature	TEXT
tpressure	Pressure	TEXT

the pressure measurements. Table 4-6 shows the schema for the database, which I named HomeWeatherStation.

I created the database and table using the same procedures shown in the previous chapters. However, I changed my approach to developing the actual classes by using the Eclipse IDE environment on a laptop in lieu of developing directly on the Pi. I had several reasons for my decision but the primary reason was that it is much faster and easier to develop and debug using a full-scale IDE than it is using the command line entries on the Pi. Using the IDE meant I had to create pseudo or simulation classes that would fill in for the real classes that depended on the native Pi4J library, specifically:

- `XferTemp`
- `XferPres`

These simulation (sim) classes were relatively easy to create as they only returned a double when called. I also included a random number generator in the sim classes to enhance the data realism somewhat and make it reflect what the real classes would provide. The XferTemp code listing that follows shows one of the sim classes:

```
package sensor;

public class XferTemp implements DataBehavior {
      public double read() {
            double temp;
            temp = (double)Math.round((Math.random() + 20)*100) /100;
            return temp;
      }
}
```

I next ran a simple adaptation of the Pi TestTemp1 program, which I named TestTemp to check if all the sim classes were functioning as I expected. This program is similar to the original TestTemp1, which ran on the Pi, but runs in the Eclipse environment with the simulated objects.

```
package sensor;

public class TestTemp {

      final static int TEMP_TYPE = 0;
      final static int PRES_TYPE = 1;

      public static void main(String[] args) throws InterruptedException {
            try {
            SEN_11824 tempSensor = new SEN_11824(TEMP_TYPE);
            SEN_11824 presSensor = new SEN_11824(PRES_TYPE);

            double tempData, presData;

            while(true) {
                        tempData = tempSensor.performData();
                        presData = presSensor.performData();


                        System.out.println("Temp = " + tempData + " degrees C");
                        System.out.println("Pressure = " + presData + " milliBars");
                        System.out.println("===================================");
                        Thread.sleep(10000);
            }
            }
            catch(InterruptedException e){ }
      } // end of main
} // end of class
```

A sample of this program's output is shown in Figure 4-22.

I next created a client class named FillDB2 that contains the main method, which starts the application as well as establishing the MySQL database connectivity and inserting the data created by the sim objects into the database records. This was a bit of a shortcut, but I was eager to test the application.

```
package sensor;

import java.sql.Connection;
import java.sql.DriverManager;
import java.sql.ResultSet;
import java.sql.SQLException;
import java.sql.Statement;
import java.util.Date;
import java.text.SimpleDateFormat;

public class FillDB2 {
      final static int TEMP_TYPE = 0;
      final static int PRES_TYPE = 1;

      // JDBC driver name and database URL
```

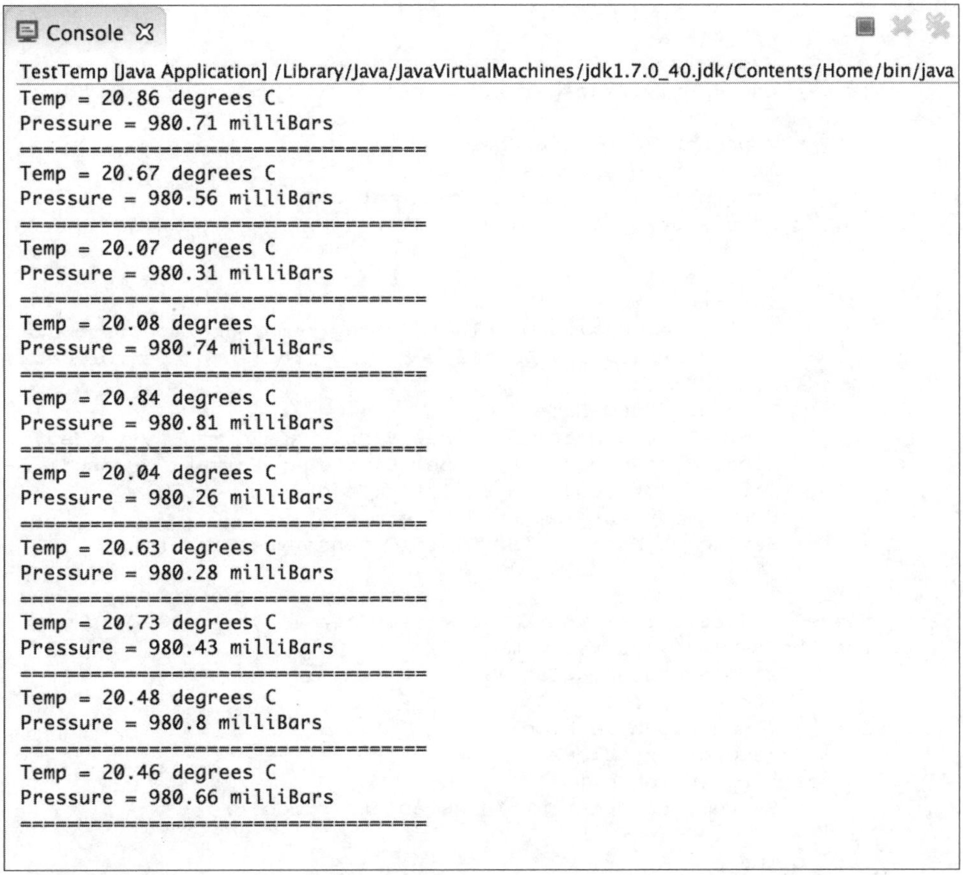

```
Console

TestTemp [Java Application] /Library/Java/JavaVirtualMachines/jdk1.7.0_40.jdk/Contents/Home/bin/java
Temp = 20.86 degrees C
Pressure = 980.71 milliBars
====================================
Temp = 20.67 degrees C
Pressure = 980.56 milliBars
====================================
Temp = 20.07 degrees C
Pressure = 980.31 milliBars
====================================
Temp = 20.08 degrees C
Pressure = 980.74 milliBars
====================================
Temp = 20.84 degrees C
Pressure = 980.81 milliBars
====================================
Temp = 20.04 degrees C
Pressure = 980.26 milliBars
====================================
Temp = 20.63 degrees C
Pressure = 980.28 milliBars
====================================
Temp = 20.73 degrees C
Pressure = 980.43 milliBars
====================================
Temp = 20.48 degrees C
Pressure = 980.8 milliBars
====================================
Temp = 20.46 degrees C
Pressure = 980.66 milliBars
====================================
```

Figure 4-22 Eclipse TestTemp program output

```
static final String JDBC_DRIVER = "com.mysql.jdbc.Driver";
static final String DB_URL = "jdbc:mysql://localhost/";

//  Database credentials
static final String USER = "root";
static final String PASS = "";

public static void main(String[] args) {
        Connection conn = null;
        Statement stmt = null;
        SEN_11824 tempSensor = new SEN_11824(TEMP_TYPE);
        SEN_11824 presSensor = new SEN_11824(PRES_TYPE);

        try {
           //Register JDBC driver
          Class.forName("com.mysql.jdbc.Driver");
```

```
                //Open a connection
               System.out.println("Creating a connection object");
              conn = DriverManager.getConnection(DB_URL, USER, PASS);

               //Connect to the database
             stmt = conn.createStatement();
             String sql = "USE HomeWeatherStation";
             stmt.executeUpdate(sql);

             // forever loop
             while(true) {
               String tempData = Double.toString(tempSensor.performData());
               String presData = Double.toString(presSensor.performData());

               Date dt = new Date();
               SimpleDateFormat sdf1 = new SimpleDateFormat("yyyyMMdd");
               SimpleDateFormat sdf2 = new SimpleDateFormat("HHmmss");
               String currentDate = sdf1.format(dt);
               String currentTime = sdf2.format(dt);
               String logData = "INSERT INTO sensorMeasurements

( ttemp, tpres,  tdate,ttime) VALUES (" + tempData + "," + presData + "," +
"," + currentDate + "," + currentTime + ")";
               stmt.execute(logData);
               Thread.sleep(10000);
             }// end of while block
           } // end of try block
           catch(Exception e){
               System.out.println("Threw an exception");
           }
       }// end of main
} // end of class
```

Figure 4-23 shows a portion of the sensorMeasurement table contents after the program had run for a while. I used the command `SELECT * FROM sensorMeasurements;` at the interactive MySQL command line to create this display.

At this point, I have demonstrated that the program architecture works as intended and all the classes operate per the design requirements. All that's left is to "port" the Eclipse design over to the Pi and refactor the classes a bit to extract the database functions from the client to their own class. The client class should normally be very small and really only designed to launch the application and call the appropriate functions such as creating the database connection, instantiating the desired sensor objects, and starting the measurement cycle. The porting and refactoring are made quite simple by my use of interfaces. I only have to substitute the three "real" classes for the sims and extract the database features into their own class and just instantiate it in the client when it starts.

Figure 4-23 Portion of the sensorMeasurement table records

```
mysql> USE HomeWeatherStation;
Reading table information for completion of table and column names
You can turn off this feature to get a quicker startup with -A

Database changed
mysql> SELECT * FROM sensorMeasurements;
+----+-------+--------+----------+--------+
| id | ttemp | tpres  | tdate    | ttime  |
+----+-------+--------+----------+--------+
|  1 | 20.35 | 980.47 | 20140414 | 212010 |
|  2 | 20.8  | 980.21 | 20140414 | 212020 |
|  3 | 20.51 | 980.21 | 20140414 | 212030 |
|  4 | 20.58 | 980.31 | 20140414 | 212040 |
|  5 | 20.05 | 980.42 | 20140414 | 212050 |
|  6 | 20.13 | 980.33 | 20140414 | 212100 |
|  7 | 20.08 | 980.98 | 20140414 | 212110 |
|  8 | 20.64 | 980.12 | 20140414 | 212120 |
|  9 | 20.82 | 980.53 | 20140414 | 212130 |
| 10 | 20.2  | 980.1  | 20140414 | 212140 |
| 11 | 20.76 | 980.34 | 20140414 | 212150 |
| 12 | 20.63 | 980.48 | 20140414 | 212200 |
| 13 | 20.69 | 980.47 | 20140414 | 212210 |
| 14 | 20.66 | 980.03 | 20140414 | 212220 |
| 15 | 20.22 | 980.27 | 20140414 | 212230 |
| 16 | 20.26 | 980.78 | 20140414 | 212240 |
| 17 | 20.72 | 980.37 | 20140414 | 212250 |
| 18 | 20.14 | 980.14 | 20140414 | 212300 |
| 19 | 20.1  | 980.1  | 20140414 | 212310 |
| 20 | 20.06 | 980.69 | 20140414 | 212320 |
| 21 | 20.48 | 980.25 | 20140414 | 212330 |
| 22 | 20.92 | 980.52 | 20140414 | 212340 |
| 23 | 20.97 | 980.05 | 20140414 | 212350 |
| 24 | 20.5  | 980.25 | 20140414 | 212400 |
| 25 | 20.84 | 980.59 | 20140414 | 212410 |
| 26 | 20.58 | 980.5  | 20140414 | 212420 |
| 27 | 20.19 | 980.24 | 20140414 | 212430 |
| 28 | 20.82 | 980.55 | 20140414 | 212440 |
| 29 | 20.99 | 980.67 | 20140414 | 212450 |
| 30 | 20.08 | 980.08 | 20140414 | 212500 |
| 31 | 20.37 | 980.18 | 20140414 | 212510 |
| 32 | 20.09 | 980.73 | 20140414 | 212520 |
| 33 | 20.78 | 980.28 | 20140414 | 212530 |
| 34 | 20.13 | 980.27 | 20140414 | 212540 |
| 35 | 20.93 | 980.41 | 20140414 | 212550 |
| 36 | 20.5  | 980.36 | 20140414 | 212600 |
+----+-------+--------+----------+--------+
36 rows in set (0.00 sec)

mysql>
```

The following listing shows the new Pi database class I named HWSDB, which naturally is short for Home Weather Station DataBase.

```
package sensor;

import java.sql.Connection;
import java.sql.DriverManager;
import java.sql.ResultSet;
import java.sql.SQLException;
import java.sql.Statement;
```

```
import java.util.Date;
import java.text.SimpleDateFormat;
public class HWSDB {
      //JDBC driver name and database URL
      static final String JDBC_DRIVER = "com.mysql.jdbc.Driver";
      static final String DB_URL = "jdbc:mysql://localhost:3306/";
      // Database credentials
      static final String USER = "root";
      static final String PASS = "<your password>";
      Connection conn = null;
      Statement stmt = null;

      public HWSDB() throws Exception {
          try {
            // Register JDBC driver
            Class.forName("com.mysql.jdbc.Driver").newInstance();
            // Open a connection
            conn = DriverManager.getConnection(DB_URL, USER, PASS);
            System.out.println("got a connection");
            // Connect to the database
            stmt = conn.createStatement();
            String sql = "USE HomeWeatherStation";
            stmt.executeUpdate(sql);
          }
          catch(Exception e) {
            System.out.println("Threw an exception " + e);
          }
        } // end of ctor

      public void store(String[] dataStr) throws SQLException {
          try {
            String tempData = dataStr[0];
            String presData = dataStr[1];
            Date dt = new Date();
            SimpleDateFormat sdf1 = new SimpleDateFormat("yyyyMMdd");
            SimpleDateFormat sdf2 = new SimpleDateFormat("HHmmss");
            String currentDate = sdf1.format(dt);
            String currentTime = sdf2.format(dt);
            String logData = "INSERT INTO sensorMeasurements
(ttemp, tpres, tdate, ttime) VALUES(" + tempData + "," + presData +
"," + currentDate + "," + currentTime + ")";
            stmt.execute(logData);
          }
          catch(SQLException se) {
            System.out.println("Threw an exception in the store method" + se);
          }
      } // end of store method
} // end of class definition
```

The client class, which is shown next, will instantiate all the required objects, collect the measurements, and pass them on to the database object for storage. It also sets the interval between measurement sets, which is fixed at 10 seconds for this example program. I named this program OperateHWS from which I know you can guess the meaning:

```
package sensor;

public class OperateHWS {
      final static int TEMP_TYPE = 0;
      final static int PRES_TYPE = 1;
      String[] dataInput = String[3];

      public static void main(String[] args) {
            SEN_11824 tempSensor = new SEN_11824(TEMP_TYPE);
            SEN_11824 presSensor = new SEN_11824(PRES_TYPE);

            HWSDB db = new HWSDB();
            try {
             // forever loop
            while(true) {
                  dataInput[0] = Double.toString(tempSensor.performData());
                  dataInput[1] = Double.toString(presSensor.performData());
                  db.store(dataInput);
                  Thread.sleep(10000);
              }// end of while block
            } // end of try block
            catch(Exception e){
                System.out.println("Threw an exception");
            }
      }// end of main
} // end of class
```

You must create the MySQL HomeWeatherStation database before attempting to run this program or an exception will be thrown. All the source code within the package must also be compiled before running this program. Once everything is compiled, you need only to enter this command to start logging weather station data:

```
sudo java -classpath .:/classes:/opt/pi4j/lib/'*' sensor/OperateHWS ↵
```

As with all the run forever programs, you will need to press the CTRL-C key combination to stop the program.

Figure 4-24 is a combined selection of the first 20 and last 20 records from the sensorMeasurement table that I retrieved using the interactive MySQL command line procedure that I discussed earlier. The time span for record storage was approximately 20 hours, and you can observe that the temperature dropped considerably while the pressure increased during this period.

```
pi@raspberrypi: /opt/pi4j/examples
File Edit Tabs Help
49 rows in set (0.03 sec)

mysql> SELECT * FROM sensorMeasurements WHERE id < 51;
+----+-------+-------------------+----------+--------+
| id | ttemp | tpres             | tdate    | ttime  |
+----+-------+-------------------+----------+--------+
|  1 | 15.0  | 1006.7226354650069 | 20140415 | 114954 |
|  2 | 15.0  | 1006.7525917504389 | 20140415 | 115005 |
|  3 | 15.0  | 1006.7226354650069 | 20140415 | 115015 |
|  4 | 15.0  | 1006.5728541547883 | 20140415 | 115026 |
|  5 | 15.0  | 1006.6627229175313 | 20140415 | 115037 |
|  6 | 15.0  | 1006.8125043446909 | 20140415 | 115048 |
|  7 | 15.0  | 1006.5428979161327 | 20140415 | 115059 |
|  8 | 15.0  | 1006.6327666554876 | 20140415 | 115109 |
|  9 | 15.0  | 1006.7525917504389 | 20140415 | 115120 |
| 10 | 15.0  | 1006.7226354650069 | 20140415 | 115130 |
| 11 | 15.0  | 1006.692679187371  | 20140415 | 115141 |
| 12 | 15.0  | 1006.7525917504389 | 20140415 | 115152 |
| 13 | 15.0  | 1006.8125043446909 | 20140415 | 115203 |
| 14 | 15.0  | 1006.842460653511  | 20140415 | 115213 |
| 15 | 15.0  | 1006.7525917504389 | 20140415 | 115224 |
| 16 | 15.0  | 1006.7525917504389 | 20140415 | 115235 |
| 17 | 15.0  | 1006.6627229175313 | 20140415 | 115246 |
| 18 | 15.0  | 1006.692679187371  | 20140415 | 115256 |
| 19 | 15.0  | 1006.692679187371  | 20140415 | 115308 |
| 20 | 15.0  | 1006.9323296267478 | 20140415 | 115318 |
| 6698 | 6.0   | 1042.166371775138  | 20140416 | 80443  |
| 6699 | 6.0   | 1042.1064407821661 | 20140416 | 80453  |
| 6700 | 6.0   | 1042.2263027992944 | 20140416 | 80504  |
| 6701 | 6.0   | 1042.2263027992944 | 20140416 | 80515  |
| 6702 | 6.0   | 1042.2263027992944 | 20140416 | 80526  |
| 6703 | 6.0   | 1042.1064407821661 | 20140416 | 80536  |
| 6704 | 6.0   | 1042.166371775138  | 20140416 | 80548  |
| 6705 | 6.0   | 1042.0764752973741 | 20140416 | 80558  |
| 6706 | 6.0   | 1042.166371775138  | 20140416 | 80609  |
| 6707 | 6.0   | 1042.2263027992944 | 20140416 | 80620  |
| 6708 | 6.0   | 1042.2263027992944 | 20140416 | 80630  |
| 6709 | 6.0   | 1042.1963372833181 | 20140416 | 80641  |
| 6710 | 6.0   | 1042.2263027992944 | 20140416 | 80652  |
| 6711 | 6.0   | 1042.1963372833181 | 20140416 | 80703  |
| 6712 | 6.0   | 1042.166371775138  | 20140416 | 80714  |
| 6713 | 6.0   | 1042.166371775138  | 20140416 | 80726  |
| 6714 | 6.0   | 1042.2263027992944 | 20140416 | 80736  |
| 6715 | 6.0   | 1042.166371775138  | 20140416 | 80747  |
| 6716 | 6.0   | 1042.316199393999  | 20140416 | 80758  |
| 6717 | 6.0   | 1042.166371775138  | 20140416 | 80808  |
+------+-------+--------------------+----------+--------+
6717 rows in set (0.43 sec)
```

FIGURE 4-24 First 20 and last 20 records from the sensorMeasurement table

Remote Access to the HWS Database

The HWS database can be remotely accessed using the procedures detailed at the end of Chapter 2. It is entirely irrelevant how the record data was collected and stored, which was by using Python in Chapter 2 and Java in this chapter. It is essential that you create a new user and password to remotely access the database as this chapter's example was based only on a root-level user. Allowing root-level access to your system over the Internet is only inviting disaster into your computing environment. Believe me—there are unscrupulous types on the Web who will wreak havoc just for the fun of it. Chapter 2 details procedures on how to add a user and create a password.

Summary

This is a very brief summary of a very long chapter. This chapter's project was the building and programming of a home weather station that sensed both temperature and pressure. A MySQL database was also created to store the measurements for later retrieval either locally or through the Web. I discussed how Java may be used to control GPIO pins using the very clever Pi4J library. A simple LED blinky project was demonstrated to further clarify the library functions.

In addition, this chapter included a simple thermostat application that proved the project could turn on a heating system if the measured temperature dropped below a set point. I showed you how to establish this control temperature set point both locally using a keyboard and also remotely using a web page served by the Pi's Apache web server.

5
CHAPTER

Webcam and Raspberry Pi Camera Projects

Remotely accessing webcams is a fairly common activity that many people perform routinely. It is my intention in this chapter to show you how to put together three projects, each with unique capabilities. The first project uses a standard USB webcam with a comprehensive open source software package named Motion. The second project deals with the specially designed Pi Camera, which only works with the Pi. The final project also uses the Pi Camera but uses the Motion software instead of the specialty applications that typically come bundled with the Pi Camera.

Conventional Webcam

The first project in this chapter uses a high-quality webcam that connects to the Pi using one of the PI's two USB ports. I used a Logitech C920, shown in Figure 5-1.

This a high-definition camera capable of producing excellent videos, but the open source software used within this project will constrain its performance. You may still use any one of the many different webcams that are available as the video requirements are modest and the current Raspian distribution will automatically detect and support many types, including older ones. If you already own a webcam, my suggestion is to plug it into one of the Pi's USB ports and then type the following into a command line:

```
lsusb↵
```

Figure 5-2 is a screenshot of the command's output. Device 012 is the webcam while the other Logitech device listed as 006 is a keyboard. If in doubt, with multiple devices from the same manufacturer, simply unplug the device and rerun the command to see which one disappears.

Note that I have found this particular webcam to be somewhat sensitive in terms of the USB port it is plugged into. It may not be detected when plugged into one of the two Pi USB ports, in which case, try plugging it into a powered USB hub that provides at least the minimum specification current of 500mA for each USB port.

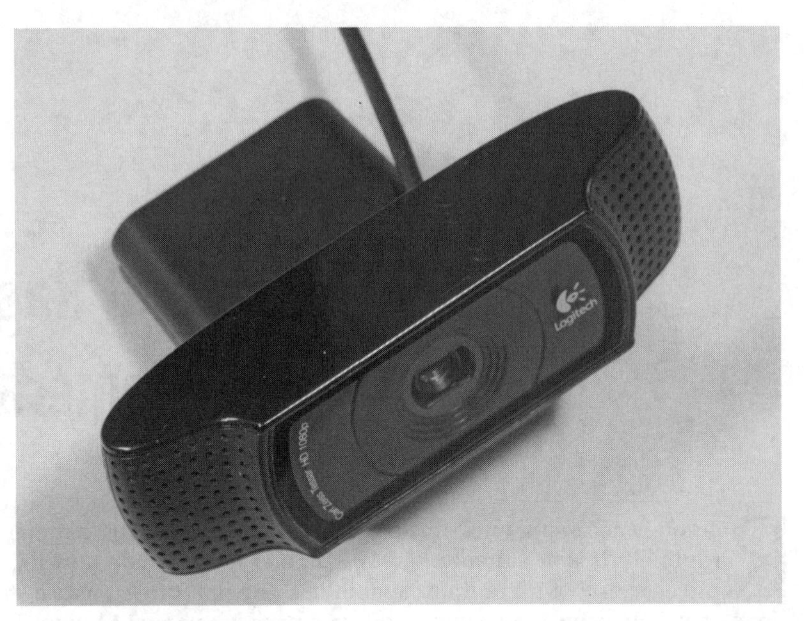

FIGURE 5-1 Logitech C920 webcam

```
pi@raspberrypi: ~
File  Edit  Tabs  Help
pi@raspberrypi ~ $ lsusb
Bus 001 Device 002: ID 0424:9512 Standard Microsystems Corp.
Bus 001 Device 001: ID 1d6b:0002 Linux Foundation 2.0 root hub
Bus 001 Device 003: ID 0424:ec00 Standard Microsystems Corp.
Bus 001 Device 012: ID 046d:082d Logitech, Inc.
Bus 001 Device 004: ID 1a40:0201 Terminus Technology Inc. FE 2.1 7-port Hub
Bus 001 Device 005: ID 1241:1177 Belkin F8E842-DL Mouse
Bus 001 Device 006: ID 046d:c326 Logitech, Inc.
Bus 001 Device 007: ID 0bda:8171 Realtek Semiconductor Corp. RTL8188SU 802.11n W
LAN Adapter
Bus 001 Device 008: ID 1a40:0101 Terminus Technology Inc. 4-Port HUB
pi@raspberrypi ~ $
```

FIGURE 5-2 `lsusb` output screenshot

Motion Software Package

I selected an open source software package named Motion to enable remote viewing of the webcam. This is a very comprehensive package containing a substantial number of features, far more than could be covered in this chapter. Creating similar software for a more traditional board, such as one from the Arduino series, would be a substantial undertaking.

The key feature that is used from the Motion package is the built-in web server. This server receives the video stream from the webcam and sends it off in TCP/IP format over a predefined port. All you need to remotely view the webcam video is a browser pointed to the Pi's IP address and port number, nothing more. This feature makes the viewing exercise extremely simple. But there is more: Motion enables you to use more than one webcam. You can set up multiple webcams, each with its own port number. Thus, you can monitor multiple locations throughout the observed area. Each webcam video feed is handled by what is known as a *thread* within the Motion software. I recognized provision for four threads in the Motion configuration file from which I presume four webcams could be handled. However, I seriously wonder if the Pi has the processing power to manage four simultaneous video feeds. In any case, this project is concerned with only one feed, which I know works very well.

Motion Features

Motion has a substantial number of features that enable it to accomplish an amazing amount of functions. The user manual is online at www.lavrsen.dk/foswiki/bin/view/Motion/WebHome, and it is over 100 pages in length. While it would take a complete book to document all the features, the configuration file itself contains many self-documenting comments that should help you explore some of the additional features of this software package.

Motion is described on its home page as follows: "Motion is a program that monitors the video signal from cameras. It is able to detect if a significant part of the picture has changed; in other words, it can detect motion." I will not be using this key feature as the project's main purpose is to simply demonstrate a straightforward approach to remotely display real-time video using a Pi as a webcam controller. However, you should keep Motion's powerful capabilities in mind if you want to expand the project beyond simple remote viewing.

Motion Setup

You will need to install the Motion package before using it. I strongly suggest that you update and upgrade your distribution prior to installing Motion. Simply type the following at a command-line prompt to update and upgrade the Raspian distribution in use:

```
sudo apt-get update ↵
sudo apt-get upgrade ↵
```

Be patient as the updates and upgrades can take a bit of time if there are many to install. Next, install Motion by typing the following:

```
sudo apt-get install motion ↵
```

Again, be a bit patient as this package is over 20MB in size and has many component parts.

Motion will be run in the "background" as a daemon, which means that it will be constantly available for service. To enable the daemon, you must edit the /etc/default/motion file. Type the following:

```
sudonano /etc/default/motion↵
```

You will see in the nano editor the line:

```
start_motion_daemon=no
```

Change the no to yes, and then save the nano buffer (CONTROL-O) and exit the editor (CONTROL-X).

Next comes Motion's configuration file. Motion has no graphics user interface (GUI) so it must be configured by making changes to its configuration file, /etc/motion/motion.conf. This is a very big text file—well over 600 lines, although much of the file content consists of comments inserted to help the user. Fortunately, only a few changes are necessary for this project. I have provided the changes in Table 5-1 to be made by configuration file section, but I do not show you step-by-step instructions as you should be fairly comfortable with how to use the nano editor.

Start the nano editor session as follows:

```
sudonano /etc/motion/motion.conf↵
```

Make the changes as shown in Table 5-1 if the contents of the table have not already been configured.

Save the changes and exit the nano editor. Now, you must start the Motion server, which is done by entering the following:

```
sudo service motion start ↵
```

One nice feature of having the Motion web server running as a daemon is that it is automatically started each time you boot the Pi. You may also stop or restart the service by typing the following:

```
sudo service motion stop ↵
sudo service motion restart ↵
```

That's it for the changes to be made in the configuration file. I do want to briefly discuss why these changes were made. The change from daemon off to daemon on is obvious as it was needed to run Motion as a daemon.

TABLE 5-1 Motion Configuration File Changes

Section	Change From	Change To
Daemon	daemon off	daemon on
Live Webcam Server	webcam_port 0	webcam_port 8081
Live Webcam Server	webcam_localhost on	webcam_localhost off
HTTP Based Control	control_port 0	control_port 8080
HTTP Based Control	control_localhost on	control_localhost off

The next change—making the port number 8081—is a bit historical as the Motion web service has traditionally been assigned to this port. It is not a required port number and you can easily change it to any number that you desire as long as it is greater than 1024 and less than 65535. This range avoids the "well known ports" and goes to the maximum possible port number. My recommendation is to leave it at 8081.

The next change concerns localhost operation. Localhost refers to the same machine that is hosting the Motion web server. No other system can access the webcam if you restrict the service to localhost so it must be turned off.

The next two changes are similar to what was discussed previously but are concerned with the remote control functions of the webcam. You do not enable any remote control functions such as webcam panning or tilting in this first project, but it is certainly doable. I suggest keeping the port number as assigned and disabling the localhost operation.

Webcam Viewing

It is now time to test the remote webcam viewing functionality. You will need a separate computer on the same network that connected to the Pi. You will also need the Pi IP address. It does not matter if the Pi is connected via Ethernet cable or by a Wi-Fi wireless adapter. My suggestion is to log in to your network router and click Attached Devices. The local IP address for the Pi should appear in the appropriate list. My Netgear router has separate lists for wired and wireless devices connecting to the network. Yours might be slightly different but should show something similar to Figure 5-3.

Attached Devices

Wired Devices

#	IP Address	MAC Address	Device Name
1	192.168.1.15	00:26:2D:03:61:9C	NESS-LLC
2	192.168.1.14	00:10:75:06:0D:6A	BA-060D6A
3	192.168.1.43	B8:27:EB:BC:74:67	RASPBERRYPI
4	192.168.1.13	00:90:A9:6F:8E:CB	MYBOOKWORLD

Wireless Devices (Wireless intruders also show up here)

#	IP Address	MAC Address	Device Name
1	192.168.1.3	00:A0:96:AA:B1:00	<unknown>
2	192.168.1.4	48:02:2A:B4:15:08	ANDROID-C7CE956E54930205
3	192.168.1.10	00:04:20:2A:89:63	SQUEEZEBOXRADIO
4	192.168.1.9	F0:4F:7C:42:F7:8F	<unknown>
5	192.168.1.7	CC:3A:61:43:8A:C7	ANDROID-817AE5448DFACA06
6	192.168.1.2	F4:CE:46:0D:A9:4A	HP0DA949
7	192.168.1.8	60:03:08:C9:C2:1E	APPLE-TV
8	192.168.1.16	20:68:9D:E1:44:2A	KARENS_LAPTOP
9	192.168.1.6	D0:E7:82:C6:9C:90	CHROMECAST
10	192.168.1.22	00:0D:4B:71:73:8D	<unknown>
11	192.168.1.5	28:CF:E9:1C:BD:8D	DONS-MBP

Figure 5-3 Router Attached Devices list

FIGURE 5-4 Video screen capture from the Pi webcam

The entry RASPBERRYPI has the IP 192.168.1.43 on the wired portion of my home network. This is all you need to remotely view the webcam. You just have to type 192.168.1.21:8081 into a browser on another networked computer to view the real-time webcam video stream. Figure 5-4 is a screen capture of the video feed from the attached Pi webcam. The object in the screen capture is an Elev-8 quadcopter, which I built as a project for my book *Build Your Own Quadcopter*.

This completes the first webcam project, which employed a high-quality webcam to provide remote video monitoring using a Pi running the open source Motion software. It was relatively easy to complete this project as it required only that you install a software package, make some minor file configuration changes, and plug in a USB webcam. The next webcam project uses the same software but has an entirely different webcam.

Raspberry Pi Camera

In this project, I will be using the relatively inexpensive Raspberry Pi camera module introduced in 2013 and shown in Figure 5-5.

This camera was specifically designed to operate solely with the Pi using the camera serial interface (CSI) connector, as shown in Figure 5-6. There are two identical 15-wire flex ribbon connectors mounted on the Pi. The CSI connector is the one located directly behind the RJ45 network connector, as shown in the figure. The other 15-wire flex connector is for

Figure 5-5 Raspberry Pi camera

the display serial interface (DSI), which is designed to work with a display device slated to be introduced by the Raspberry Pi Foundation in 2014.

Note *The 15-wire flex cable is plugged into the CSI connector with the printed side facing the RJ45 connector, as shown in Figure 5-6. Please be careful when inserting the ribbon cable into the connector as it is quite easy to misalign the contacts, which will cause problems.*

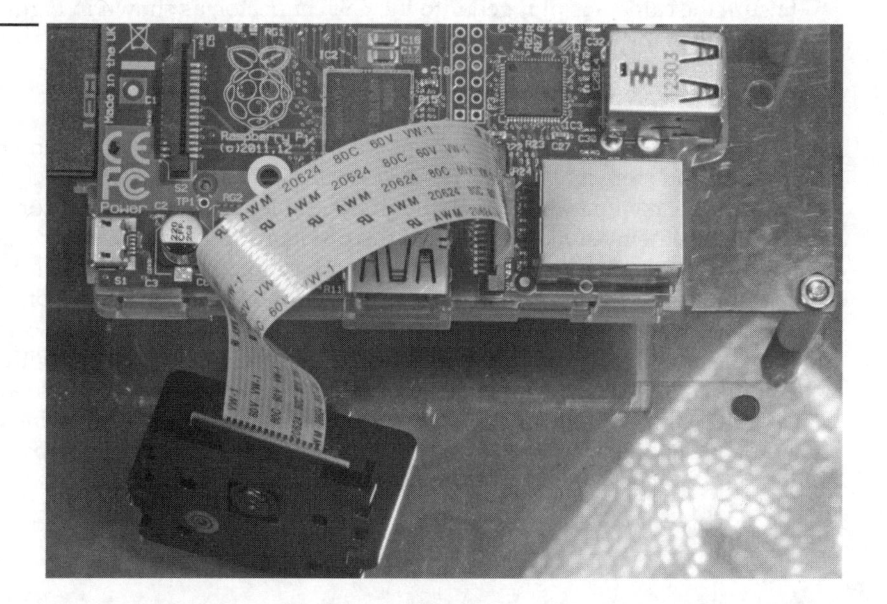

Figure 5-6 CSI connector

Table 5-2 Pi Camera Video and Still Modes

Resolution	Max Frame Rate (fps)	Remarks
2592 × 1944	15	Stills
1920 × 1080	30	1080p30 mode
1296 × 972	42	Stills, 4:3 aspect
1296 × 720	49	Stills, 720p30 mode, 16:9 aspect
640 × 480	60	VGAp60 mode
640 × 480	90	VGAp90 mode

I would like to now discuss the Pi camera hardware and its properties before discussing the camera software. The camera incorporates a 5-megapixel CMOS sensor along with an automatic control system to generate both real-time video streams and stills. The video stream is also compliant with the H264 codec standard. Table 5-2 details the various video and still modes possible with this camera.

Raspberry Pi Camera Software

It turns out that there are a variety of different software packages available that will control the Pi Camera. I will demonstrate three packages, beginning with the command line version that essentially takes still photographs or videos. This package is installed when you click OK on the RasPi camera option during the RasPi-Config menu selection. If you did not select it at that time, simply type

```
sudo raspi-config↵
```

and go to the menu selection number 5 to enable the camera.

Ensure the camera is plugged into the CSI connector, as shown in Figure 5-6, and type the following at the command line:

```
raspistill -v -o test.jpg ↵
```

The console screen will show a 5-second preview from the camera and then take a still photograph and store it as a JPEG file named test.jpg in the current directory from which you entered the command. Recheck the ribbon cable connection if this command does not work as that is the most likely source of the problem. You can obviously change the filename to avoid overwriting the existing photograph. Figure 5-7 shows a sample photograph taken of a Beaglebone Black development board using the preceding command.

The photograph is a bit fuzzy, which I suspect occurred because the camera was a bit too close to the subject being photographed. That is always a problem with a fixed-focus, wide-angle camera lens like the one you find in the Pi Camera. I had also attached the Pi Camera on a simple stand, as shown in Figure 5-8, which greatly helps in steadying the camera while taking photographs. The Pi Camera itself is held in place by a two-piece holder that I purchased from an online supplier.

Figure 5-7
Test photograph

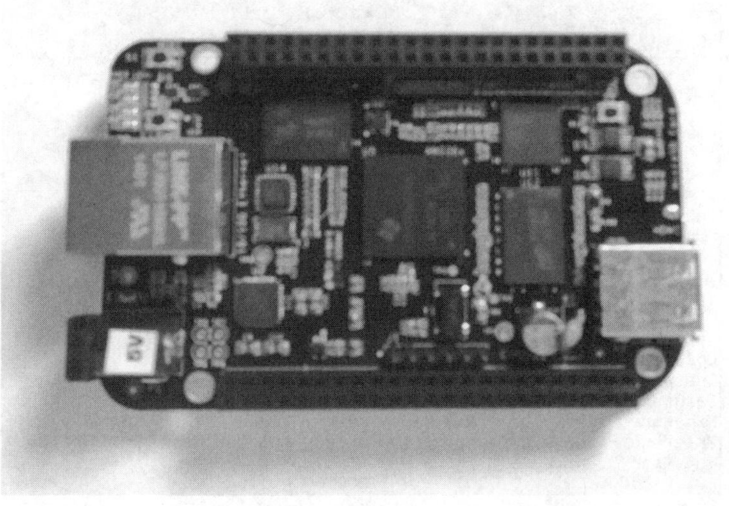

There is also a substantial amount of metadata shown on the Pi console screen, (see Figure 5-9). Several process descriptions are also shown in the figure, which clearly illustrates the steps needed to take the photograph.

The raspistill application has many more options, which I have not delved into but are further explained at the Raspberry Pi Foundation's camera website, http://www.raspberrypi.org/help/camera-module-setup/.

Figure 5-8
Pi Camera stand with a camera holder support

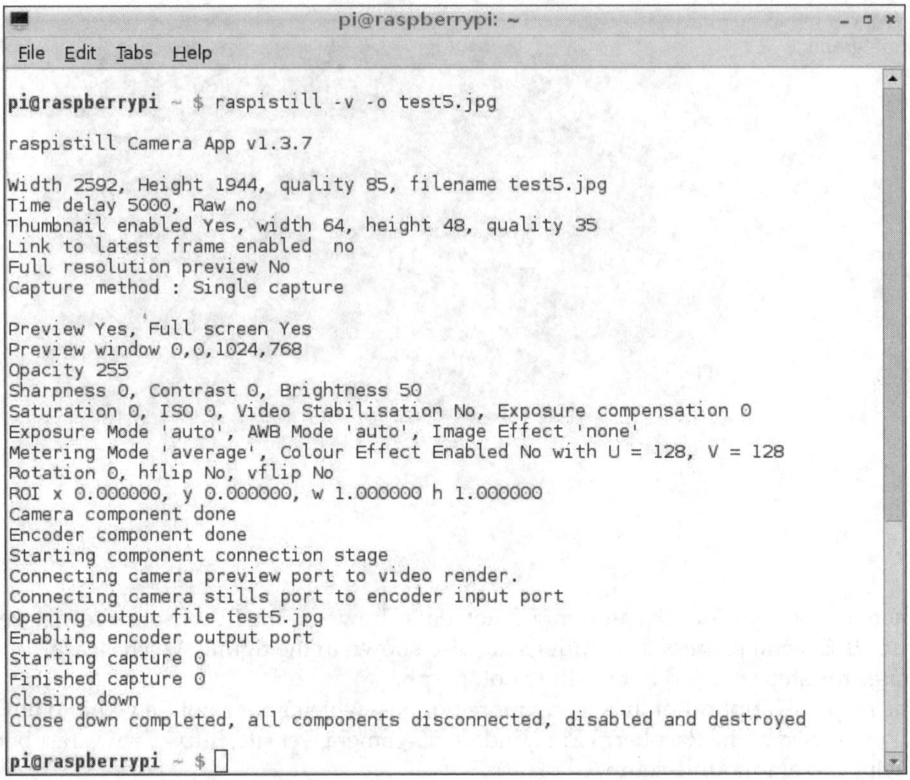

```
pi@raspberrypi: ~
File  Edit  Tabs  Help

pi@raspberrypi ~ $ raspistill -v -o test5.jpg

raspistill Camera App v1.3.7

Width 2592, Height 1944, quality 85, filename test5.jpg
Time delay 5000, Raw no
Thumbnail enabled Yes, width 64, height 48, quality 35
Link to latest frame enabled  no
Full resolution preview No
Capture method : Single capture

Preview Yes, Full screen Yes
Preview window 0,0,1024,768
Opacity 255
Sharpness 0, Contrast 0, Brightness 50
Saturation 0, ISO 0, Video Stabilisation No, Exposure compensation 0
Exposure Mode 'auto', AWB Mode 'auto', Image Effect 'none'
Metering Mode 'average', Colour Effect Enabled No with U = 128, V = 128
Rotation 0, hflip No, vflip No
ROI x 0.000000, y 0.000000, w 1.000000 h 1.000000
Camera component done
Encoder component done
Starting component connection stage
Connecting camera preview port to video render.
Connecting camera stills port to encoder input port
Opening output file test5.jpg
Enabling encoder output port
Starting capture 0
Finished capture 0
Closing down
Close down completed, all components disconnected, disabled and destroyed

pi@raspberrypi ~ $
```

FIGURE 5-9 Test photograph metadata

Video capture is also relatively easy to accomplish by entering the following command:

```
raspivid -t 5000 -o video.h264 ↵
```

Executing this command will result in a 1080p 5-second video clip displayed on the console display. The `5000` in the command refers to the number of milliseconds that the video runs and the `video.h264` refers to the codec mode to be used. Please go to the Foundation's website just mentioned to learn about the many other options that are available for use with the raspivid application.

Recording a video is also very easy and can be done with this command:

```
raspivid -t 15000 -o > test_vid.h264 ↵
```

Running this command will result in the creation of a 15-second 1080p video in the current directory with the name test_vid.h264. Linux-savvy readers will recognize that the greater than operator, >, used in the command redirects the video stream to the file. The console display is also shown while the video data is streamed to the file. Incidentally, Linux automatically creates the file if it doesn't already exist in the directory. Additionally, if you rerun the command without changing the filename, it will overwrite the existing video record. Another word of caution is that video files can be quite large, so be prudent in how long your videos are. The 15-second video I created using the preceding command resulted in a 25MB file, which means that about 100MB will be used for every minute of recording. You can easily "crash" your Pi if you attempt to create a 30-minute video with a 4GB SD card installed. I highly recommend that you install an external USB hard drive with at least a 1TB capacity if you wish to experiment with long-duration videos. Also, plug the hard drive into a high-quality, externally powered USB hub because the Pi will not support one plugged directly into it.

One of the limitations of using the raspistill and raspivid applications is that you are not able to easily use them in a programmed manner. This limitation is overcome using the approach that I discuss in the next section.

Using Python with the Raspberry Pi Camera

I will use a clever Python library named python-picamera, created by Dave Jones (Github), which will permit the Pi Camera to both be operated in an interactive mode as well as programmed for specific operations. You must have Python 2.7 installed for this library to function. I believe the latest Raspian download version contains it, but if not, enter this command to install Python:

```
sudo apt-get install python python-dev↵
```

Once Python is ready, enter the following command to load the picamera Python library:

```
sudo apt-get install python-picamera↵
```

Start Python and enter the following commands to test the library and see how well it functions with the Pi camera (the arrow prompts indicate this is an interactive Python session):

```
-->importpicamera
-->camera = picamera.PiCamera()
-->camera.start_preview()
```

After entering the third command, you should see a full screen console display showing whatever the Pi Camera is facing. Enter CTRL-D (^d) to terminate the preview display. This will kick you out of the interactive Python session and back to the normal command line prompt.

Creating a script that does a series of camera functions is also quite easy. The following Python script takes a series of pictures at 10-second intervals and stores them in consecutive numbered image files located in the current directory. The script is named TestCamera.py and is available on this book's companion website.

```
importpicamera
import time

camera = picamera.PiCamera()
str1 = 'image0'
```

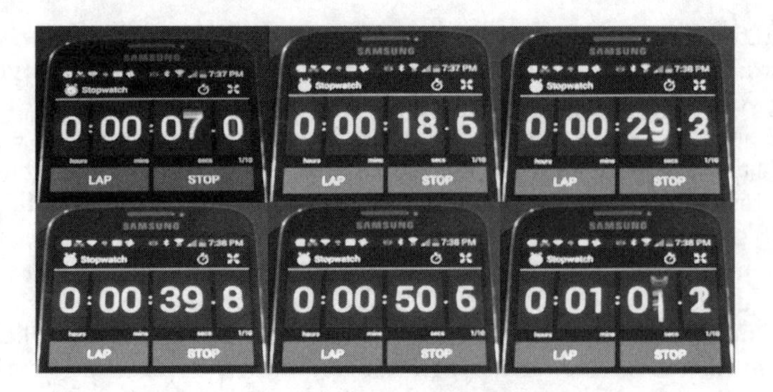

FIGURE 5-10 Composite image of six sequential images

```
for i in range(1, 12):
    str2 = str1 + str(i) + '.jpg'
camera.capture(str2)
    time.sleep(10)
```

The script causes a series of pictures to be taken with approximately 10 seconds between pictures and lasting two minutes overall. The captured images are stored in a series of incrementally labeled files starting with image01.jpg to image011.jpg. I have created a composite image of the first six images taken of a smartphone stopwatch application. This composite is shown in Figure 5-10.

You will quickly realize that the 10-second interval between image captures is only approximate as it takes about .5 to .6 seconds for the Pi Camera to take a photograph and have the application store it on the SD card. In most cases, this additional time will be of little consequence but you should keep it in mind for very tightly timed photographic situations. You can also easily change the image names by modifying the string `str1` in the Python script.

There are 15 camera properties that can be adjusted within the Python library to suit your particular setup. I have listed these properties in Table 5-3 along with their default values. In most cases, the default values will be suitable, but it is nice to have direct control over some of the camera attributes to help in difficult or unusual photographic setups.

I recognize that many readers will not know what many of these properties mean and how to set them appropriately. I thus highly recommend that you go to the website http://picamera.readthedocs.org/en/release-1.3/, which contains a wealth of information regarding the Pi Camera, its properties, operational modes, and a lot more. Best of all, just about everything can be controlled using Python, which means the python-picamera library opens up a tremendous opportunity to apply the Pi Camera in many innovative ways. It truly is a great open source resource that all developers should appreciate.

By the way, changing a property in a Python script is as simple as entering the following in the script, assuming the Pi Camera has been instantiated as an object named `camera`:

```
camera.brightness = 70
```

Recording a video with the library is again very easy to do using the following command:

```
camera.start_recording('video.h264')
```

TABLE 5-3 Pi Camera Properties with Default Values

Property	Default Value
Brightness	50
Sharpness	0
Contrast	0
Saturation	0
ISO	0
Video_stabilization	0
Exposure_mode	Auto
Meter_mode	Average
Awb_mode	Auto
Image_effect	None
Color_effect	None
Rotation	0
Hflip	False
Vflip	False
Crop	(0.0, 0.0, 1.0, 1.0)

This command starts a 1080p high-definition (HD) video stream to a file named video.h264. You stop the recording with this command:

```
camera.stop_recording()
```

My previously stated caution about storage capacity still holds true for recordings made using the Python library. It is just too easy to create a huge video file, which will crash your Pi once it runs out of SD card memory unless you are using an external hard drive. The documentation web page does discuss more limited video resolution modes that can be used, which will drastically cut back on storage requirements at the expense of video resolution. Of course, lower video resolution means diminished viewing quality, but that is a tradeoff that you must constantly consider.

Remote Raspberry Pi Camera Viewing

I will show you how to install an application named MJPG Streamer, which will allow the video generated by the Pi Camera to be viewed on a browser. This software installation is based on instructions I found on Miguel Grinberg's blog, which can be read at http://blog.miguelgrinberg.com/post/how-to-build-and-run-mjpg-streamer-on-the-raspberry-pi. Please follow these steps very closely as some commands are complex and quite lengthy:

1. Install the necessary libraries by entering the following:

```
sudo apt-get install libjpeg8-dev imagemagick libv4l-dev ↵
```

NOTE *The last entry is* `libv4l-dev`, not `libv41-dev`. This caused me some problems until I realized it was not the number 41 after `libv` but a lowercase *L*.

2. Add a new link for a required header file by entering the following:

```
sudo ln -s /usr/include/linux/videodev2.h  /usr/include/linux/
videodev.h↵
```

3. Download the compressed MJPG-Streamer file from sourceforge.net:

```
sudowgethttp://sourceforge.net/code-snapshots/svn/m/mj/mjpg-streamer/
code/mjpg-streamer-code-182.zip↵
```

4. Unzip the download. I used my Home directory, but you can use a temp directory if you so choose:

```
sudo unzip mjpg-streamer-code-182.zip ↵
```

5. You now need to compile some of the many downloaded and extracted files. Just a few of the files, which are plugins, are needed for this streamer application. Enter the following commands in the sequence shown:

```
cd mjpg-streamer-code-182/mjpg-streamer ↵
sudo make mjpg_streamer  input_file.so  output_http.so ↵
```

6. It is time to install the key MJPG-Streamer files into the appropriate directories by entering these commands in sequence:

```
sudocpmjpg_streamer  /usr/local/bin ↵
sudocp output_http.so  input_file.so  /usr/local/lib ↵
sudocp -R www  /usr/local/www ↵
```

7. The Pi Camera's still photo application is now started with certain preset options.

NOTE *You may alter the options as you desire using the raspistill options discussed earlier in the chapter. Enter the following to start the camera in a VGA mode:*

```
sudomkdir /tmp/stream
raspistill --nopreview -w 640 -h 480 -q 5 -o /tmp/stream/pic.jpg -tl
100 -t 9999999 -th 0:0:0 &↵
```

Entering the preceding command will start the camera taking stills at about a 10-second interval but does not start the streamer. The next step does that.

8. Start the streamer by entering the following:

```
sudo LD_LIBRARY_PATH=/usr/local/lib  mjpg_streamer  -i  "input_file
.so  -f  /tmp/stream -n  pic.jpg"  -o  "output_http.so  -w  /usr/
local/www" ↵
```

9. The still picture stream should now be available on your local network. The easiest way to view the stream is to open a browser on a local, networked computer and go to the Pi's IP address with port 8080 appended. In my case it was:

```
192.168.1.26:8080.
```

Figure 5-11 illustrates this site.

The last paragraph in Figure 5-11 states:

"The image displayed here was grabbed by the input plugin. The HTTP request contains the GET parameters *action=snapshot*. This requests one single picture from the image-input. To display another example, just click on the picture."

I clicked on the image and received another image, which is shown in Figure 5-12.

So far, so good. What I really wanted was to see a continuing stream of still pictures. This was easily accomplished by clicking on the here link, as shown in Figure 5-12. The result was the picture shown in Figure 5-13, which was updated about every 10 seconds, as indicated by the elapsed time shown on the stopwatch smartphone app.

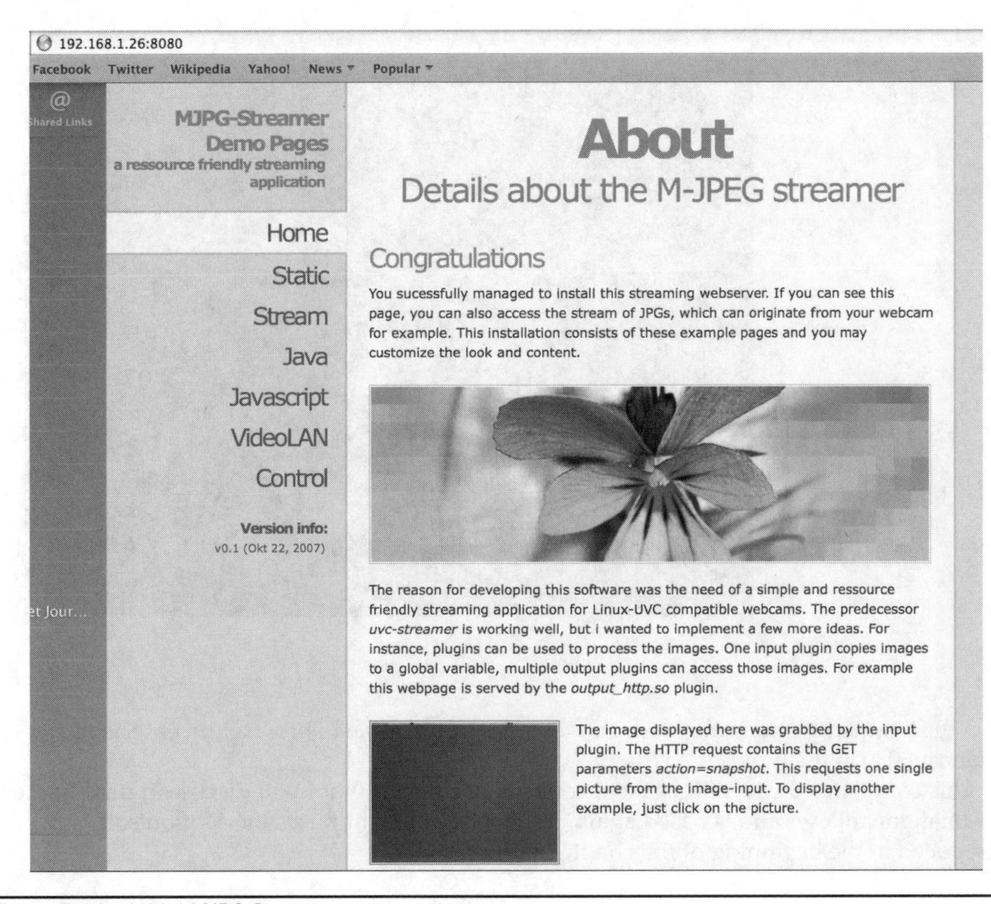

Figure 5-11 Initial MJPG-Stream screen capture

Static
A static snapshot
Hints
This example shows a static snapshot. It should work with any browser. To see a simple example click here.
Source snippet
<img src="/?action=snapshot" />

0:00:48.3
LAP
STOP

FIGURE 5-12 Result of clicking on the initial snapshot

The streamed pictures may also be viewed locally on the Pi browser by entering http://localhost:8080 in the browser's URL box.

This completes the second of the three camera projects that I will discuss in this chapter. The final one follows and is based again on the Pi Camera but uses the Motion software introduced at the beginning of the chapter.

Figure 5-13 One of the continuing streamed pictures from MJPG-Streamer

Raspberry Pi Camera with Motion Software

The procedures for setting up a Pi Camera to work with the Motion software are based on the steps on this Instructables website: www.instructables.com/id/Raspberry-Pi-as-low-cost-HD-surveillance-camera/?ALLSTEPS.

This Instructable project was created by the user scavix, who apparently lives in Germany. I do wish to acknowledge scavix's fine project. The following is a series of steps I created based largely on scavix's Instructable article.

1. Install Motion per the instructions detailed at the beginning of this chapter.
2. Enter this very long apt-get install command. Take your time to ensure that it is entered exactly as shown here:

```
sudo apt-get install -y libjpeg62 libjpeg62-dev libavformat53
libavformat-dev libavcodec53 libavcodec-dev libavutil51-dev libc6-dev
zlib1g-dev libmysqlclient13 libmysqlclient-dev libpq5 libpq-dev↵
```

3. Get a specially modified Motion build that was created to work with the Pi Camera. Enter the following:

```
sudo wget https://www.dropbox.com/s/xdfcxm5hu71s97d/motion-mmal.tar
.gz ↵
```

Note *The compressed file is about 148KB in size.*

4. Copy the compressed file to the /tmp directory by entering the following:

```
sudocp motion-mmal.tar.gz /tmp/motion-mmal.tar.gz↵
```

5. Change to the /tmp directory and extract the compressed file by entering these commands:

```
cd /tmp↵
sudo tar zxvf motion-mmal.tar.gz↵
```

6. Move the extracted files as follows:

```
sudo mv motion /usr/bin/motion↵
sudo mv motion-mmalcam.conf  /etc/motion.conf↵
```

7. Edit the standard Motion daemon file to allow daemon operation by editing it as follows:

```
sudo nano /etc/default/motion↵
Change start_motion_daemon=no to start_motion_daemon=yes
```

   Save the file and exit the nano editor.

8. Replace the motion.conf file found in the /etc directory to this one, which you can download from www.scavix.com/files/raspberry_surveillance_cam_scavix.zip. This file is compressed and must be uncompressed to retrieve the desired file. Enter the following to unzip this file:

```
sudo unzip -lv raspberry_surveillance_cam_scavix.zip ↵
```

   Next, move the extracted motion.conf file to the /etc directory:

```
sudo mv motion.conf  /etc/motion.conf ↵
```

   The series of steps that I detailed from the Instructables website enables you to manually edit the motion.conf instead of downloading, extracting, and overwriting the existing motion.conf file. I just found it much easier to download the altered configuration file.

9. Reboot the Pi so that all the configuration changes can take place, and start the Motion daemon. Enter this to reboot:

```
sudo reboot ↵
```

   If all went well, you should see the red LED on the Pi Camera light after the boot sequence completes.

10. You can view the Pi camera remotely in exactly the same manner as described in the first project, which used a webcam with the Motion software. Simply open a browser on a networked computer that the Pi is connected with and navigate to the Pi's IP address with the 8080 port appended, which in my case was

```
http://192.168.1.26:8080
```

Figure 5-14 is a screen capture from my laptop of the now infamous smartphone stopwatch app.

This last project concludes this chapter's discussion on how to use both a webcam and the Pi Camera with the Pi to achieve both local and remote viewing.

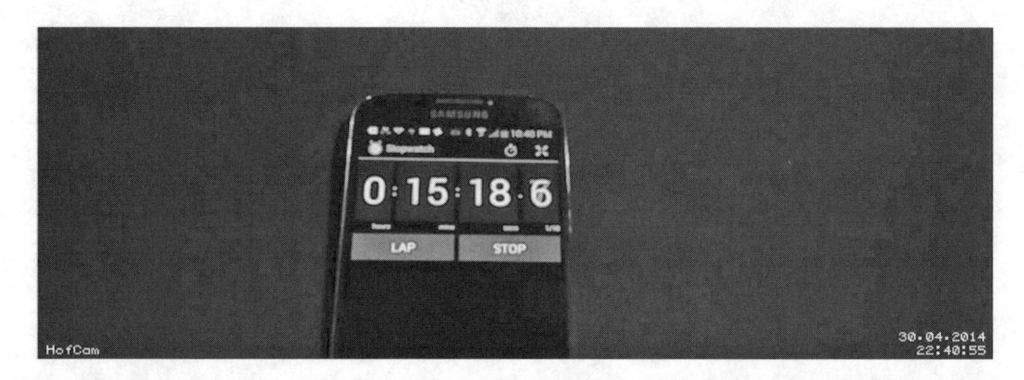

Figure 5-14 MacBook screen capture of the Safari browser connected with the Motion web server

Summary

This chapter included three projects, each designed to illustrate how to effectively use both a webcam and the Pi Camera to enable video and still pictures to be taken and viewed both locally and remotely over a network.

The first chapter project focused on setting up a traditional USB webcam to function with an open source software package named Motion. I explained some of Motion's key features and pointed out that I would be using very few of these features for the demonstration project. I went through all the steps to install and set up Motion for use with the Pi. This project concluded with a discussion of how to view both still pictures and videos taken with the webcam both locally and from a remote networked computer.

The next project in the chapter introduced the Pi Camera, which was designed and made available by the Raspberry Pi Foundation specifically for the Pi. I covered the various still picture and video modes, as well as the two bundled binary applications, raspistill and raspivid, which enable the camera to take stills and videos, respectively.

I also showed you how to use a Python library named python-picamera to create programmed scripts that can automate the Pi Camera's functions. I demonstrated a simple program that automatically takes a series of time-lapsed still pictures and stores them on the SD card.

The next section showed you how to install and set up the MJPG Streamer application, which enables the Pi Camera and Pi to stream both still pictures and video over the network. I demonstrated how to access these streams using a browser operating wirelessly from a remote computer connected to the same network to which Pi was connected.

The final project in this chapter demonstrated how to install and set up the Motion software to function with the Pi Camera. Using Motion with the Pi Camera opens up many capabilities that would be unavailable with either one of the two binary applications, or would be too hard to develop with the Python library. Remote viewing the Motion-created pictures was shown to be identical to the way it was done with the first webcam project.

6

CHAPTER

Internet-Enabled, Arduino Powered Garage Door Opener

This chapter's project will use the classic Arduino Uno development board with an Internet connection to provide a means to open an electrically powered garage door using a web browser, smartphone, or tablet. This project complements the use of a conventional radio control transmitter that is ordinarily used by most homeowners to remotely operate their garage doors. The Arduino door opener project will also make use of passwords to provide a strong security to the project.

Arduino Hardware

There are two hardware boards used in this project that I will separately discuss. The first is the Arduino Uno development board, which contains the microcontroller, and the second is the Ethernet Shield board, which connects the Uno board to an Ethernet network.

Arduino Uno Development Board

I will start with a brief overview of the Arduino Uno development board as I suspect most readers will already be familiar with it. If not, I would highly recommend reading Simon Monk's excellent book on programming the Arduino boards, *Programming Arduino: Getting Started with Sketches* (McGraw-Hill Education, 2011). The word "Sketches" in the book title refers to the name the Arduino Project gives to programs written for Arduino development boards. I will discuss Sketches and many other related programming elements in the software section, but first I want to provide a brief tour of the Arduino hardware I will be using in this project.

The Arduino Uno board I used is shown in Figure 6-1. It is a rev 3 board, which is important to note as the pin sockets changed slightly between the board revisions.

You can quickly identify rev 3 boards as the reset button was relocated from the center right-hand side on earlier versions to the upper left-hand side on rev 3 boards. The key specifications of the Uno board are specified in Table 6-1.

Don't be concerned if you do not understand some of the specification abbreviations in the preceding table as I will explain them if they are needed for the project. I recommend

FIGURE 6-1 Arduino Uno rev 3 board

looking at Atmel's ATMEGA328P datasheet if you want to learn more about the detailed microprocessor specifications.

The single most important item to be mindful of regarding the Uno is that it is a microcontroller board and not a fully operational computer such as a Raspberry Pi. The significant difference is the Uno has no capability of hosting an operating system and cannot support any programming development using only the board. It must be connected to an external computer in order to be programmed. This does not make the Uno inferior to the Pi; it just is designed for a different approach for controlling embedded projects as compared to the Pi or the Beaglebone Black, which I discuss in upcoming chapters.

Item	Value	Remarks
Microcontroller	8-bit Atmel ATMEGA328P	28-pin DIP socket
Operational voltage	5V	Input range: 7–12V
Digital GPIO	14	6 capable of PWM
Analog IO	6	10-bit
Program memory	Flash 32KB, EEPROM 1KB	SRAM 2KB
Clock speed	16 MHz	
USB	Type B socket	
Programmer	In-system firmware	USB-based
Serial communications	SPI, I2C	Software UART
Other	RTC, watchdog, interrupts	

TABLE 6-1 Arduino Uno Key Specifications

The open-source Arduino Project may be accessed at http://arduino.cc, the home page that contains many links to other pages that I know you will find very informative. In fact, I highly recommend that you stop reading this book for a while and go to this site and become acquainted with the Arduino concept as it will help you comprehend the software underlying the Arduino boards.

Ethernet Shield Board

Figure 6-2 shows the Ethernet Shield board that I used to establish a wired network connection between the Uno board and my local area network (LAN).

The word "shield" in the board title simply refers to any one of a number of boards that are designed to plug directly on top of an Arduino microcontroller board such as the Uno. This one contains all the hardware required to establish a wired connection between the Uno and the LAN. This board is an older version of the Arduino Ethernet Shield and is based on the Wiznet W5100 chip. Figure 6-3 is the W5100 block diagram where you can see all the component parts that make up the built-in networking functions.

It is important to point out that while the W5100 chip does contain 16KB for memory buffers, it does not have any provision to store a program and is totally dependent upon the Uno for its programmed instructions in order to function. This means a separate Ethernet library must be loaded into the Uno's flash memory in order for this shield to work properly. This approach generally holds true for most of the shield boards that are designed for use with the Arduino series of microcontrollers.

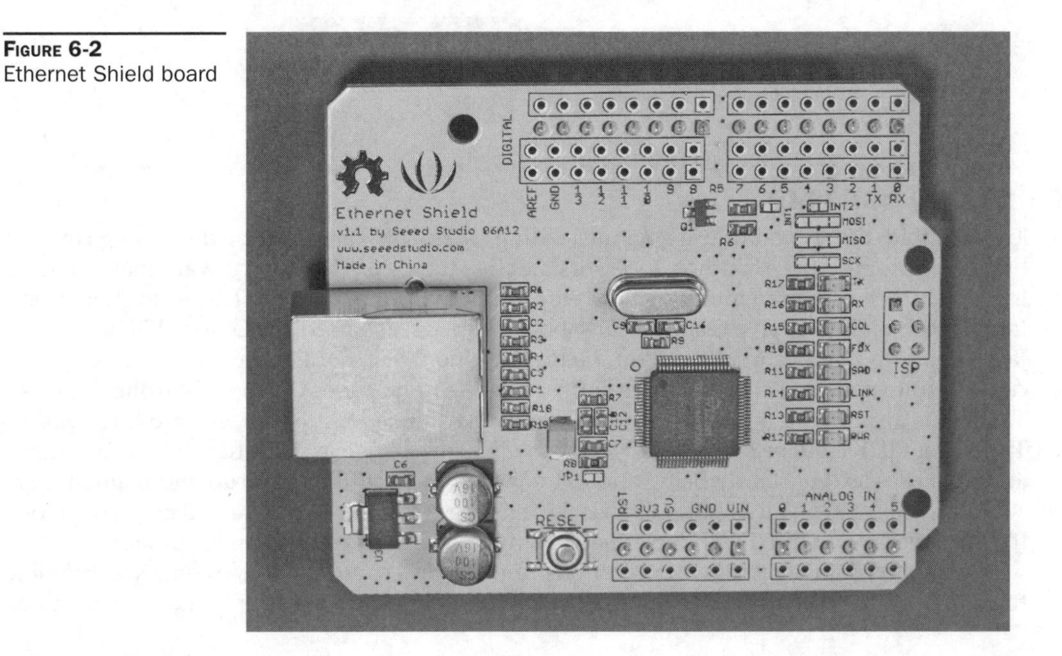

FIGURE 6-2 Ethernet Shield board

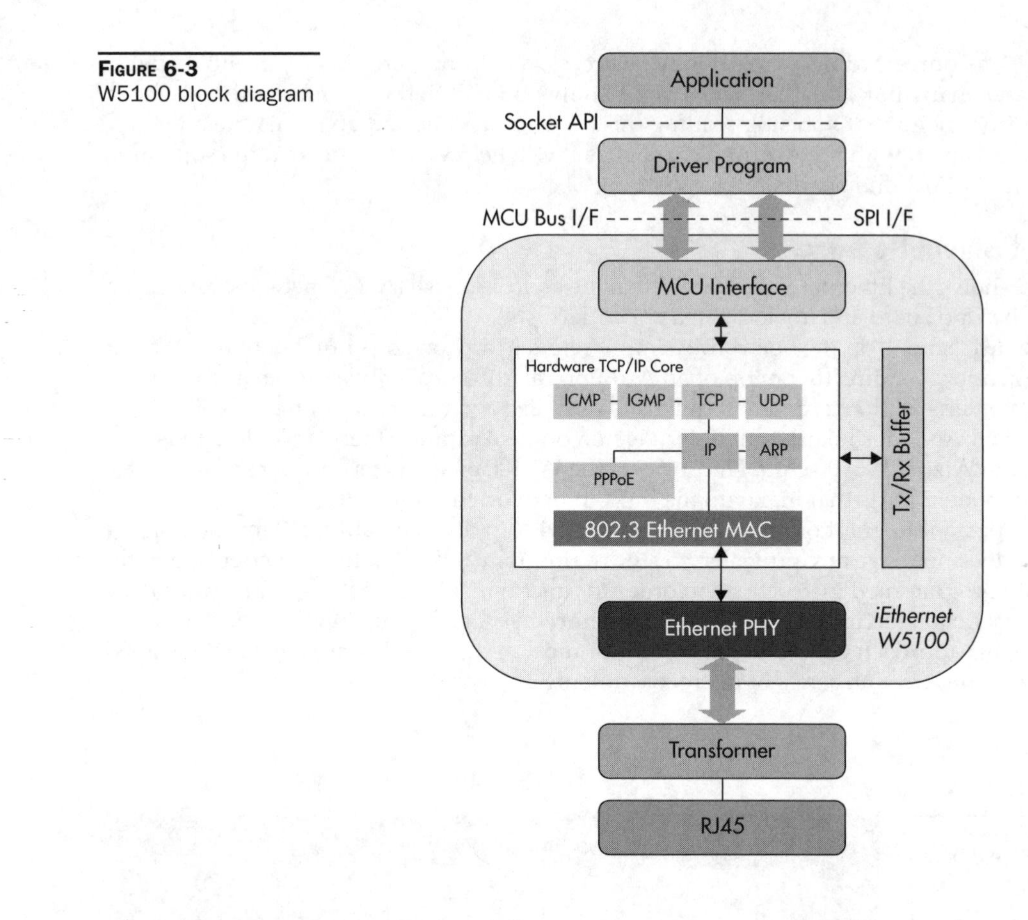

FIGURE 6-3 W5100 block diagram

Arduino Uno Software

Truth be told, the section title is a bit misleading as the software I will be discussing covers a broad spectrum of the Arduino processors, not just the Uno. The key software that you need to program the Uno is an integrated development environment (IDE). This term should be familiar to you if you have read the early book chapters, as I used the Eclipse IDE to test Java programs that eventually ran on the Pi. The Uno IDE is not Eclipse, but it is a fully capable suite designed specifically to support the Arduino series of microcontroller boards. The IDE is available as a free download from the Arduino website that was provided earlier. The current IDE I will be using is 1.05, which will likely change in the future as improvements and upgrades are constantly being added by the very smart folks that run and maintain the Arduino Project. One nice feature is that the existing Arduino hardware will always run on the latest version of the IDE. No planned or unplanned obsolescence in this arena.

I would recommend that you power on your Uno and connect it to the computer running the IDE using a standard USB cable. Almost any "wall wart" power supply that uses a 2.1 mm

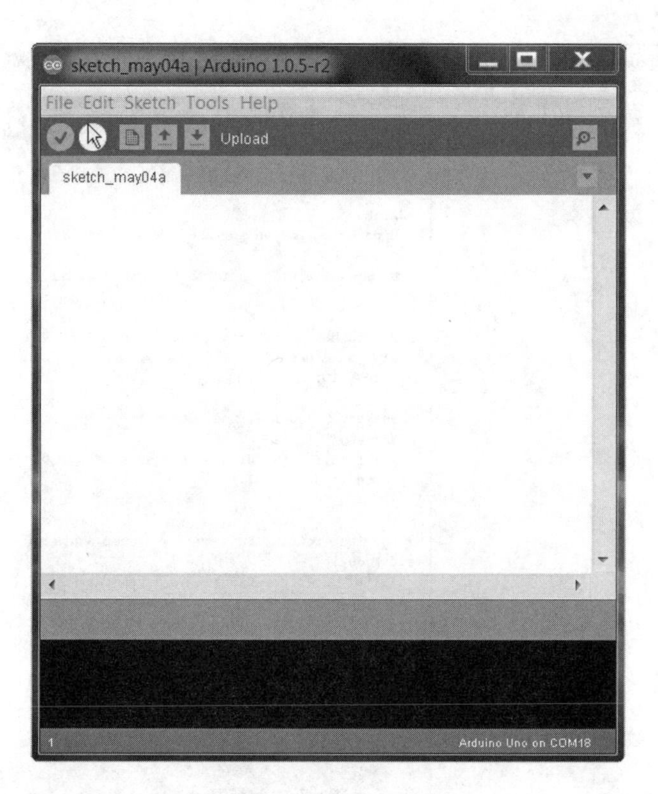

Figure 6-4 Arduino IDE startup screen

outer barrel with a positive center connector will work. Remember that supply must be between 7 and 12VDC. I used a surplus power supply that provides 7.5V at 1.5A, which is more than ample for this project. Your computer should show a dialog that a driver is being installed after the Uno is plugged into the computer. Wait until the driver has been installed before starting the IDE.

Figure 6-4 is a screenshot of the Arduino v1.05 Start screen on a Windows laptop.

The IDE automatically created a default sketch entitled sketch_may04a, which obviously contains the date that I ran the IDE program. You would normally use this blank sketch to create a program and then rename it to whatever suits your application. I will not be creating a sketch for this demonstration but will instead load a pre-stored example to demonstrate the classic LED blink program. There are many example programs automatically loaded into the computer during the IDE download. The program I opened was aptly named "blink" and was loaded by following this sequence: Select progressively File | Open | Examples | 01.Basics | Blink | Blink.

Figure 6-5 shows the loaded Blink program, which appears in its own window. Note that the original window for the sketch_may04a is still open in the background. This makes working on multiple programs very easy and convenient as all you need to do is select the desired window to resume development in that program.

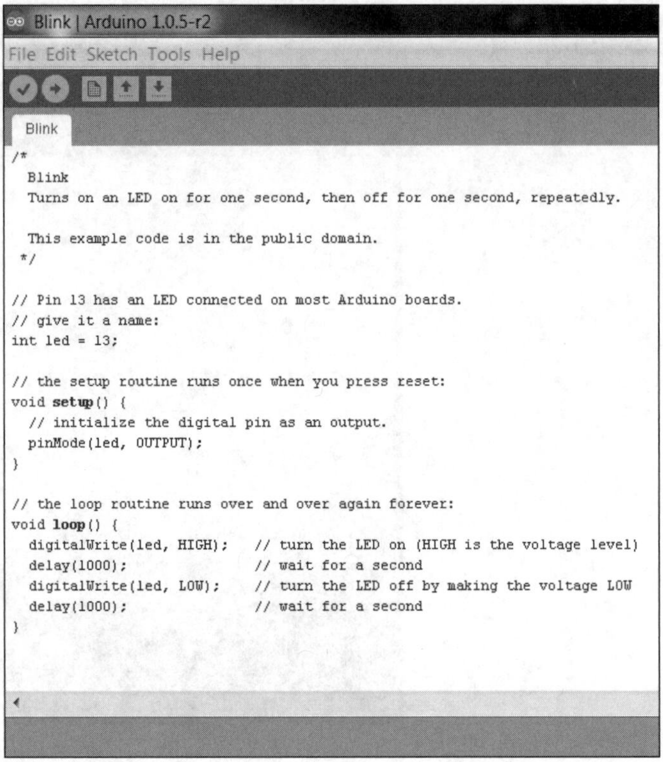

```
Blink | Arduino 1.0.5-r2
File Edit Sketch Tools Help
Blink
/*
  Blink
  Turns on an LED on for one second, then off for one second, repeatedly.

  This example code is in the public domain.
 */

// Pin 13 has an LED connected on most Arduino boards.
// give it a name:
int led = 13;

// the setup routine runs once when you press reset:
void setup() {
  // initialize the digital pin as an output.
  pinMode(led, OUTPUT);
}

// the loop routine runs over and over again forever:
void loop() {
  digitalWrite(led, HIGH);   // turn the LED on (HIGH is the voltage level)
  delay(1000);               // wait for a second
  digitalWrite(led, LOW);    // turn the LED off by making the voltage LOW
  delay(1000);               // wait for a second
}
```

Figure 6-5 Blink code

I'm including the following Blink code in order to point out some key program parts for this introductory example. I will not normally list example program code, as it is easy to load and examine by yourself.

```
/*
  Blink
  Turns on a LED for one second, then off for one second.
 /This example code is in the public domain.
 */
 // Pin 13 has a LED connected on most Arduino boards.
// give it a name:
int led = 13;
// the setup routine runs once when you press reset:
void setup() {
  // initialize the digital pin as an output.
  pinMode(led, OUTPUT);
}
// the loop routine runs over and over again forever:
void loop() {
  digitalWrite(led, HIGH);   // turn the LED on (HIGH is the voltage level)
  delay(1000);               // wait for a second
  digitalWrite(led, LOW);    // turn the LED off by making the voltage LOW
  delay(1000);               // wait for a second
}
```

This sketch has two methods: `setup` and `loop`. The `setup` method is always run first, followed by the `loop` method. The `setup` method provides the logical name "led" to the LED attached to the Uno's pin 13. It also makes GPIO pin 13 an output.

The `loop` method is a forever loop that alternately turns on the LED for one second and then turns it off for one second. The `digitalWrite` method is the means by which the Uno controls pin 13 and, ultimately, the LED.

You should note that I didn't mention any physical wiring was required for this demonstration as the Uno board already has a yellow LED permanently connected to pin 13. You can easily see this LED in Figure 6-1 as it is labeled with an "L" and is located just to the left and above the ARDUINO silkscreen name.

Selecting the right-facing arrow in the toolbar shown in Figure 6-5 will cause the program to be compiled and uploaded to the Uno. The Blink program will start immediately and continue indefinitely. You also might be a bit confused as the LED probably was already blinking before you uploaded the Blink program. That blinking was due to the default "heartbeat" that runs when no program was previously loaded. You can prove to yourself that the Blink program functions as expected by changing the delay time and observing the new blink rate matches whatever you entered. Simply enter new values for the delay time, say 2000ms, which will make the LED blink at a two-second rate. Compile and upload the changed program by pressing the right-facing arrow and watch the LED slowly blink every two seconds.

Testing the Ethernet Connection

In this section, I will demonstrate how to use the Ethernet Shield board with the Uno. The first step is to unplug the Uno's power and USB cables. You never should attempt to attach or remove a shield board while the Uno is powered on as you might damage it. There is only one way that the shield pins will align with the Uno sockets. Carefully align the boards and firmly press them together. Figure 6-6 shows the Ethernet Shield attached to the Uno. Notice that the shield's Ethernet connector is on the same side as the Uno's power and USB sockets. The shield board is identical in size to the Uno board and should precisely cover it.

Next, connect the Uno's power and USB cables and also connect an Ethernet patch cable from the shield to your network's router, hub, or switch, depending upon how you have configured your LAN. In my case, I connected to a 10/100 switch that is on my development workbench. You should immediately see four or five green LEDs light up on the shield indicating both power is being applied to the board and network activity is being detected via the patch cable. Figure 6-7 shows these active LEDs.

If you do not see these LEDs light, immediately remove the power from the Uno and recheck how the shield and Uno are connected together. The most likely cause is that you have shifted the pins over one or even two spaces. Reposition the shield and reattach it to the Uno and you should be fine.

The Ethernet library, which you will need to access, is available from the IDE's menu selections: Open | Examples | Ethernet menu.

All the current Ethernet library menu selections are detailed in Table 6-2.

Open the DhcpAddressPrinter program and upload it into the Uno. After it uploads, select the spyglass icon near the upper-right corner of the IDE to open a serial terminal. You should see the Uno's IP address displayed, as shown in Figure 6-8.

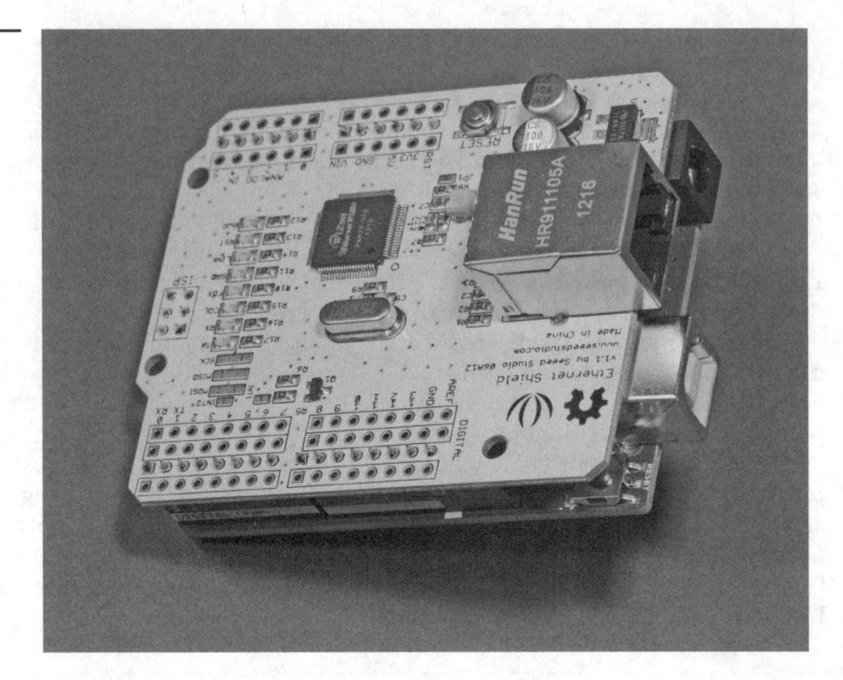

FIGURE 6-6 Ethernet Shield attached to the Uno

My network router assigned a 192.18.1.29 IP address to the Uno/Ethernet Shield. Your address would most certainly be different, however; please make note of it, as you will need it for the next program.

Next, load the program named WebServer, located in the Ethernet library, into the Uno. This program creates an active server that responds to HTTP requests at the IP address you

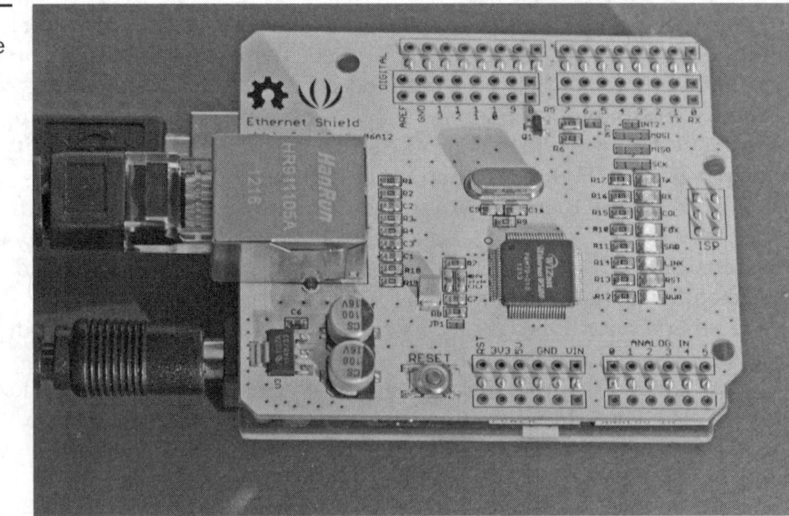

FIGURE 6-7 Ethernet Shield active LEDs

Program Name	Brief Description
ChatServer	Set up a simple chat server.
ChatClient	Set up a simple chat client to interact with the chat server.
WebClient	Make an HTTP request.
WebServer	Host a simple HTML page displaying six analog values.
PachubeClient	Now uses Xively.com, a free data logging site.
PachubeClientString	Send strings to Xively.com.
BarometricPressureWebServer	Outputs the values from a barometric pressure sensor.
UDPSendReceiveString	Send and receive text strings via UDP.
UdpNtpClient	Query a Network Time Protocol (NTP) server using UDP.
DnsWebClient	DNS and DHCP-based Web client.
DhcpChatServer	A simple DHCP Chat Server.
DhcpAddressPrinter	Get an IP address via DHCP and print it out.
TelnetClient	A simple Telnet client.

Table 6-2 The Ethernet Library

determined with the previous program, and it returns the raw data from the Uno's six analog input channels. WebServer uses the default HTTP port 80 so all you will have to do is enter the IP address on a remote, networked computer to see the results. However, you must make the following change in WebServer in order for it to work with the Uno's preassigned IP address. Just type in your IP address where it is shown in the following code snippet from the WebServer sketch:

```
// Enter a MAC address and IP address for your controller below.
// The IP address will be dependent on your local network:
byte mac[] = {
  0xDE, 0xAD, 0xBE, 0xEF, 0xFE, 0xED };
IPAddress ip(192,168,1,29); <-- This is where you will enter your own ip address
```

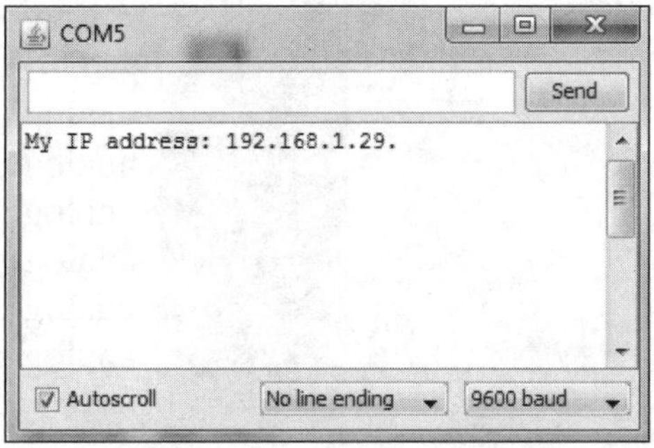

Figure 6-8 IP address displayed

Note that the IP numbers are separated (delimited) by commas and not periods as they normally would be. Incidentally, for those readers who have some background in network protocols, I would point out that a dummy mac address is being used for this and all the other Arduino Ethernet programs for this project. There are two reasons; first it doesn't make any real-world difference what the actual mac address is as long as it is unique within the network. The second reason is that this older Ethernet Shield did not come with any sticker showing the actual firmware mac address assigned to the chip. I think this situation holds true for most of the old revision Ethernet Shields. I also understand newer versions do have a sticker on the board showing the actual address. If that's the case, all you need to do is change the preset mac address to match the sticker address. Strangely, it should also work if you do not make the change but I believe it is good practice to always match the mac addresses if possible.

You should now open a browser on a networked computer and enter the Uno's IP address, which in my case was 192.168.1.29. It is purely optional to enter the "http://" prefix as all modern browsers assume that is the default protocol to use. Figure 6-9 shows the output on MacBook Pro when I entered that address.

In the figure, you can see numbers ranging from a low of 282 to a high of 321. These represent the voltage "floating" on the six unconnected Uno analog input channels. Each channel is 10 bits in resolution and has a full-scale voltage input of 5.0V, which is the same as the Uno's operating voltage. This means that a count of 1023 would represent a 5.0V input to a particular analog channel. Consequently, numbers averaging around a 300 count represent an approximate 1.47V level, which is determined by a simple proportional calculation:

Floating input voltage = 5.0 * (300 / 1023) = 1.47V

I soldered a jumper wire between ground and the analog input A0, as you can see in Figure 6-10.

My purpose in this effort was to further confirm the analog inputs were properly operating as I expected to see a 0V level when I next opened a browser to the WebServer page. My expectations were met as evidenced by Figure 6-11, which is the WebServer operating with the A0 input grounded.

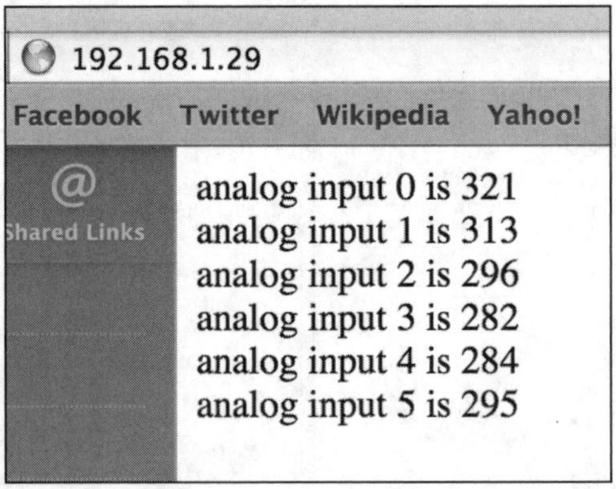

FIGURE 6-9 The WebServer program on a remote browser

192.168.1.29
Facebook Twitter Wikipedia Yahoo!
Shared Links
analog input 0 is 321
analog input 1 is 313
analog input 2 is 296
analog input 3 is 282
analog input 4 is 284
analog input 5 is 295

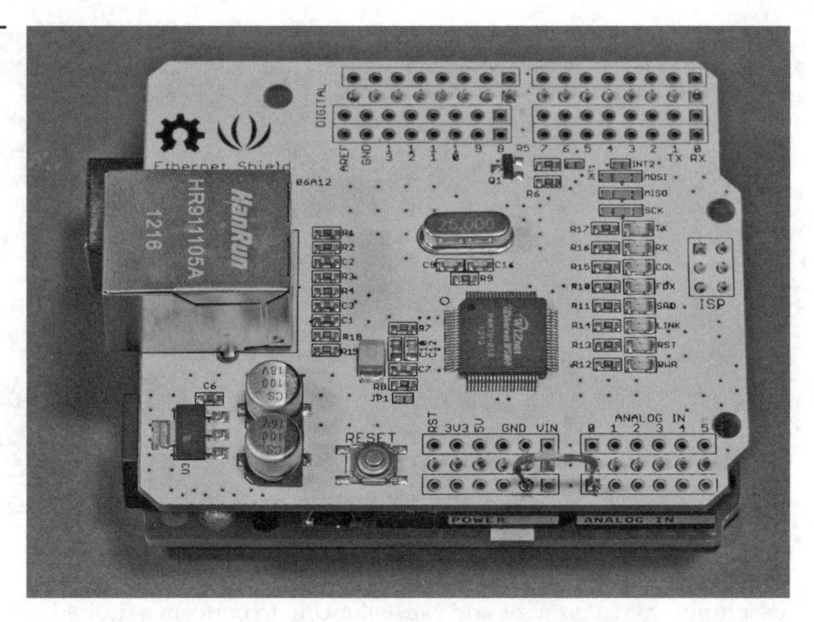

FIGURE 6-10 Jumper wire soldered between ground and analog input A0

Interestingly, you can see that not only does the A0 input have a 0 value but all the others are much lower than they were before when nothing was connected to the inputs. I can only infer that the inputs must have been at very high impedances, which are easily influenced by stray and nearby electrical fields. This also explains why, when I attempted to measure the open voltages on the analog inputs with a standard multimeter, I recorded much lower voltages than I had earlier calculated. My moderate impedance multimeter simply loaded down the high impedance inputs and showed voltages that were a result of the multimeter shunting the high impedance input. This is a good example of recognizing when your test equipment can have an unexplained effect upon the device under test (DUT). By the way, DUT is a standard electronics industry description for any electrical/electronic equipment being tested. This possible effect should always be kept in mind when testing any device.

At this point, I believe I have fairly well demonstrated how to connect an Uno to a network and have it perform useful tasks. It is now time to demonstrate a program that will accept

FIGURE 6-11 WebServer page with A0 grounded

analog input 0 is 0
analog input 1 is 89
analog input 2 is 149
analog input 3 is 174
analog input 4 is 207
analog input 5 is 198

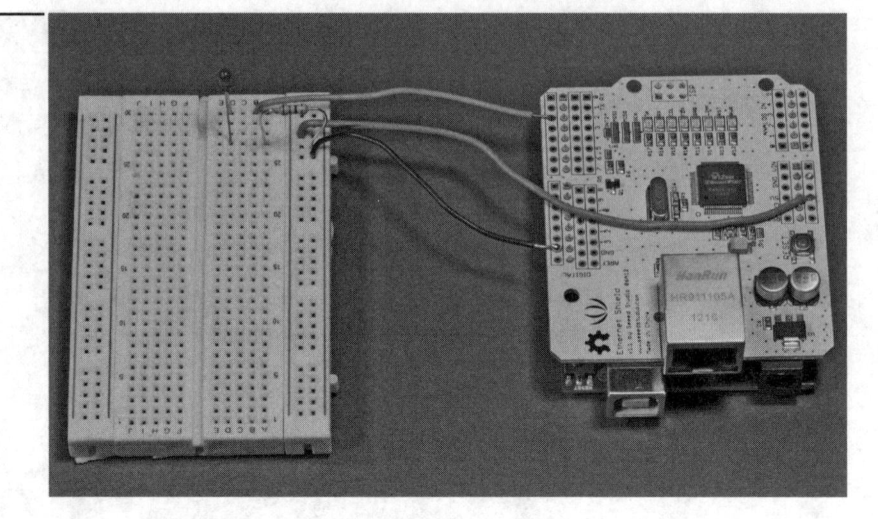

Figure 6-12 Physical LED test setup

a user input from a browser and cause the Uno to perform a desired operation. This program will show you the fundamental concepts involved in controlling a microprocessor over a networked connection. This program will also be the basis for the simplified garage door opener project. For this initial demonstration, I will only be lighting a LED when a checkbox is selected.

A simple hardware setup should be done before discussing the code. This simply entails connecting a LED's anode to the Uno's pin 2 and 220Ω resistor in series from the LED's cathode to ground. Figure 6-12 shows the physical setup. Note that I soldered wires directly to the Ethernet Shields pins that directly connect with pin 2 and one of the ground pins. I also connected to the 5V supply, which I use in a later example.

```
The program that controls this demonstration is named eth_websrv_LED and all the
program authors are credited in the beginning comments section. The code is listed
here in its entirety. I'll break it down section by section in the material that
follows. /*--------------------------------------------------------------
Program: eth_websrv_LED

Description: Arduino web server that serves up a web page allowing the user to
control a LED.

Hardware: - Arduino Uno and official Arduino Ethernet Shield. Should work with
other Arduinos and compatible Ethernet Shields.  A LED and a 220Ω resistor are put
in series between Arduino pin 2 and ground.

Software: Developed using Arduino 1.0.3 software.  Should be compatible with
Arduino 1.0 +

References: - WebServer example by David A. Mellis and modified by Tom Igoe
            - Ethernet library documentation: http://arduino.cc/en/Reference/
Ethernet
```

```
Date:         11 January 2013

Author:       W.A. Smith, http://startingelectronics.com
------------------------------------------------------------*/

#include <SPI.h>
#include <Ethernet.h>

// MAC address from Ethernet Shield sticker under board
byte mac[] = { 0xDE, 0xAD, 0xBE, 0xEF, 0xFE, 0xED };
IPAddress ip(10, 0, 0, 20); // IP address, may need to change depending on network
EthernetServer server(80);  // create a server at port 80

String HTTP_req;              // stores the HTTP request
boolean LED_status = 0;  // state of LED, off by default

void setup()
{
    Ethernet.begin(mac, ip); // initialize Ethernet device
    server.begin();                 // start to listen for clients
    Serial.begin(9600);         // for diagnostics
    pinMode(2, OUTPUT);  // LED on pin 2
}

void loop()
{
    EthernetClient client = server.available();// try to get a client

    if (client) {                       // got a client?
        boolean currentLineIsBlank = true;
        while (client.connected()) {
            if (client.available()) {// client data available to read
                char c = client.read(); // read 1 byte (character) from client
                HTTP_req += c;// save the HTTP request 1 char at a time
                // last line of client request is blank and ends with \n
                // respond to client only after last line received
                if (c == '\n' && currentLineIsBlank) {
                    // send a standard http response header
                    client.println("HTTP/1.1 200 OK");
                    client.println("Content-Type: text/html");
                    client.println("Connection: close");
                    client.println();
                    // send web page
                    client.println("<!DOCTYPE html>");
                    client.println("<html>");
                    client.println("<head>");
                    client.println("<title>Arduino LED Control</title>");
                    client.println("</head>");
                    client.println("<body>");
                    client.println("<h1>LED</h1>");
                    client.println("<p>Select to switch LED on and off.</p>");
                    client.println("<form method=\"get\">");
                    ProcessCheckbox(client);
                    client.println("</form>");
                    client.println("</body>");
                    client.println("</html>");
```

```
                    Serial.print(HTTP_req);
                    HTTP_req = "";// finished with request, empty string
                    break;
                }
                // every line of text received from the client ends with \r\n
                if (c == '\n') {
                    // last character on line of received text
                    // starting new line with next character read
                    currentLineIsBlank = true;
                }
                else if (c != '\r') {
                    // a text character was received from client
                    currentLineIsBlank = false;
                }
            } // end if (client.available())
        } // end while (client.connected())
        delay(1);      // give the web browser time to receive the data
        client.stop(); // close the connection
    } // end if (client)
}

// switch LED and send back HTML for LED checkbox
void ProcessCheckbox(EthernetClient cl)
{
    if (HTTP_req.indexOf("LED2=2") > -1) {//see if checkbox was selected
        // the checkbox was selected, toggle the LED
        if (LED_status) {
            LED_status = 0;
        }
        else {
            LED_status = 1;
        }
    }

    if (LED_status) {    // switch LED on
        digitalWrite(2, HIGH);
        // checkbox is checked
        cl.println("<input type=\"checkbox\" name=\"LED2\" value=\"2\" \
        onselect=\"submit();\" checked>LED2");
    }
    else {              // switch LED off
        digitalWrite(2, LOW);
        // checkbox is unchecked
        cl.println("<input type=\"checkbox\" name=\"LED2\" value=\"2\" \
        onselect=\"submit();\">LED2");
    }
}
```

The code starts with two `include` statements that refer to all the dependencies necessary for this program to function. These are the SPI and basic Ethernet libraries.

A "dummy" mac address assignment follows, which you can use as is or else enter your own if known. The allocated Uno IP address is next. This must be entered to match the real IP address or the Ethernet connection cannot be established, as I discussed in an earlier example.

Next are the two standard Arduino methods named `setup` and `loop` that I have already mentioned. This particular `setup` method performs these initializations:

- Instantiates an Ethernet connection object
- Starts listening for client requests
- Starts the serial terminal at 9600 baud
- Sets the Uno GPIO pin 2 as an output

The `loop` method follows the `setup` and is much more extensive than the previous example's `loop` method. The main reason for this is that the Uno does not use an operating system and consequently cannot set up and maintain a file system. This means that code that interacts with the remote browser (client) must be dynamically created each time the web server program is run. That's the purpose of all the `client.println` statements contained within the `loop` method. This necessity should be compared with the way the Raspberry Pi web service functions. The Pi has a full Linux file system, which means the client code can be stored in a predefined directory from which clients can access it. Normally, it is in /etc/www directory and the client file is typically named index.htm or index.html for the HTTP protocol using port 80.

The first statement in the `loop` method is

```
EthernetClient client = server.available();
```

which will assign a non-null value to the client reference if an HTTP request is detected. Next, all the `println` statements will be executed, resulting in a web page being displayed on the client. The page displayed in Figure 6-13 appears when the client, again being a browser on a remote network computer, connects with the Uno's web server program.

The `loop` method will now loop, essentially waiting until the checkbox is selected, thus indicating that the user wants to turn on the connected LED. Checking the status of the checkbox is the function of the third method contained in the ProcessCheckbox method. This method takes a client argument and will either cause the LED to turn on if it was previously off or it will turn it off if it was previously on. The LED status, whether on or off, is stored in a program variable aptly named `LED_status`. If you carefully examined the `ProcessCheckbox` code, you should notice that the HTML call to the browser is exactly the same whether turning the LED on or off. This call simply toggles the checkmark in the checkbox and has nothing to do with actually turning the LED on or off. That's all handled by the Uno in the web server program.

I ran the program and observed the LED turning on and off as expected when the checkbox was selected. I urge you to duplicate this demonstration as it should reaffirm what you know about how the Uno interacts with a web page to control a GPIO pin. The next step is to modify the LED program to control a relay that in turn controls a garage door.

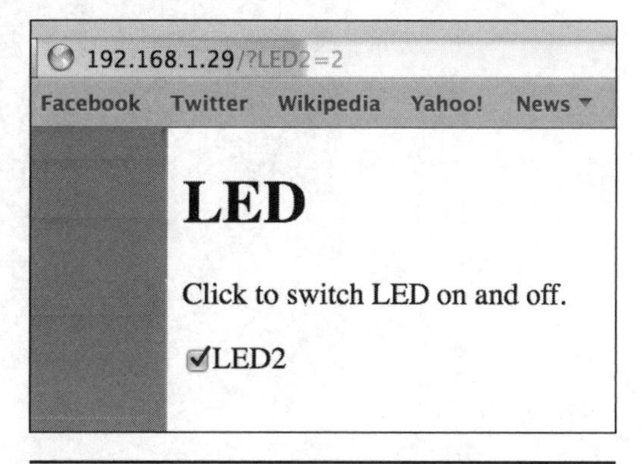

FIGURE 6-13 The client after it has connected to the Uno's web server program

Simplified Garage Door Opener

This portion of the project uses a modification of the previous LED demonstration where a relay is being controlled in lieu of a LED. The relay in turn switches the control power to a garage door opener. I will need to discuss my specific garage door opener system before proceeding with the actual project.

Actual Garage Door Opener

In this section I will show you how to build a basic, no-frills, remotely activated garage door controller. It is designed to operate with an existing residential style opener, which is shown in Figure 6-14.

I believe this opener is fairly typical of the garage door openers that are used for residential service. Either the press of a wall-mounted push button or the press of a button on a radio control transmitter that I have attached to the driver side visor activates it. The plan is simply to connect two wires in parallel to those wires, which are currently connected to the wall-mounted push button. These two new wires will be connected to a set of normally open relay contacts such that when the relay is energized by the Uno, which will then close the contacts and act the same as if the wall-mounted push button was pressed. The relay is needed because there is a 16VDC potential on the wires going to the push button, which far exceeds the Uno GPIO voltage specifications.

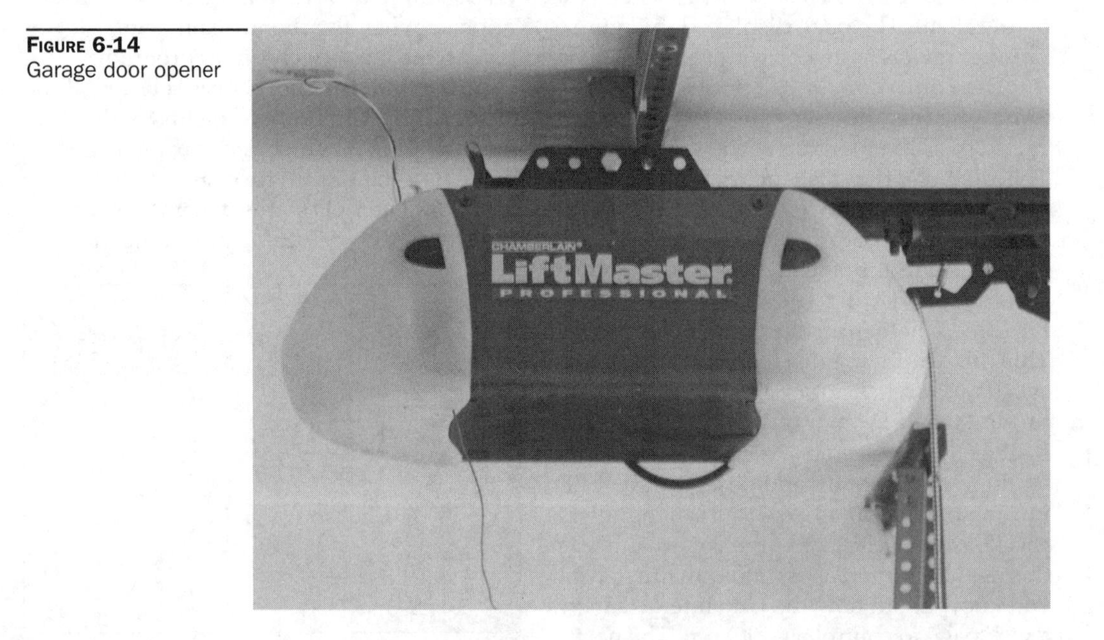

FIGURE 6-14 Garage door opener

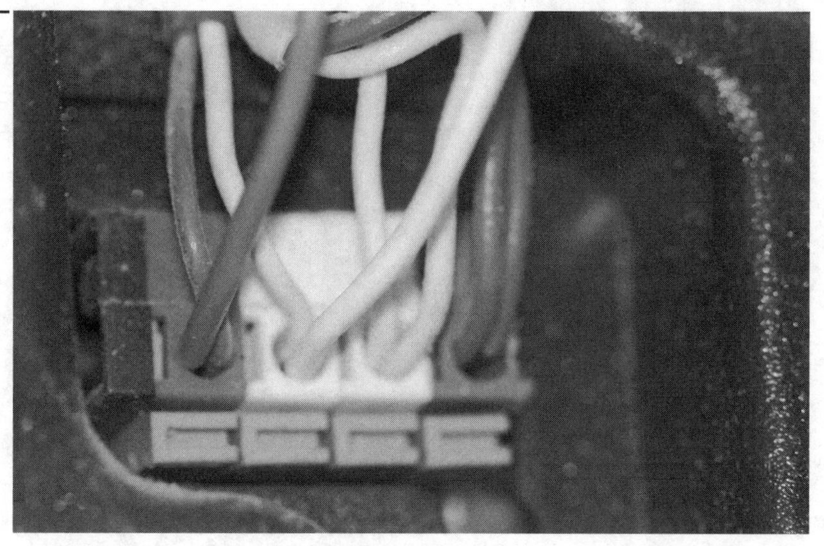

Figure 6-15 Garage door opener connection terminals

Figure 6-15 is a closeup of the connection terminals where I made the push button wires parallel. They are the two left-most wires shown in the figure. You may have to check your own opener's connection diagram to determine which set of wires goes to the wall-mounted push button.

If you refer to Figure 6-14, you can see these two new wires protruding from the left side of the opener. I used #20 gauge, solid-core, twisted bell wire for the new wires as that was exactly the same type that was installed going to the wall-mounted push button. I also briefly touched the newly installed wires together at the remote end where I was installing the Uno to confirm the door did operate as expected.

Modified LED Program to Open Garage Door

I modified the existing LED control program so that it will operate a transistor that will, in turn, control a relay, which has the garage door contact wires connected to a set of normally open contacts. Figure 6-16 is a schematic for this straightforward control circuit. Almost any common NPN switching transistor can be used to switch the relay. I used a 2N3904 in the circuit shown in the schematic.

The relay switching circuit was wired on a solderless breadboard for convenience and easy modification if necessary. Both the breadboard and the Uno board were mounted in a plastic case near the garage door that was being controlled. Figure 6-17 shows the Uno and relay circuit mounted in the case.

I ran both power and an Ethernet cable to the case along with the control wire pair connected to the garage door opener mechanism. You can easily see all the connections within the case, which was mounted on the garage wall.

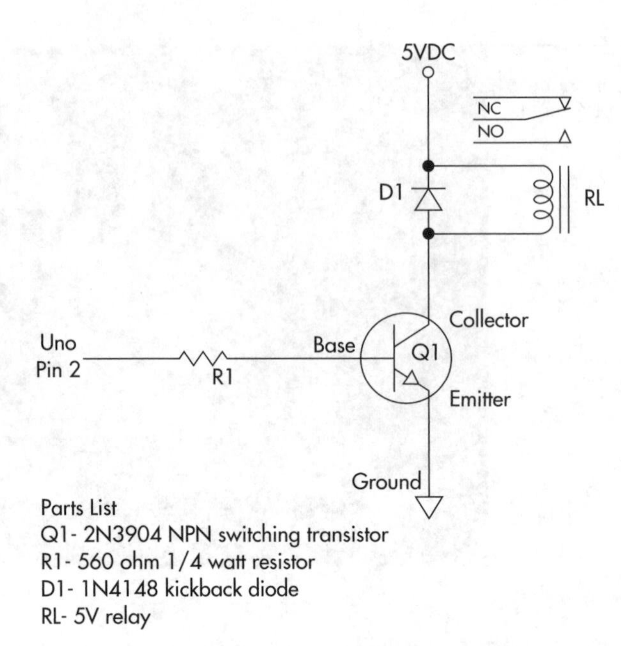

FIGURE 6-16 Relay switching circuit schematic

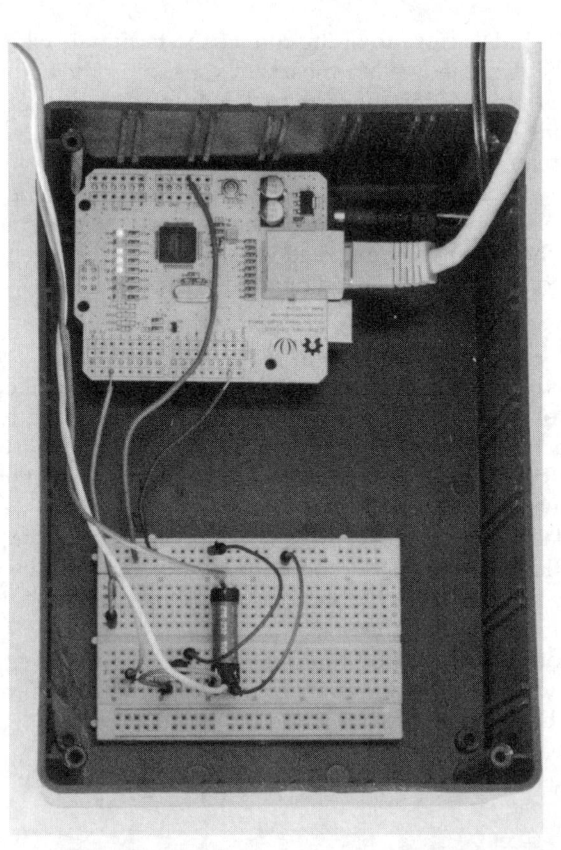

FIGURE 6-17 Uno and relay switcher mounted in plastic case

The following is the modified LED control program. I didn't change much except to operate the relay for one second every time the checkbox is selected. It doesn't matter if there is a checkmark in the box; all you need do is select it to operate the door. This modified code is named Garage_Door_Open.ino and is available on this book's companion website.

```
/*--------------------------------------------------------------
Program: Garage_Door_Open

Description: Arduino web server that serves up a web page allowing the user to
control a garage door.

Hardware: - Arduino Uno and official Arduino Ethernet Shield. Should work with other
            Arduinos and compatible Ethernet Shields.
          - Arduino pin 2 controls a transistor that in turn operates a relay

Software: Developed using Arduino 1.0.3 software.  Should be compatible with
Arduino 1.0 +

References: - eth_websrv_LED modified by D. J. Norris
            - WebServer example by David A. Mellis and modified by Tom Igoe
            - Ethernet library documentation: http://arduino.cc/en/Reference/
Ethernet

Date: 10 May 2014

Author: original W.A. Smith, http://startingelectronics.com
--------------------------------------------------------------*/

#include <SPI.h>
#include <Ethernet.h>

// MAC address from Ethernet Shield sticker under board
byte mac[] = { 0xDE, 0xAD, 0xBE, 0xEF, 0xFE, 0xED };
IPAddress ip(192, 168, 1, 29); // IP address, may need to change depending on network
EthernetServer server(80);  // create a server at port 80

String HTTP_req;          // stores the HTTP request
boolean LED_status = 0;   // state of LED, off by default

void setup()
{
    Ethernet.begin(mac, ip);  // initialize Ethernet device
    server.begin();           // start to listen for clients
    Serial.begin(9600);       // for diagnostics
    pinMode(2, OUTPUT);       // to transistor on pin 2
}

void loop()
{
    EthernetClient client = server.available();  // try to get client

    if (client) {  // got client?
        boolean currentLineIsBlank = true;
        while (client.connected()) {
```

```
            if (client.available()) {   // client data available to read
                char c = client.read(); // read 1 byte (character) from client
                HTTP_req += c;  // save the HTTP request 1 char at a time
                // last line of client request is blank and ends with \n
                // respond to client only after last line received
                if (c == '\n' && currentLineIsBlank) {
                    // send a standard http response header
                    client.println("HTTP/1.1 200 OK");
                    client.println("Content-Type: text/html");
                    client.println("Connection: close");
                    client.println();
                    // send web page
                    client.println("<!DOCTYPE html>");
                    client.println("<html>");
                    client.println("<head>");
                    client.println("<title>Arduino Garage Door Control</title>");
                    client.println("</head>");
                    client.println("<body>");
                    client.println("<h1>Garage Door Opener</h1>");
                    client.println("<p>Select to activate door.</p>");
                    client.println("<form method=\"get\">");
                    ProcessCheckbox(client);
                    client.println("</form>");
                    client.println("</body>");
                    client.println("</html>");
                    Serial.print(HTTP_req);
                    HTTP_req = "";     // finished with request, empty string
                    break;
                }
                // every line of text received from the client ends with \r\n
                if (c == '\n') {
                    // last character on line of received text
                    // starting new line with next character read
                    currentLineIsBlank = true;
                }
                else if (c != '\r') {
                    // a text character was received from client
                    currentLineIsBlank = false;
                }
            } // end if (client.available())
        } // end while (client.connected())
        delay(1);// give the web browser time to receive the data
        client.stop(); // close the connection
    } // end if (client)
}

// turn on relay and send back HTML for LED checkbox
void ProcessCheckbox(EthernetClient cl)
{
    if (HTTP_req.indexOf("DOOR=2") > -1) {//see if checkbox was selected
        // the checkbox was selected, toggle the LED
        if (LED_status) {
            LED_status = 0;
        }
```

```
        else {
            LED_status = 1;
        }
    }
    //kept the same logic but it now does the same action no matter when selected
    if (LED_status) {     // turn relay on for 1 second
        digitalWrite(2, HIGH);
        delay(1000);
        digitalWrite(2, LOW);
        // checkbox is checked
        cl.println("<input type=\"checkbox\" name=\"DOOR\" value=\"2\" \
        onselect=\"submit();\" checked>DOOR");
    }
    else {              // turn relay on for 1 second
        digitalWrite(2, HIGH);
        delay(1000);
        digitalWrite(2, LOW);
        // checkbox is unchecked
        cl.println("<input type=\"checkbox\" name=\"DOOR\" value=\"2\" \
        onselect=\"submit();\">DOOR");
    }
}
```

Figure 6-18 shows the browser connected to the Uno web server for this door opener program.

Selecting the checkbox either opened or closed the garage door depending upon its previous position.

NOTE *This program in no way compromises the inherent safety features of the garage door opener. It will still stop and reverse its path if it encounters an obstacle when closing. In addition, the safety infrared beams at the door bottom still operate normally.*

I also wanted to confirm that I could control the garage door using a smartphone. To do this, I simply entered the Uno's IP address into the phone's browser and subsequently connected to the web server. Figure 6-19 shows my smartphone's browser display.

The door operated precisely in the same manner as when I used the laptop's browser. Of course, there is no extra security when operating in this manner other than the WPA passkey used in my network's secure Wi-Fi connection. I would not be overly concerned as WPA should be sufficient to prevent unauthorized garage door operation. Definitely do not use this garage door software with an open Wi-Fi connection. You might as well hang your house keys next to your door if you choose to operate in this manner.

FIGURE 6-18 The browser display for the garage door opener

Garage Door Opener

Click to activate door.

DOOR

FIGURE 6-19 Smartphone browser display connected to the garage door web server

It is now time to take a slight deviation and introduce another IDE that will be used along with the Arduino IDE to create an enhanced garage door opener project.

Visual Studio 2012 IDE

It is both possible and useful to the Microsoft Visual Studio IDE to develop software for use with the Arduino board series. Using this IDE creates an opportunity to have access to a wide array of programming tools and libraries not found in the more limited Arduino libraries. You will need three software packages to develop with this software:

- Visual Studio 2012
- Arduino version 1.05
- MegunoLink Pro

The Microsoft Visual Studio 2012 Express edition will be used to develop the garage door software. This IDE, which will be referred to as simply VS2012, is available free of charge for non-commercial purposes at Microsoft's download website: www.microsoft.com/en-us/download/details.aspx?id=34673.

This is a Windows program and it's fairly large so it will take a while to download. Once it is installed, you will have to register online in order to obtain a free product key. Otherwise, the software "expires" after 30 days. Again, it is free as long as you certify that you are using it for personal, non-commercial use. The latest commercial version is expensive if bought directly from Microsoft. Figure 6-20 shows the VS2012 application without any projects loaded.

The second software item is the Arduino IDE, which you have likely already downloaded. Just confirm that it is version 1.05 or later as earlier versions may have compatibility issues with the VS2012.

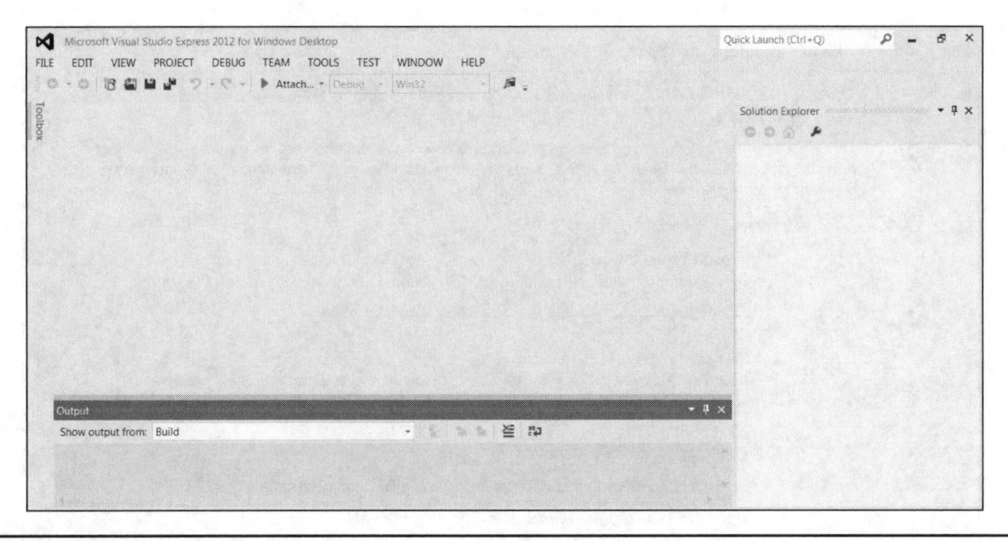

Figure 6-20 Visual Studio 2012 IDE

The last piece of software is named MegunoLink Pro and serves as a bridge between VS2012 and Arduino boards. It is impossible to upload any compiled VS2012 into an Arduino board without the use of the MegunoLink Pro software. There is a free seven-day trial period; you must purchase a license if you want to use it once the trial period has ended. The license fee for personal use is very reasonable and in my opinion well worth it for the significant functionality it provides in allowing the use of VS2012 with the Arduino boards as well as providing some very professional appearing graphical user interfaces (GUIs). MegunoLink Pro is available at www.megunolink.com.

Follow the website installation instructions to initially set up the MegunoLink Pro software. You should see the build-tool installation screen, as shown in Figure 6-21, when you select the gear icon on the MegunoLink Pro toolbar. Select Setup Arduino Build Tool for Visual Studio 2012 to install the software module within VS2012, which permits compiled programs to be uploaded into Arduino boards.

You will need to enter the directory location for the arduino.exe binary executable. This may vary somewhat depending on the options you selected during the Arduino software installation. In my case, the location was simply C:\Program Files\Arduino, as you can see in the figure.

The current serial port connecting the Uno to the laptop must also be selected in the MegunoLink Pro Connection Manager Visualizer module. The Uno must be powered on and connected to the laptop via a USB cable in order to have the serial port recognized by the Connection Manager. Figure 6-22 shows my Connection Manager establishing a connection via COM46 at 9600 baud.

Completing all of the preceding steps should now allow programs created in VS2012 to be compiled and uploaded to the Uno. I will next demonstrate how to create, compile, and upload an LED blink program to an Uno using VS2012 and MegunoLink Pro.

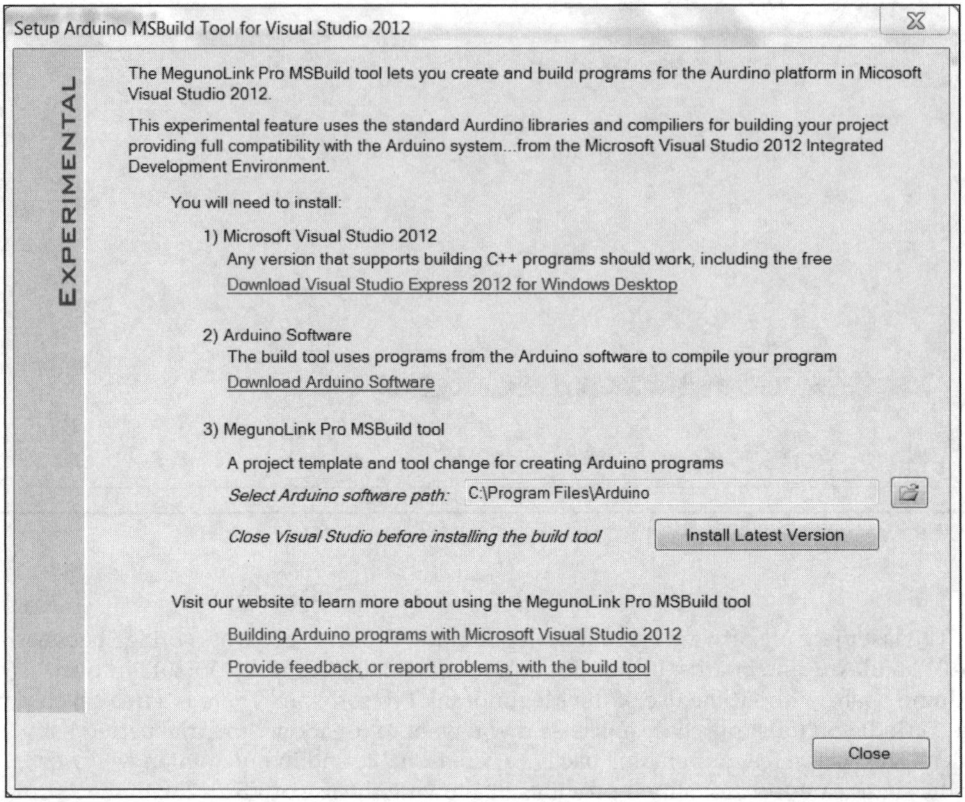

Setup Arduino MSBuild Tool for Visual Studio 2012

EXPERIMENTAL

The MegunoLink Pro MSBuild tool lets you create and build programs for the Aurdino platform in Micosoft Visual Studio 2012.

This experimental feature uses the standard Aurdino libraries and compilers for building your project providing full compatibility with the Arduino system...from the Microsoft Visual Studio 2012 Integrated Development Environment.

You will need to install:

1) Microsoft Visual Studio 2012
Any version that supports building C++ programs should work, including the free
Download Visual Studio Express 2012 for Windows Desktop

2) Arduino Software
The build tool uses programs from the Arduino software to compile your program
Download Arduino Software

3) MegunoLink Pro MSBuild tool
A project template and tool change for creating Arduino programs
Select Arduino software path: C:\Program Files\Arduino
Close Visual Studio before installing the build tool Install Latest Version

Visit our website to learn more about using the MegunoLink Pro MSBuild tool
Building Arduino programs with Microsoft Visual Studio 2012
Provide feedback, or report problems, with the build tool

Close

Figure 6-21 MegunoLink Pro build-tool installation

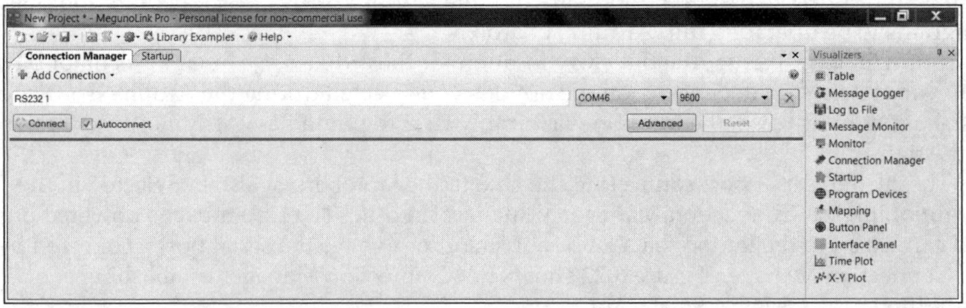

New Project * - MegunoLink Pro - Personal license for non-commercial use
Library Examples • Help •
Connection Manager | Startup
Add Connection •
RS232 1 | COM46 | 9600
Connect | Autoconnect | Advanced | Reset
Visualizers
Table
Message Logger
Log to File
Message Monitor
Monitor
Connection Manager
Startup
Program Devices
Mapping
Button Panel
Interface Panel
Time Plot
X-Y Plot

Figure 6-22 Connection Manager Visualizer

VS2012 LED Blink Program

Any program created with VS2012 must be part of a project. Create a project by following these three steps:

1. Select File | New | Project | Templates | Visual C++ | Arduino Program.
2. Provide a name for the new project. For this example, I chose "HelloArduino." Use the default Location to create the project directory. The solution name will automatically be the same as the project name except for the file extension.
3. Select OK.

Figure 6-23 shows the HelloArduino project screen with the source code for Program .cpp shown in the main editor window. Program.cpp is the default name automatically provided by VS2012 as part of the VC++/Arduino template. It also contains the working code for a LED blink application that uses a LED connected to pin 13, as was the case for the first blinking program shown earlier in this chapter. I found that having sample code immediately in the editor does make initial program development easier as it is always better to start with modifications to a working program.

Select Build | Build Solution to compile the source code. You can also simply press the F7 function key to do an immediate build. The compiled code that is first created uses the project name with a .elf file extension. This file type is a CodeWarrior ELF Debug Executable type and for my setup was automatically stored at C:\Users\Don\My Documents\Visual Studio 2012\Projects\HelloArduino\HelloArduino\bin\debug.

The MegunoLink Pro software automatically creates a companion hex file, which is the only type that can be uploaded into the Uno. The new hex file is also stored in the same directory as the original ELF file.

The actual hex file program upload is accomplished using the MegunoLink Pro Program Device Visualizer. You need to select the Program Device Visualizer from the Visualizer list and add in the location of the hex file and the serial port that was previously set up. Also ensure that the Arduino Uno is selected from the device drop-down menu. Then simply select the Program button. It should only take a few seconds to upload the newly created hex file into the Uno. A graphical progress bar shows what is currently happening in the process. Figure 6-23 shows this Program Device Visualizer screen.

Please note that after every program modification, you must recompile with VS2012 and upload using the MegunoLink Pro software. I can assure you that the process becomes almost second nature as you repeatedly use the software.

I did observe a blinking LED after I compiled and uploaded the binary to the Uno. Now, let's move on to the enhanced garage door project.

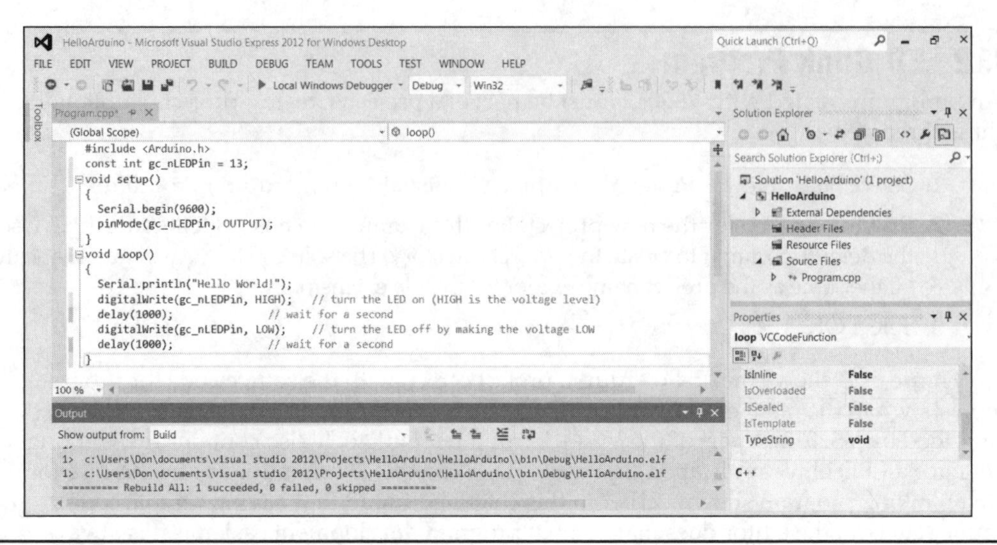

FIGURE 6-23 Program Device Visualizer screen

Enhanced Garage Door Project

This section is called "Enhanced Garage Door Project" simply because I will be demonstrating some features beyond the basic functionality that was shown in the first version of this project. Password set and retrieve features will be shown that will allow the garage door to be operated from an open Wi-Fi connection without the risk of allowing unauthorized door activations. This revised project will make use of the VS2012 IDE as well as the MegunoLink Pro software. This project is the creation of Paul Martinsen and may be downloaded in zip form from GitHub at https://github.com/Megunolink/GarageDoorOpener.

Go to that website and select the Download zip button located in the lower-right corner of the opening page. The archive file is medium sized at about 5.6MB and must be extracted and stored somewhere on the laptop's hard drive. I put all the extracted files into a subdirectory named GarageDoor located in the Downloads directory. There is a VS2012 solutions file in the archive, which you will use to re-create the original project on your laptop. This file is named Garage Door Opener.sln and is located in the GarageDoorOpener-master directory that was automatically created during the archive extraction. Incidentally, there is another file in that same directory named Garage door opener config interface.mix that you will also need later when setting up the password(s).

The Garage Door Opener project is easily created in VS2012 by following these steps:

1. Select File | Open Project.
2. Browse to the solution file wherever you extracted it using the Open Project dialog box.
3. Select the solution file (Garage Door Opener.sln).
4. Select OK.

```
#include <Arduino.h>
#include "HardwareConfiguration.h"
#include "PasswordManager.h"
#include "CommandProcessing.h"
#include "Webserver.h"
#include "AccessControl.h"

// Global password manager object.
CPasswordManager PasswordManager(PWD_BASE_ADDRESS, PWD_RECORD_LENGTH, PWD_MAX);
void setup()
{
  Serial.begin(9600);
  InitializeCommandProcessing();
  InitializeGarageDoorWebserver();
  SetupAccessControl();
  Serial.println(F("Garage door opener ready."));
}
void loop()
{
  ProcessSerialInput();
  ProcessGarageDoorWebserver();
  PasswordManager.CheckGracePeriod();
}
```

FIGURE 6-24 VS2012 Garage Door Opener project

Figure 6-24 shows this newly created project with the main application file program.cpp displayed in the source code editor window.

You should notice that Program.cpp contains only the same two methods, `setup` and `loop`, that you have seen in all of the other Arduino sketches. There are, however, a substantial number of additional files that provide additional functionality required to support web access as well as password protection. A list of these additional files is shown in the Solution Explorer in the VS2012 right-hand pane. I will not offer a detailed explanation of all these files as that would be well beyond the scope of this chapter and book. However, I do need to discuss one particular file, as some configuration changes must be made within this file in order to have a functional project. This file is Configuration.h and is classified as an include file. Files of this type contain supplemental information that is crucial to support application files, i.e., the ones ending with the .cpp filename extension. Simply double-click on the filename, which is shown in the include file list shown in Figure 6-25.

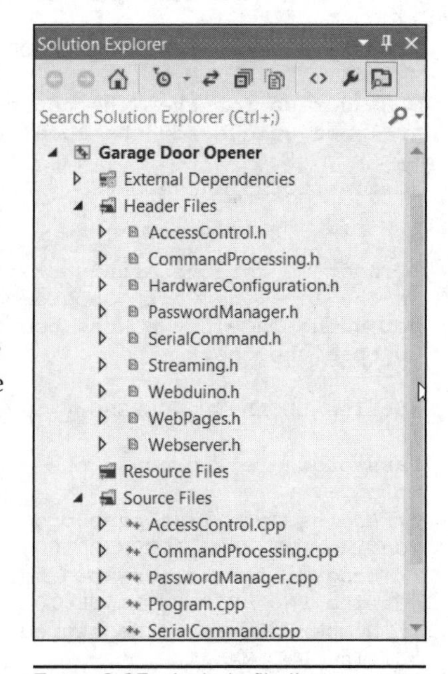

FIGURE 6-25 Include file list

The source code will be displayed in the VS2012 main editor window. The altered Configuration.h source code is shown here for your information:

```
/* **************************************************************************
*****************
** Pin assignments
** **************************************************************************
***************** */
#pragma once
#include <Arduino.h>
/*
Digital output connected to a relay that activates the door
*/
#define PIN_DOOR 2
/*
Digital output for LED that signals door activation.
*/
#define PIN_INDICATOR 13
/*
Duration to close the switch on the door opener. This should be long
enough for the mechanism to start; typically it doesn't need to remain
activated for the door to complete its motion. It is the same as the
time you'd hold down the button to start the door moving.
*/
#define DOOR_ACTIVATION_PERIOD 600 // [ms]
/*
MAC address for the Ethernet controller. Preferably globally unique, but
at least unique for the local network.
*/
#define LOCAL_MAC_ADDRESS { 0xDE, 0xAD, 0xBE, 0xEF, 0xFE, 0xDE }

/*
IP address for the Arduino. Either globally unique, or at least unique
on the local network. Should be a statically allocated address (that is
not using DHCP so we know how to connect to the Arduino when we want
to open the door.
*/
#define LOCAL_IP_ADDRESS { 192, 168, 1, 29};
/*
Passwords are stored in the eeprom in fixed size chunks.
*/
// Address of first password stored in the eeprom.
#define PWD_BASE_ADDRESS 100
// Length of each password record
#define PWD_RECORD_LENGTH 10
// Number of passwords stored.
#define PWD_MAX 10
```

I only changed the constant LOCAL_IP_ADDRESS to match the Uno's assigned IP address as I have previously discussed. There is also an opportunity to assign the unique mac address if you know it; otherwise use the default address. You can also reset the maximum number of passwords possible, which is set at 10, and the maximum password length, which is set at 10 characters. I do not recommend changing either one of these parameters. The passwords are set using a special configuration application that I discuss shortly. They are permanently stored in the Uno's eeprom memory to allow instantaneous access when the main garage door application is run. Save the altered Configuration.h file before proceeding with the compilation.

Next, select Build | Build Solution to create the hex file that will be uploaded into the Uno. The MegunoLink Pro Program Device Visualizer must next be used to upload the hex file into the Uno using the procedures that I discussed earlier.

The next step is to set at least one password in order to test this project. The passwords are set using the MegunoLink Pro configuration file I mentioned previously. All you need to do is double-click on the file Garage door opener config interface.mix, which is located in the same directory as the project solution file. Figure 6-26 shows the MegunoLink Pro screen that results when this file is run.

You need first to select the initialize button, which prepares the Uno's eeprom to accept new passwords. Next, select one of ten slots in which you will store the password being set. The first slot should be preselected. In addition, there is a default password "test" appearing in the Enter new password textbox. I used that as it was sufficient to proceed with the initial test. Select the Add/Update button to store the password into the selected slot. You can repeat this process until you have reached the ten-password limit. Note that you can always overwrite any password by simply assigning a new password to a given slot and selecting the Add/Update button. Do not select the Initialize button as it will likely erase any stored passwords that are in the eeprom.

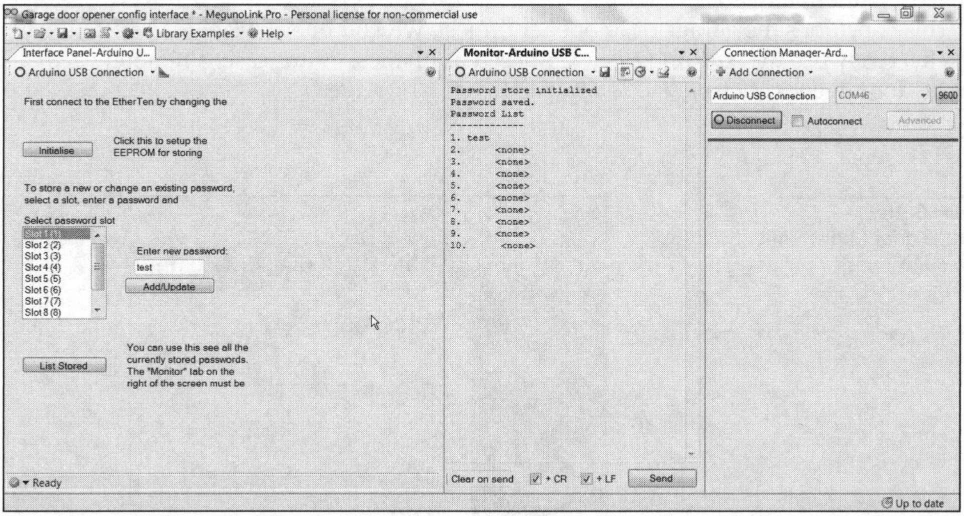

FIGURE 6-26 MegunoLink Pro password configuration

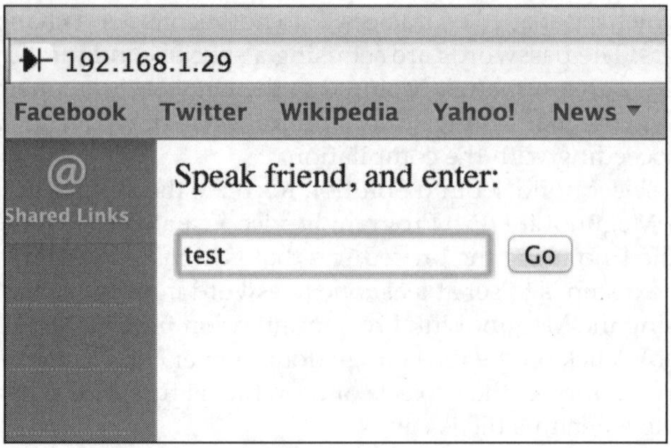

FIGURE 6-27 Initial web page for the enhanced garage door opener project

Testing the Enhanced Garage Door Opener

The newly programmed Uno with the relay switching circuit was reinstalled in the same box that held the original project circuits. A browser on my MacBook Pro was set to the Uno's IP address and I was greeted with the web page shown in Figure 6-27. Enter any of the stored passwords and you should see Figure 6-28, which is the next web page to activate the garage door.

Selecting the Activate Door button either opened or closed the garage door as expected. I also accessed the door opener using my iPad, as you can see in Figure 6-29. It worked perfectly without any issues.

This concludes the garage door opener project, which hopefully showed you how you can effectively use the Arduino as a web-based appliance, in this case, a remotely activated garage door opener/closer.

FIGURE 6-28 Garage door activation web page

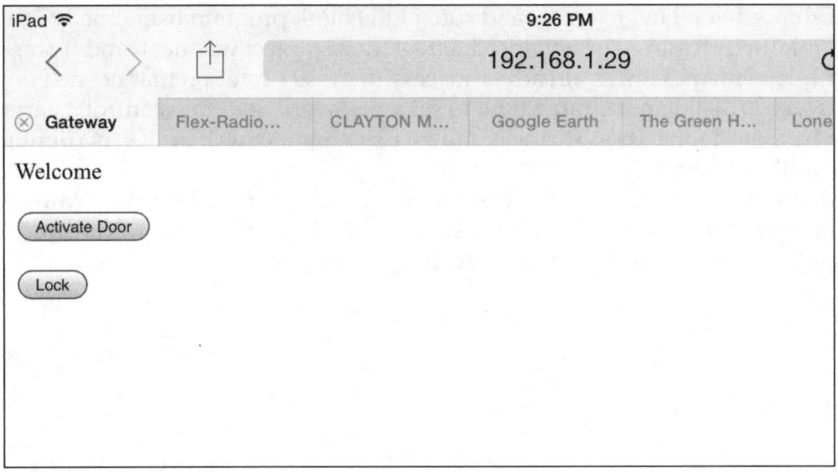

Figure 6-29 iPad access to garage door opener

Summary

I started this chapter with an overview of the Arduino Uno development board and explained key features that you should know when using it. I also discussed the Ethernet Shield board as that is the means by which the Uno connects to a network using the Ethernet.

The Arduino integrated development environment (IDE) was discussed next as that is required to create the software that controls the Uno functions. I used the customary and traditional LED "blinking" program to show you how to create and upload a program into the Uno. Incidentally, no additional parts were required for this demonstration as the program makes use of the Uno's built-in LED connected to GPIO pin 13.

Next, I demonstrated a network-enabled program that allows a browser running on a remote, networked computer to control and receive data from the Uno. This program displayed the data from the Uno's six analog-to-digital (ADC) converter channels in real time to the browser client.

I included a discussion on how the Uno, lacking an operating system, can run a web server application. I went into some detail about how this was accomplished and compared it to how similar boards, such as the Raspberry Pi, approach implementing web server applications.

I then demonstrated a simplified garage door opener project that used a modified version of an LED web control program that I showed earlier in the chapter. This project lacks security, save the inherent security present using a WPA Wi-Fi network with the opener hardware. I used both a remote client web browser and a smartphone to successfully control a garage door.

I next discussed how the Arduino software may be developed using Microsoft's Visual Studio 2012 (VS2012) IDE in conjunction with the MegunoLink Pro software package. This combination is very powerful as it allows you to have full access to a comprehensive C/C++ development environment if you so choose. I used a completed project solution to demonstrate an enhanced garage door opener that employed passwords for better security.

I first showed you how to create and run a LED blink program using the VS2012 and MegunoLink Pro software. The enhanced garage door project was next, and it was very quickly implemented as it was already a successful project solution that needed only to be loaded into VS2012. I demonstrated how to set a password and then control a garage door using the password for access. This was shown using remote web browsers running a laptop as well as a tablet.

You should feel comfortable after completing this chapter in using the Arduino Uno for web-enabled projects. The next chapter builds on this chapter's content to create a slightly more complex project: a web-enabled home irrigation system.

7

CHAPTER

Arduino Irrigation Control System

This chapter's project will use the Uno development board with an Internet connection to remotely operate a home irrigation system. This project will be split into two phases. The first phase will be a simple implementation where a homeowner can remotely start watering using any one of up to eight irrigation zones. An irrigation zone typically consists of two or more pop-up sprinkler heads.

The second phase will report back soil moisture content using an Uno web page so the homeowner can determine if immediate irrigation is needed. The soil moisture sensor subsystem will communicate with the main Uno controller using an XBee digital transceiver.

I will also be using the same Ethernet Shield that was introduced in the last chapter to connect the Uno with the home network. I recommend that you read Chapter 6 if you happened to skip straight to this chapter. I covered the fundamentals of how the Uno web server functions, which is quite different than a typical server installation.

Irrigation System Design

This irrigation project uses my own home's installed irrigation system as the initial platform. It has six zones that are controlled by a controller installed in the basement. Figure 7-1 is a system block diagram, which shows the basic system components.

This system is relatively uncomplicated wherein the controller operates water solenoids connected to individual zones. Only one zone in this system can be activated as my home water supply volume will not support operating more than one zone. The controller can be programmed to operate any zone for a predetermined time, time of day, and number of days per week. It can also manually operate a single zone or all the zones. Figure 7-2 shows the irrigation controller.

In the figure you can also see a wireless rain sensor module located above the controller, which sends a digital signal to the controller when the rain level collected in the sensor reaches a predetermined point. Unfortunately, I found that this sensor was neither accurate nor reliable, which is why it is turned off. I decided to incorporate the new moisture sensor into this project because I had no reliable way to test the lawn moisture.

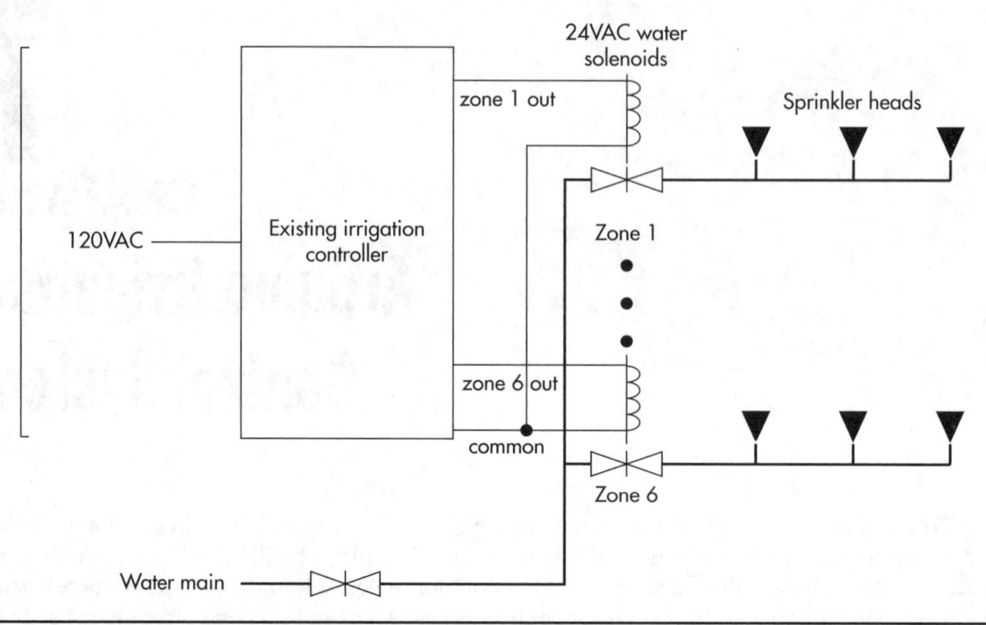

Figure 7-1 Home irrigation block diagram

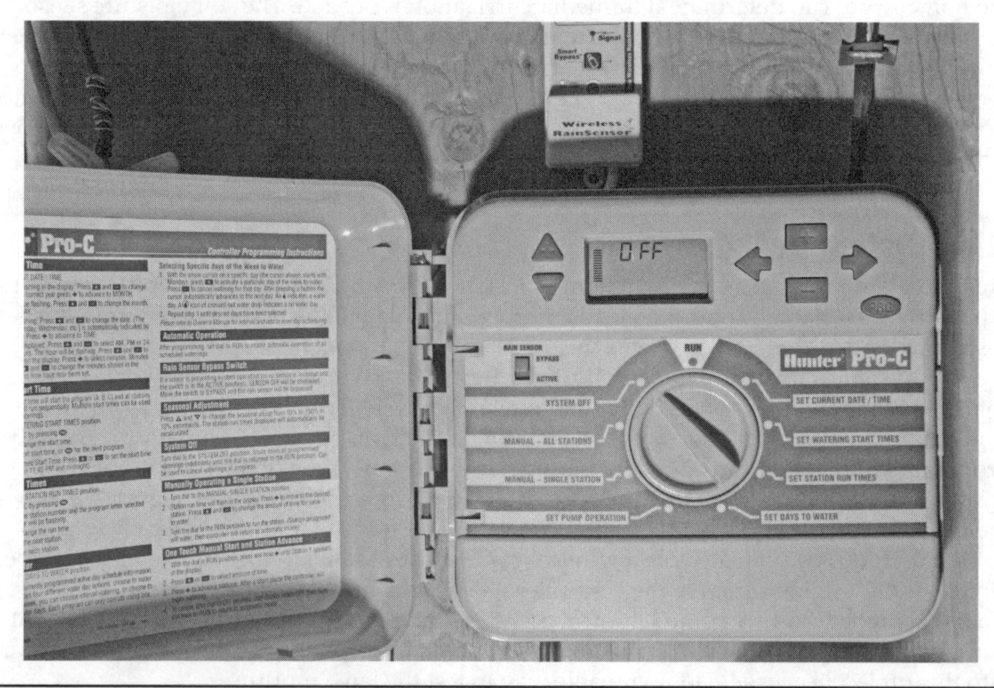

Figure 7-2 Home irrigation system controller

I had to disable the original controller as I didn't want to accidentally enable two zones at the same time, which would disrupt my home water supply and likely cause a domestic disruption with my spouse.

Consequentially, I had to provide an alternate 24VAC supply, which is needed to operate the water solenoids because the main controller was offline. The new solenoid voltage was applied using new relay control circuits that I implemented with two 4-channel relay modules. Figure 7-3 shows the Arduino development board, consisting of two relay modules along with the Uno, Ethernet Shield, and a solderless breadboard that I used to interconnect eight of the Uno's GPIO pins to the relay board inputs. I did connect all eight, but I only needed six of the eight pins.

I purchased two inexpensive, 4-channel relay-switching modules from SainSmart through Amazon.com. I was impressed by the build quality as well as the ease of interfacing it to the Uno. Figure 7-4 is a close-up photograph of one of the modules where you can easily see the digital input pins located at the board's bottom-right side.

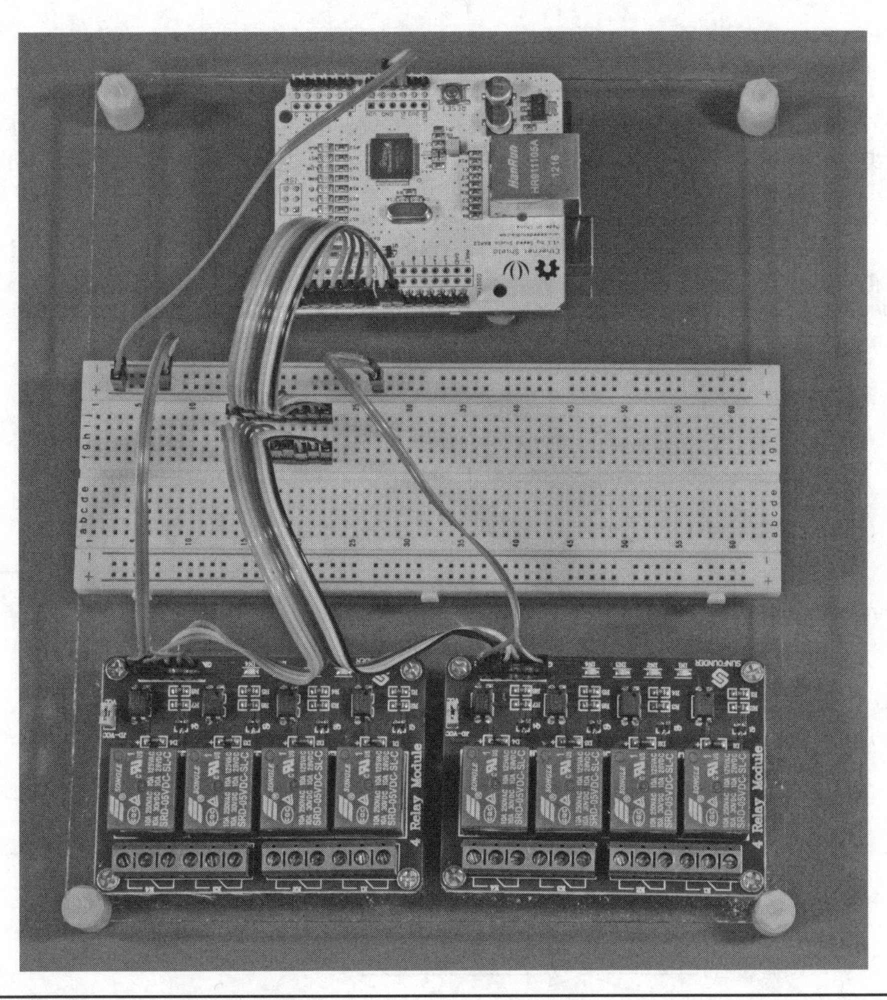

FIGURE 7-3 Arduino development board

FIGURE 7-4 SainSmart 4-channel relay module

The inputs are all FET driven, which lessens the current drive requirements from the Uno's GPIO pins. There are also LED indicators on each of the digital inputs, which makes it very easy to determine the status of every channel relay. Just be aware that the LEDs light on the GPIO LOW level, not HIGH level and will be on for all inactivated channels. Finally, connections to individual relay contacts are done using the screw terminal strips located at the bottom edge of the relay module. Each contact type is also silk-screened onto the board, which helps with making the connections to the irrigation solenoid terminal strip.

Figure 7-5 is a block diagram of the new controller with relay modules interfaced to the existing irrigation system.

I need to discuss the new Uno control program, which you should upload before installing the development board into the existing irrigation system.

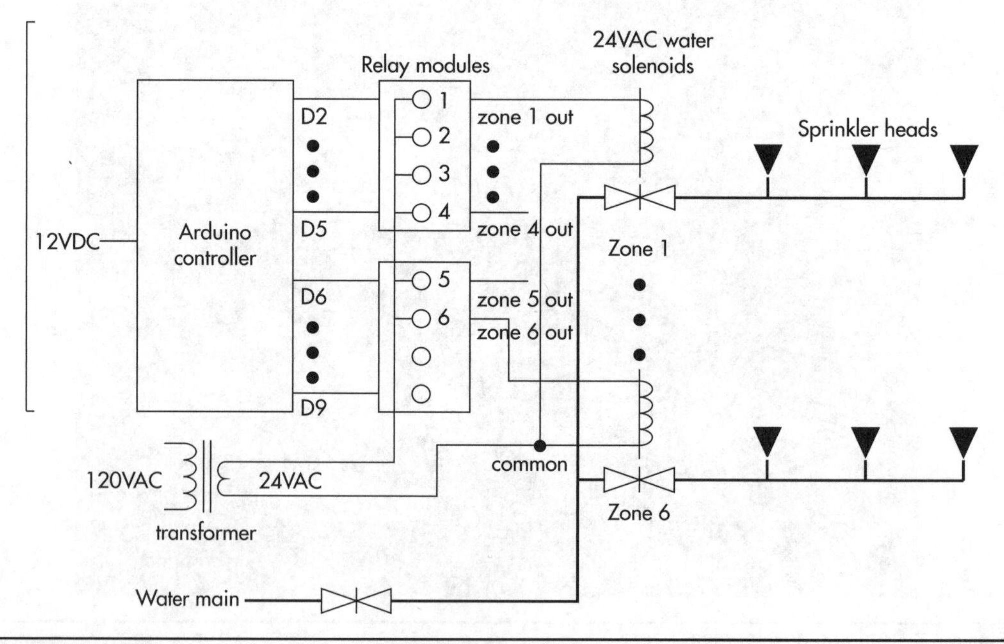

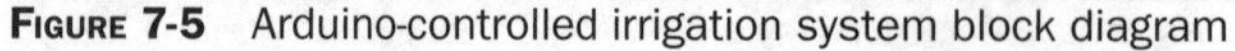

FIGURE 7-5 Arduino-controlled irrigation system block diagram

Irrigation Control Program

The new program uses the same Ethernet connectivity logic that was used in the garage door opener project but with new logic developed to function with an irrigation system. As was the case for the garage door opener project, you must determine the Uno's IP address and change the IP address in the program code to match the actual address. You will not be able to connect with the web server if you fail to make this change.

The user interface is quite simple, consisting of a series of "radio" buttons that select any one or none of the irrigation zones. Figure 7-6 shows a screen capture of the web browser connected to the Uno server program. In my case, the Uno IP address was 192.168.1.26, which is visible in the URL text box.

Clicking first on a button and then clicking on the Submit button will turn on the water supply to the zone represented by the selected radio button. Only one button may be activated at a time, which is the desired feature of the radio button interface. This ensures only that one watering zone is actuated, which matches the single operating zone requirement that I discussed earlier in the chapter.

The complete code listing, named Irrigation_Control, is shown here and, as in previous chapters, this code is available on the book's companion website.

```
/*---------------------------------------------------------------
  Program: Irrigation_Control

  Description: An Arduino web server that serves up a web page allowing users to
               control a residential, six-zone irrigation system

  Hardware: Arduino Uno and official Arduino Ethernet Shield.  Should work with
            other Arduinos and compatible Ethernet shields

  Software: Developed using Arduino 1.0.5 software.  Should be compatible with
            Arduino 1.0 + software

  References: eth_websrv_LED modified by D. J. Norris
              WebServer example by David A. Mellis and modified by Tom Igoe
              Ethernet library documentation:
              http://arduino.cc/en/Reference/Ethernet
```

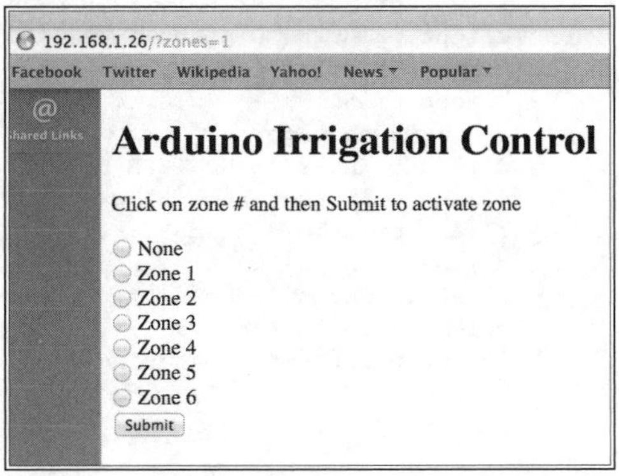

Figure 7-6 Web browser screenshot

192.168.1.26/?zones=1

Facebook Twitter Wikipedia Yahoo! News Popular

Arduino Irrigation Control

Click on zone # and then Submit to activate zone

- None
- Zone 1
- Zone 2
- Zone 3
- Zone 4
- Zone 5
- Zone 6

Submit

```
  Date:          21 May 2014
 --------------------------------------------------------------*/

#include <SPI.h>
#include <Ethernet.h>

// MAC address from Ethernet shield sticker under board (use this if no sticker)
byte mac[] = { 0xDE, 0xAD, 0xBE, 0xEF, 0xFE, 0xED };
IPAddress ip(192, 168, 1, 26); // ip address; change to match Uno's network ip
EthernetServer server(80);      // create a server at port 80 (HTTP)

String HTTP_req;            // stores the HTTP request

void setup()
{
    Ethernet.begin(mac, ip);      // initialize Ethernet device
    server.begin();                          // start to listen for clients
    Serial.begin(9600);               // for diagnostics
   // six zones, numbered 1 to 6, controlled by GPIO pins 2 to 7, respectively
    pinMode(2, OUTPUT);       // to relay on pin 2
    pinMode(3, OUTPUT);       // to relay on pin 3
    pinMode(4, OUTPUT);       // to relay on pin 4
    pinMode(5, OUTPUT);       // to relay on pin 5
    pinMode(6, OUTPUT);       // to relay on pin 6
    pinMode(7, OUTPUT);       // to relay on pin 7
}

void loop()
{
    EthernetClient client = server.available();  // try to get client

    if (client) {  // got client?
        boolean currentLineIsBlank = true;
        while (client.connected()) {
            if (client.available()) {   // client data available to read
                char c = client.read(); // read 1 byte (character) from client
                HTTP_req += c;  // save the HTTP request 1 char at a time
                // last line of client request is blank and ends with \n
                // respond to client only after last line received
                if (c == '\n' && currentLineIsBlank) {
                    // send a standard http response header
                    client.println("HTTP/1.1 200 OK");
                    client.println("Content-Type: text/html");
                    client.println("Connection: close");
                    client.println();
                    // send web page
                    client.println("<!DOCTYPE html>");
                    client.println("<html>");
                    client.println("<head>");
                    client.println("<title>Arduino Irrigation Controller</title>");
                    client.println("</head>");
                    client.println("<body>");
                    client.println("<h1>Arduino Irrigation Control</h1>");
```

```
                    client.println("<p>Click on zone # and then Submit to activate
                    zone</p>");
                    client.println("<form method=\"get\">");
                    ProcessCheckbox(client);
                    client.println("</form>");
                    client.println("</body>");
                    client.println("</html>");
                    Serial.print(HTTP_req);
                    HTTP_req = "";    // finished with request, empty string
                    break;
                }
                // every line of text received from the client ends with \r\n
                if (c == '\n') {
                    // last character on line of received text
                    // starting new line with next character read
                    currentLineIsBlank = true;
                }
                else if (c != '\r') {
                    // a text character was received from client
                    currentLineIsBlank = false;
                }
            } // end if (client.available())
        } // end while (client.connected())
        delay(10);      // give the web browser time to receive the data
        client.stop(); // close the connection
    } // end if (client)
}

// turn on selected zone relay
void ProcessCheckbox(EthernetClient cl)
{
        cl.println("<input type=\"radio\" name=\"zones\" value=\"0\" > None   <br>");
        cl.println("<input type=\"radio\" name=\"zones\" value=\"1\" > Zone 1 <br>");
        cl.println("<input type=\"radio\" name=\"zones\" value=\"2\" > Zone 2 <br>");
        cl.println("<input type=\"radio\" name=\"zones\" value=\"3\" > Zone 3 <br>");
        cl.println("<input type=\"radio\" name=\"zones\" value=\"4\" > Zone 4 <br>");
        cl.println("<input type=\"radio\" name=\"zones\" value=\"5\" > Zone 5 <br>");
        cl.println("<input type=\"radio\" name=\"zones\" value=\"6\" > Zone 6 <br>");
        cl.println("<input type=\"submit\" value=\"Submit\" >");

        if(HTTP_req.indexOf("zones=0") > -1){
          digitalWrite(2, LOW);
          digitalWrite(3, LOW);
          digitalWrite(4, LOW);
          digitalWrite(5, LOW);
          digitalWrite(6, LOW);
          digitalWrite(7, LOW);
        }
        else if(HTTP_req.indexOf("zones=1") > -1){
          digitalWrite(2, HIGH);
          digitalWrite(3, LOW);
          digitalWrite(4, LOW);
```

```
      digitalWrite(5, LOW);
      digitalWrite(6, LOW);
      digitalWrite(7, LOW);
     }
     else if(HTTP_req.indexOf("zones=2") > -1){
       digitalWrite(3, HIGH);
       digitalWrite(2, LOW);
       digitalWrite(4, LOW);
       digitalWrite(5, LOW);
       digitalWrite(6, LOW);
       digitalWrite(7, LOW);
     }
    else if(HTTP_req.indexOf("zones=3") > -1){
      digitalWrite(4, HIGH);
      digitalWrite(2, LOW);
      digitalWrite(3, LOW);
      digitalWrite(5, LOW);
      digitalWrite(6, LOW);
      digitalWrite(7, LOW);
     }
    else if(HTTP_req.indexOf("zones=4") > -1){
      digitalWrite(5, HIGH);
      digitalWrite(2, LOW);
      digitalWrite(3, LOW);
      digitalWrite(4, LOW);
      digitalWrite(6, LOW);
      digitalWrite(7, LOW);
     }
    else if(HTTP_req.indexOf("zones=5") > -1){
      digitalWrite(6, HIGH);
      digitalWrite(2, LOW);
      digitalWrite(3, LOW);
      digitalWrite(4, LOW);
      digitalWrite(5, LOW);
      digitalWrite(7, LOW);
     }
    else if(HTTP_req.indexOf("zones=6") > -1){
      digitalWrite(7, HIGH);
      digitalWrite(2, LOW);
      digitalWrite(3, LOW);
      digitalWrite(4, LOW);
      digitalWrite(5, LOW);
      digitalWrite(6, LOW);
    }
}
```

The logic to find which zone is selected is to match the beginning portion of the HTTP request string with the zone and value. For instance, a user selecting zone 3 will cause the HTTP request to have `"zones=3"` in the text. The statement

```
else if(HTTP_req.indexOf("zones=3") > -1)
```

in the chain of if/else statements will consequently be evaluated as True and all the following statements in that specific block will be executed.

Irrigation System Physical Installation

The new Arduino development boards along the 24VAC transformer were mounted on the same plywood board that supports the existing irrigation system. Figure 7-7 shows the new installation along with the ribbon cable that interconnects the relay modules with the existing solenoid terminal strip.

The relay modules were connected to the zone water solenoids by simply inserting the respective ribbon cable wire in parallel with the existing terminal connection. Figure 7-8 is a close-up photograph of this wiring installation.

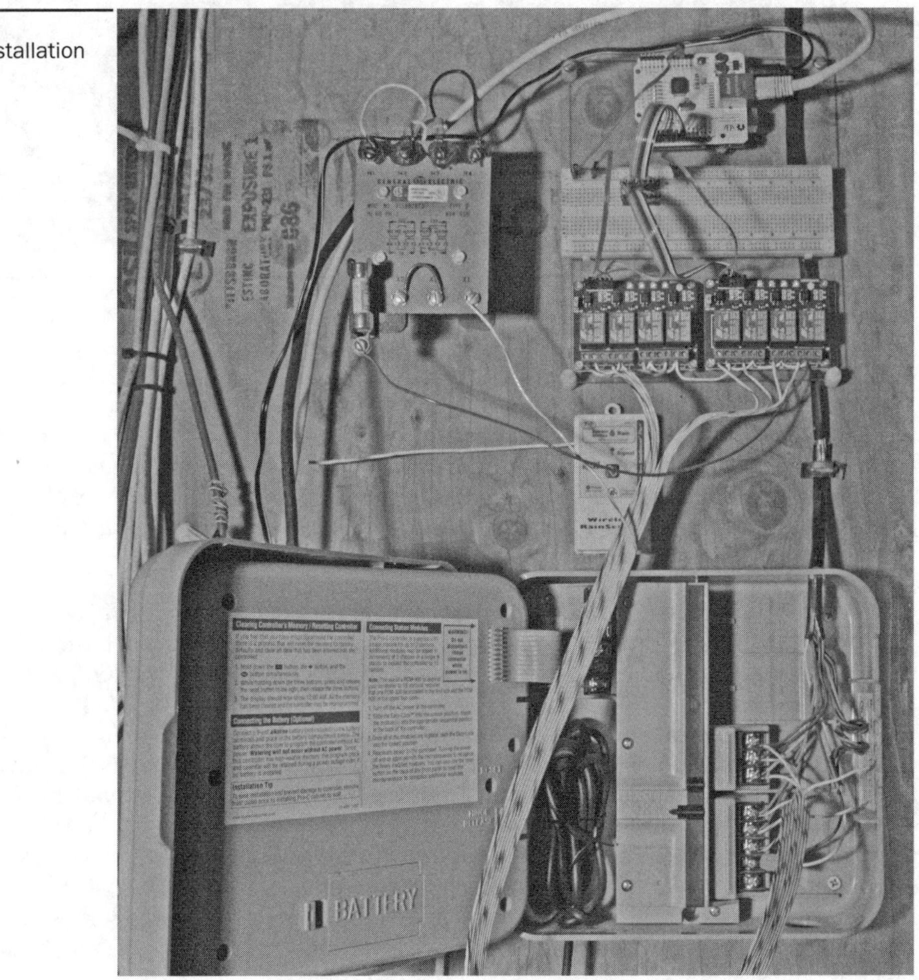

FIGURE 7-7 Physical installation

Figure 7-8 Zone water solenoid interconnections

Caution *The wire currently in the existing solenoid cable connected to the common solenoid terminal strip must be disconnected from the strip. This wire must then be directly connected to the common terminal of the new 24VAC transformer. Failure to do so will cause the secondaries of both the existing and new transformers to be connected in parallel, which will likely damage both units. It is okay to leave the existing, individual solenoid wires connected to the terminal strip as long as the common or the ground has been disconnected.*

Figure 7-9 is a close-up of the ribbon cable connections made to the relay modules. The hot side of the 24VAC transformer is connected to each relay's common terminal while each individual zone solenoid connections are made to the normally open (NO) pole for each zone relay.

The connections made to the 24VAC transformer are easily seen in Figure 7-10. I used a surplus GE control transformer that I had in my collection of junk parts. However, the particular transformer style is not critical and you can use whatever is available from an electrical supply store or home improvement center.

You can even connect two 12VAC bell transformers in series to achieve the 24VAC, which is needed to operate the solenoids.

The Uno must also be connected to a 7-12V DC supply along with an Ethernet cable; both of which can be seen in Figure 7-7. Once everything is in-place, you will be all set to test this new system.

FIGURE 7-9 Ribbon cable to relay module connections

FIGURE 7-10 24VAC transformer connections

Operating the New Irrigation System

To operate the system, you simply need to browse to the Uno's IP address; you should see the zone selection screen I presented in Figure 7-6. Click on a zone and then click on the Submit button and voilà, you should observe water being sprayed from the selected zone. Note that there is no time limitation imposed on this activation and it is your responsibility to reselect the None zone to turn off the zone. Failure to do so will simply increase your water bill and likely result in a soggy lawn.

The next phase in this project, as I mentioned at the beginning of the chapter, is to sense the moisture content and report that sensor reading to the user so that a reasoned decision can be made regarding activating the irrigation zone(s).

Moisture Sensing Subsystem

I have labeled this part of the project as a moisture sensing subsystem as it is an optional add-on to the basic Arduino irrigation system. It is not required but should prove a handy addition for making an informed choice if you should remotely activate the irrigation system. This subsystem block diagram is shown in Figure 7-11.

You should immediately notice from the block diagram that the moisture sensor portion is completely separate and remotely located from the main Arduino development board, which in my case is mounted adjacent to my existing irrigation controller. The moisture sensor communicates using a radio frequency (RF) link, which is implemented with XBee modules. I have included the following section for those readers interested in what makes up the XBee communications link and how it seamlessly integrates with the two Uno boards.

XBee Technology

Let me begin with a shameless promotion of my recent book, *Build Your Own Quadcopter.* I adapted much of the material in this section from a chapter in that book, but I've covered the Arduino Uno board in lieu of the Parallax Propeller chip, which is discussed in the quadcopter book. XBee transceivers were selected to implement the RF link because they are small, lightweight, inexpensive, and totally compatible with the Uno boards. XBee is

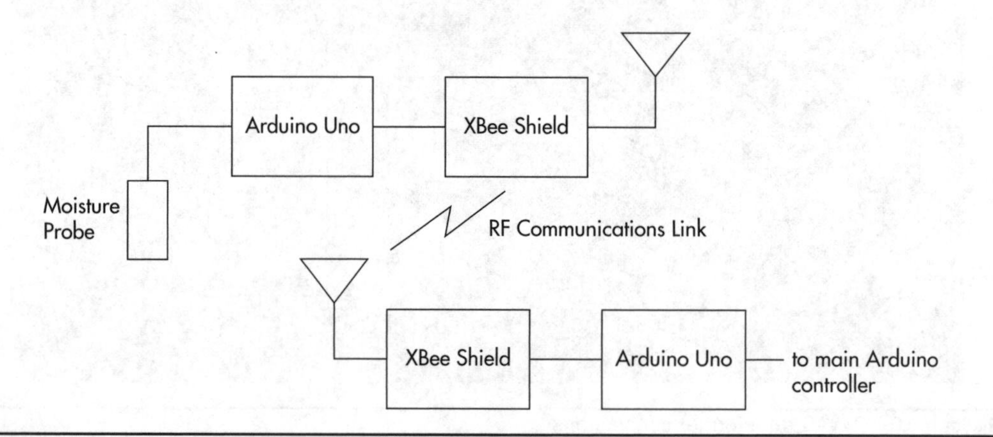

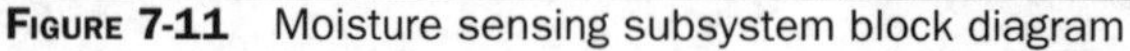

FIGURE 7-11 Moisture sensing subsystem block diagram

FIGURE 7-12 XBee Pro transceiver

the brand name for a series of digital RF transceivers manufactured by Digi International. Figure 7-12 shows one of the XBee Pro transceivers that I used.

There are two rows of 10 pins on each side of the module. These pins are spaced at 2 mm between each one, which is incompatible with the standard 0.1-inch spacing used on solderless breadboards. This means that a special connector socket must be used with the XBee module to interconnect it with the Uno. This special socket is part of an XBee Arduino Shield that is shown in Figure 7-13.

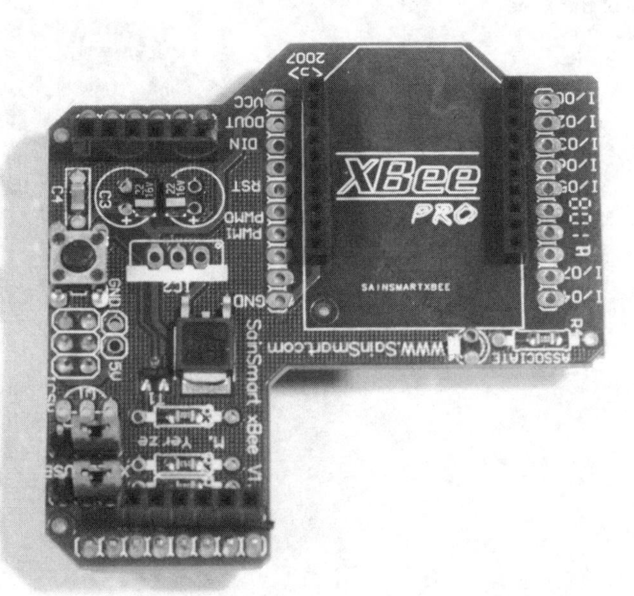

FIGURE 7-13 XBee Arduino Shield

This shield contains all the functionality needed to effectively interface an Arduino style board such as the Uno with an XBee module. The shield and accompanying software makes it very easy to create a useful RF communications link with very little effort.

I will next examine the XBee hardware to show how this clever design makes wireless transmission so easy. All the electronics in the XBee hardware, except for the antenna, are contained in a slim metal case located on the bottom side of the module, as you can see in Figure 7-14.

If you look closely at the figure, you should see the bottom of the antenna wire, which is located near the top-left corner of the case. While Digi International is not forthcoming regarding what makes up the electronic contents of the case, I did determine that the earlier versions of the XBee Pro transceivers used the Freescale model MC13192 RF transceiver. This chip is a hybrid type, meaning that it is made up of both analog and digital components. The analog components make up the RF transmit-and-receive circuits while the digital components implement all the other chip functions. It is a complex chip, which is the reason why the XBee module is so versatile and able to automatically perform a remarkable number of networking functions. Table 7-1 shows a select number of features and specifications for the MC13192.

The XBee module implements a full network protocol suite, but from a hardware perspective, it means that there must also be a microprocessor present in the electronics case. From my research, I cannot determine which type of microprocessor it is, but I am willing to make an educated guess that it would be a Freescale chip, based on the reasonable assumption that the MC13192 would be designed to be highly compatible with the company's own line of microprocessors. One other factor supporting my guess is that Digi International has recently introduced a line of programmable XBee modules named XBee Pro SB that use the 8-bit Freescale S08 microprocessor.

FIGURE 7-14 Close-up of the XBee electronics case

Features/Specifications	Description
Frequency/modulation	O-QPSK data in 5.0 MHz channels and full spread-spectrum encode and decode (modified DSSS) Operates on one of 16 selectable channels in the 2.4 GHz ISM band
Maximum bandwidth	250 Kbps (compatible with the 802.15.4 standard)
Receiver sensitivity	< -92 dBm (typical) at 1.0 percent packet error rate
Maximum output power	0 dBm nominal, programmable from –27 dBm to 4 dBm
Power supply	2.0 to 3.4 V
Power conservation modes	< 1 μA Off current 1 μA Typical Hibernate current 35 μA Typical Doze current (no CLKO)
Timers/Comparators	Four internal timer comparators available to supplement MCU resource
Clock outputs	Programmable frequency clock output (CLKO) for use by MCU
Number of general-purpose input/output (GPIO) pins	7
Internal oscillator	16 MHz with onboard trim capability
Operating temperature range	–40 to 85°C
Package size	QFN-32 Small Form Factor (SFF)

TABLE 7-1 Freescale MC13192 Features and Specifications

The XBee pins are detailed in a logical arrangement in Figure 7-15 for your information. Just be aware that only four of the pins are needed for this project, and they are shown with an asterisk next to the pin number.

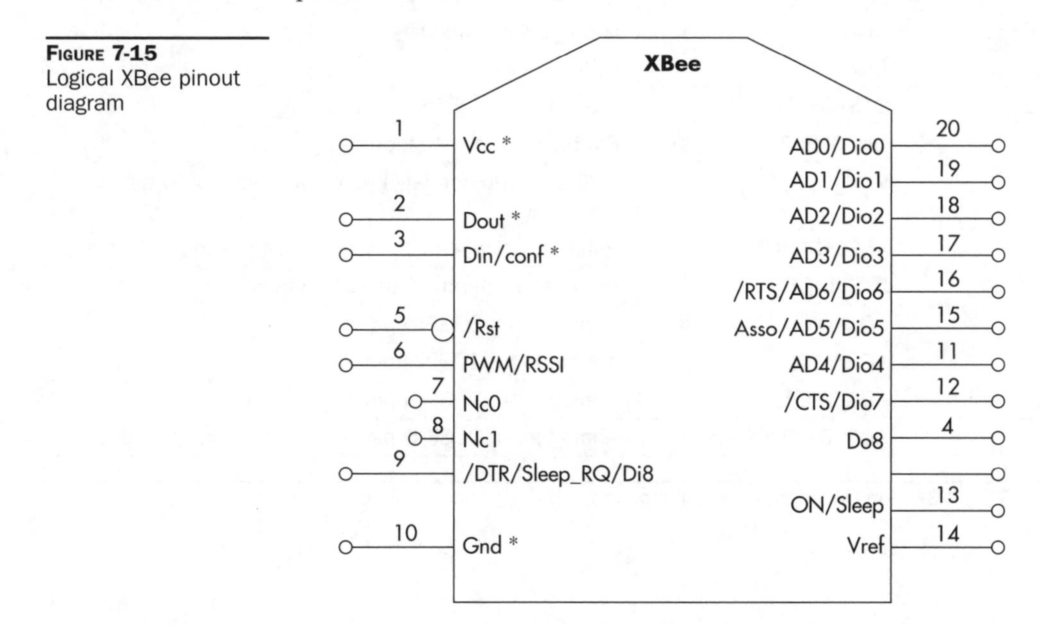

FIGURE 7-15 Logical XBee pinout diagram

All the pin and function descriptions are shown in Table 7-2.

A considerable number of functions are available to you if needed; however, this project requires only the most minimal functions for simple and reliable data transfers. Thankfully, the two XBee modules automatically connect and establish reliable communications when power is applied to them. A red, blinking LED on the XBee shield is your indication that a communications link has been established.

I will finish this section by mentioning that the XBee uses a highly capable networking protocol name ZigBee, which is also called a *Personal Area Network* (PAN). Please refer to my quadcopter book for more detailed information about ZigBee and the network protocol used with the XBees.

Pin Number	Name(s)	Description
1*	V_{cc}	Power supply, 3.3 V
2*	D_{out}	Data out (TXD)
3*	D_{in}	Data in (RXD)
4	DIO12	GPIO pin 12
5	Reset	XBee module reset, pin low
6	PWM0/RSSI/DIO10	Pulse Width Modulation (PWM Analog 0), Received Signal Strength Indicator (RSSI), GPIO pin 10
7	DIO7	GPIO pin 7
8	Reserved	Do Not Connect (DNC)
9	DTR/SLEEP_RQ/DIO8	Data Terminal Ready (DTR), GPIO Sleep Assertion (pin low), GPIO pin 8
10*	GND	Ground or common
11	DIO4	GPIO pin 4
12	CTS/DIO7	Clear To Send (CTS), GPIO pin 7
13	ON/SLEEP	Pin high when not sleeping
14	V_{ref}	Voltage reference level (used with analog to digital conversion)
15	ASSOC/DIO5	Pulse signal when connected to a network, GPIO pin 5
16	RTS/DIO6	Request To Send (RTS), GPIO pin 6
17	AD3/DIO3	Analog Input 3, GPIO pin 3
18	AD2/DIO2	Analog Input 2, GPIO pin 2
19	AD1/DIO1	Analog Input 1, GPIO pin 1
20	AD0/DIO0/COMMIS	Analog Input 0, GPIO pin 0, Commissioning Button

TABLE 7-2 XBee Pin Descriptions and Functions

Soil Moisture Sensor

This subsystem would be useless unless there were some way of sensing the amount of moisture or water present in the soil that is being irrigated. I took advantage of the fundamental property of Earth Conductivity (EC) where a measurement of how well soil conducts an electrical current is indicative of how much water or moisture is present. Soil that contains absolutely no water will not conduct an electrical current and will act as a perfect insulator, which is equivalent to zero conductivity. This situation does not normally exist as all soil on the Earth's surface normally does contain some water. Perhaps in extremely arid, desert regions you might find nearly zero water content. I would guess that the moon and Mars would also be "nearby" places where no surface water is in-place and "soil" conductivity would definitely be zero. Thus, EC was my rationale to design a very simple, yet highly effective, moisture sensor. Figure 7-16 is a photograph of this sensor.

It is a simple design consisting of two, one-foot-long, one-quarter-inch, soft copper tubes attached to a pressure treated (PT) wooden block. Notice that I flatten the ends of the copper tubes in a vise to make it easier to screw them onto the wooden block ends. I also soldered a wire to each tube, which has a sharpened end that is designed to smoothly pierce into the soil. The soil will act as a "resistor" to any electrical current that flows between the two tubes. I used a standard multimeter to measure the resistance between the two tubes, which is the inverse of the actual soil conductivity. Figure 7-17 is a photograph of my somewhat

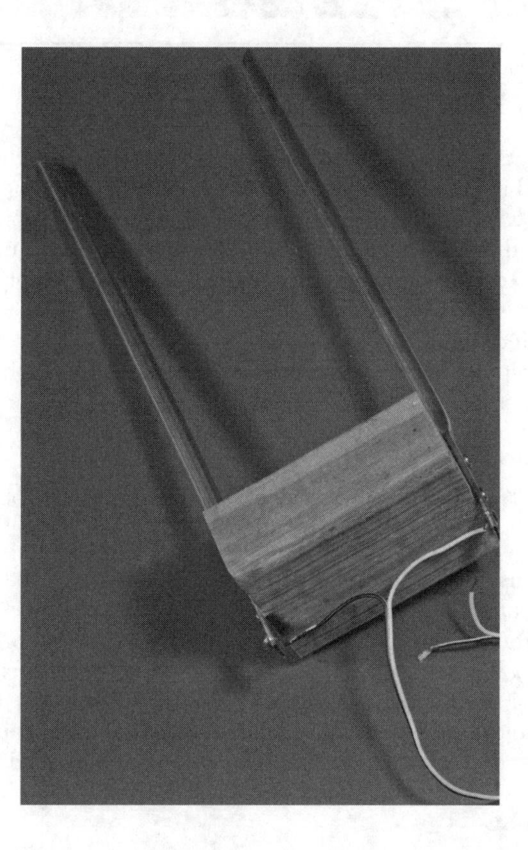

Figure 7-16 Moisture sensor

FIGURE 7-17 Soil moisture calibration setup

crude calibration setup where the sensor was measuring the conductivity of a sand and clay mixture held in a five-gallon pail.

I added water to the pail to change the soil composition from dry to a saturated condition and recorded the resistance for each stage. Figure 7-18 is the result of this calibration experiment, which serves the purpose of establishing a "trigger" point at which the irrigation system should be started.

I estimated that irrigation should be started once the sensor resistance level is measured at 40KΩ or higher. I also used the 40KΩ resistance level when I designed the rest of the moisture sensor electronics.

Moisture Sensor Design

The sensor design was based on the realization that current flow was the key parameter to be measured. This is really quite evident when you consider that absolutely dry soil will not conduct any current, while saturated soil will have the maximum conductivity. The only issue with this line of thinking is that the Uno has only analog-to-voltage (ADC) inputs and no direct means to measure current. This leads to my decision to use a fixed resistance in series with the moisture sensor and to measure the voltage drop across this fixed resistance. Now the problem is how to determine the value of this series resistor. That is actually fairly easy to do when you realize that the maximum voltage that can be handled by an Uno's ADC channel is 5VDC, which is represented by a 1023 count. The maximum voltage would be generated when the maximum current is passed through the fixed resistor under

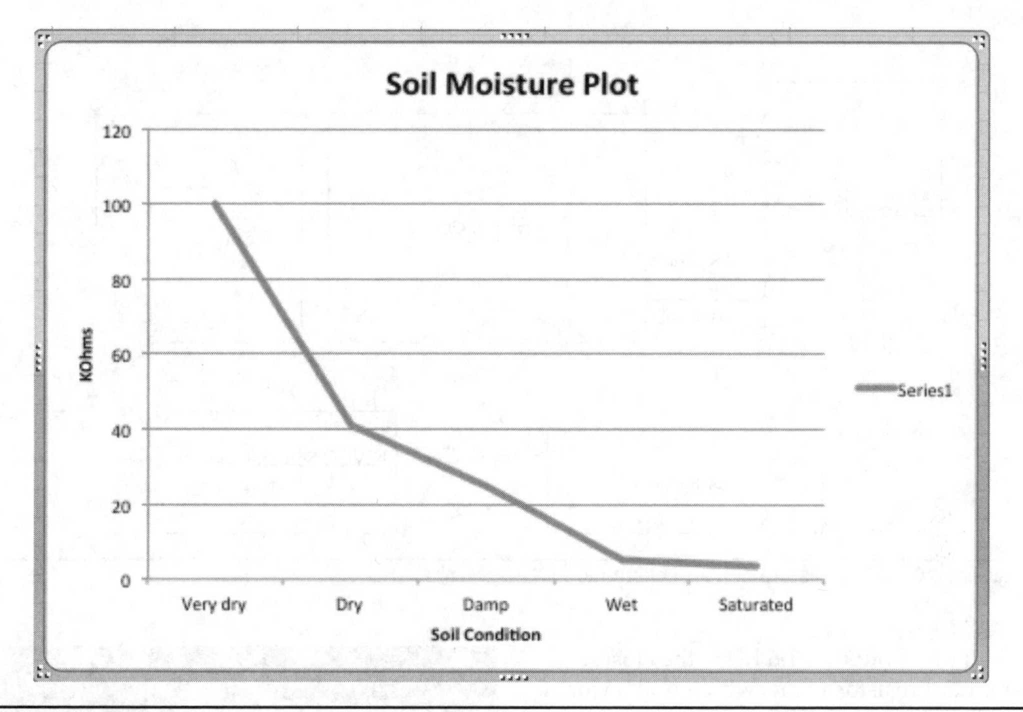

FIGURE 7-18 Soil resistance versus soil condition chart

a fully saturated soil condition. I measured the soil resistance to be 2.4 KΩ for that condition. Figure 7-19 is the equivalent circuit, which I used as a basis for the following calculations, assuming a 12VDC battery supply for the sensor:

Max current through the sensor = (12 – 5) / 2.4 = 2.92ma

Fixed resistor value (Rx) = 5 / 2.92 = 1.71 KΩ

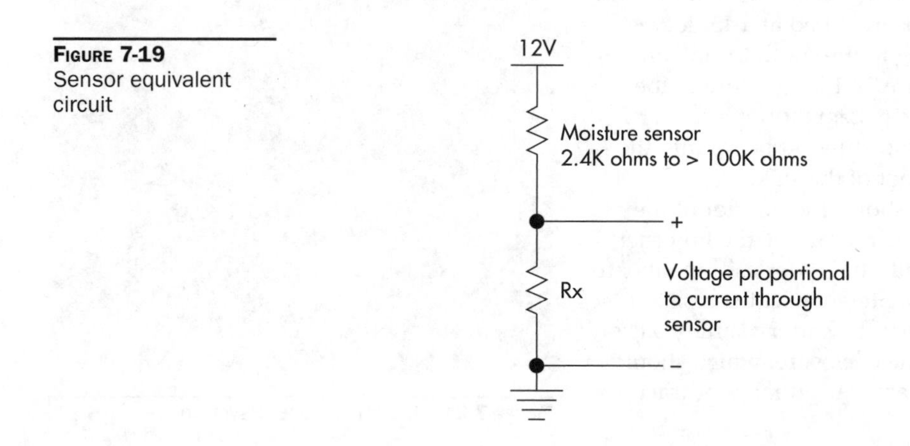

FIGURE 7-19 Sensor equivalent circuit

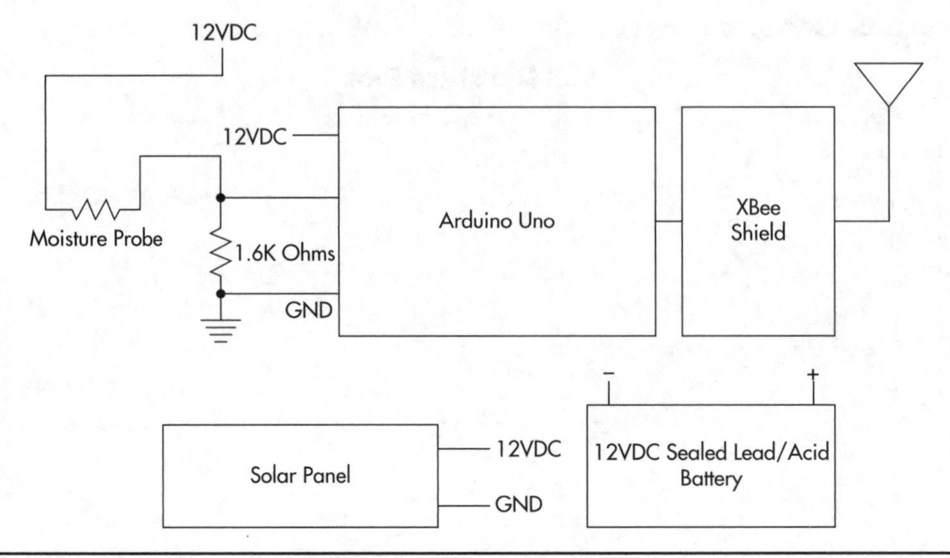

FIGURE 7-20 Soil moisture sensor schematic

It turns out that 1.6 KΩ is the closest standard resistor value, which will be just fine for this circuit. The complete moisture sensor schematic is shown in Figure 7-20.

Note that I included a 12VDC solar panel in the design to trickle charge the sealed, lead-acid battery, which should make the whole moisture sensor subsystem almost maintenance free. I mounted the sensor system in an 8×8×4-inch heavy-duty plastic enclosure that should easily withstand the outdoor environment. Obviously, the solar panel was mounted on top of the enclosure, tilted at a 45-degree angle and facing to the south for maximum solar gain. Figure 7-21 is a picture of the moisture sensor system mounted on a PT stake with the moisture probe set into the ground at the foot of the stake.

Figure 7-22 shows the interior of the enclosure where you can see the Ethernet Shield along with the solderless breadboard, which I used to interconnect the resistive divider. Note that the battery simply sits on the bottom of the enclosure, which should be fine barring any earthquakes or tractor accidents.

FIGURE 7-21 Moisture sensor system

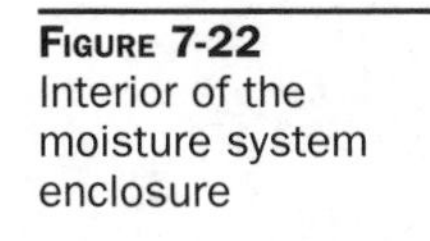

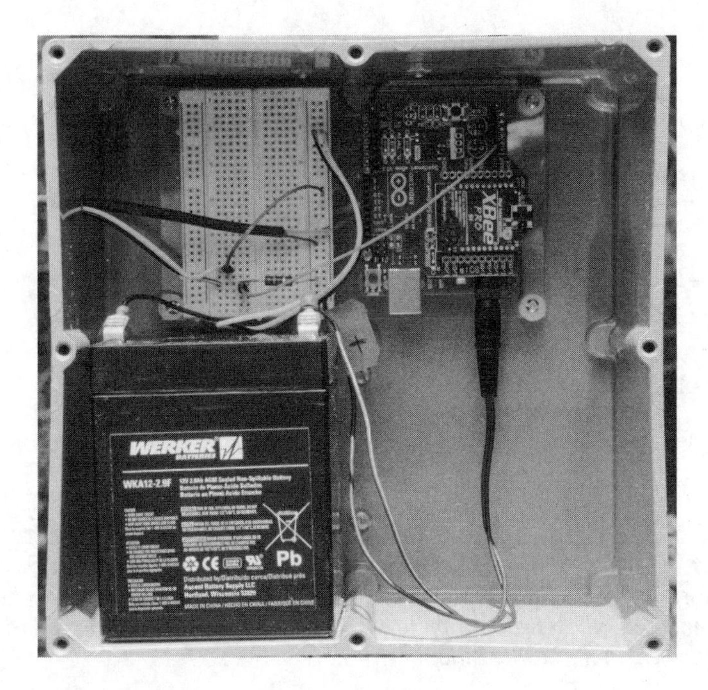

FIGURE 7-22 Interior of the moisture system enclosure

NOTE *I did not install the system outdoors until I successfully completed all the calibration testing discussed in the following section. I would strongly suggest you do the same and test all the outdoor components on a bench top before proceeding to install them outdoors.*

Moisture Sensor Software

The software, which controls the moisture sensor subsystem is straightforward as I relied on the ADC library functions as well as some XBee library functions. The XBee wireless link uses a 9600 baud rate, which is more than adequate to handle the sensor voltage readings. There are two programs involved, one for the transmitter and the other for the receiver. The transmitter code listing XbeeXmit is shown here and is available on the book's companion website, as is the receiver code:

```
/* XmitXbee
Simple program to read an ADC value from channel 0 and send it
over to another XBee.  Ensure both XBees are jumpered into the
XBee mode.
Author D. J. Norris
May, 2014
*/

int sensorData;  // a variable to read incoming serial data

void setup() {
  // initialize serial communication:
```

```
  Serial.begin(9600);
}

void loop() {
    sensorData = analogRead(A0);
    if(sensorData < 200)
      Serial.println(1);
    else if((sensorData >= 200) & (sensorData < 400))
      Serial.println(2);
    else if((sensorData >= 400) & (sensorData < 700))
      Serial.println(3);
    else if((sensorData >= 700) & (sensorData < 900))
      Serial.println(4);
    else if(sensorData >=900)
      Serial.println(5);
    delay(250);
}
```

NOTE *There are two jumpers on the XBee module, which can be in either the USB or XBee positions. They must be in the USB position in order to use the serial monitor feature. They also must be in the XBee position to enable the XBee communications link. Figure 7-23 shows these jumpers in the USB position.*

This program reads the ADC value and then selects a number from 1 to 5 depending upon the soil's moisture content. I set up five zones to encompass all of the ADC values. These are shown in Table 7-3.

FIGURE 7-23 XBee jumpers in the USB position

Table 7-3 Soil Moisture Zones

Zone Number	ADC Range	Soil Condition
1	0 to 199	Very dry
2	200 to 399	Dry
3	400 to 699	Damp
4	700 to 899	Wet
5	900 to 1023	Very Wet

I do recognize that these categories may not be very precise, but they should be ample to provide useful feedback to the user regarding the actual soil conditions. The zone number is the only data actually transmitted to the XBee receiver. The XBeeRecrTest program, which I discuss later on, deciphers the zone number's meaning and displays the actual soil condition on the browser screen.

I set up a calibration test to check on how well the XbeeXmit functions when sending the zone data. Figure 7-24 shows a potentiometer to which I connected the A0 ADC input to the tap and 5VDC and GND to the other two leads.

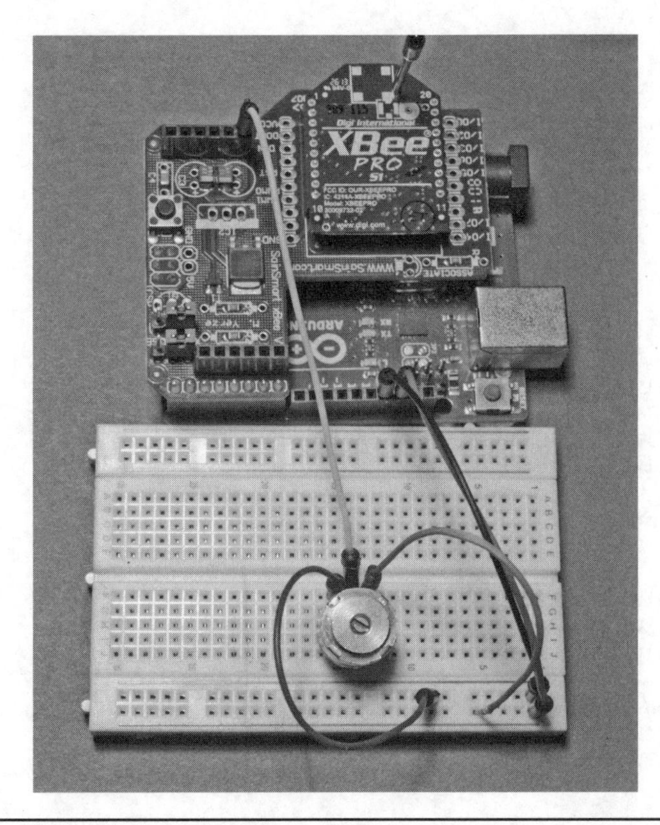

Figure 7-24 Transmitter calibration setup

This allowed me to vary the ADC input voltage, simulating the total range for the moisture probe. Of course, I also needed the XBee receiver portion running to fully carry out the calibration test. I named the following code XBeeRecvrTest to indicate that it was designed to be part of a test phase. This program receives the XBee transmitter's zone number and lights an LED corresponding to that zone. The code to accomplish this follows:

```
/* XBeeRecvrTest
Simple program to read a value sent from a transmitting XBee.
Ensure both XBees are jumpered into the XBee mode.
Author D. J. Norris
May, 2014   */

int sensorData;       // a variable to read incoming serial data

const int led1 = 2;
const int led2 = 3;
const int led3 = 4;
const int led4 = 5;
const int led5 = 6;

void setup() {
  pinMode(led1, OUTPUT);
  pinMode(led2, OUTPUT);
  pinMode(led3, OUTPUT);
  pinMode(led4, OUTPUT);
  pinMode(led5, OUTPUT);
  // initialize serial communication:
  Serial.begin(9600);
}

void loop() {
    sensorData = Serial.read();

    if(sensorData == '1'){
        digitalWrite(led1, HIGH);
        digitalWrite(led2, LOW);
        digitalWrite(led3, LOW);
        digitalWrite(led4, LOW);
        digitalWrite(led5, LOW);
    }
    else if(sensorData == '2'){
        digitalWrite(led1, LOW);
        digitalWrite(led2, HIGH);
        digitalWrite(led3, LOW);
        digitalWrite(led4, LOW);
        digitalWrite(led5, LOW);
    }
    else if(sensorData == '3'){
        digitalWrite(led1, LOW);
        digitalWrite(led2, LOW);
```

```
            digitalWrite(led3, HIGH);
            digitalWrite(led4, LOW);
            digitalWrite(led5, LOW);
        }
        else if(sensorData == '4'){
            digitalWrite(led1, LOW);
            digitalWrite(led2, LOW);
            digitalWrite(led3, LOW);
            digitalWrite(led4, HIGH);
            digitalWrite(led5, LOW);
        }
        else if(sensorData == '5'){
            digitalWrite(led1, LOW);
            digitalWrite(led2, LOW);
            digitalWrite(led3, LOW);
            digitalWrite(led4, LOW);
            digitalWrite(led5, HIGH);
        }
}
```

Figure 7-25 is a schematic for this receiver's calibration test circuit, which must be used with the preceding program.

Before running the calibration test, ensure that both XBee mode jumpers are in the XBee position as I mentioned earlier, or the test will not work.

Testing both the transmitter and receiver modules simply becomes a matter of turning the potentiometer and observing that the LEDs light sequentially as it goes through its full rotation. Recheck your receiver wiring if you do not see the LEDs blink on and then off as the potentiometer is turned from minimum to maximum voltage.

You are now ready to connect the XBee receiver to the Arduino development board, as long as the calibration test went smoothly. The hookup is straightforward in that the five wires going to the LEDs are disconnected and reconnected to the five Arduino boards' digital inputs 14 to 18, as shown in the Figure 7-26 schematic.

The physical interconnections between the Xbee receiver module and the Arduino development board are easily seen in Figure 7-27.

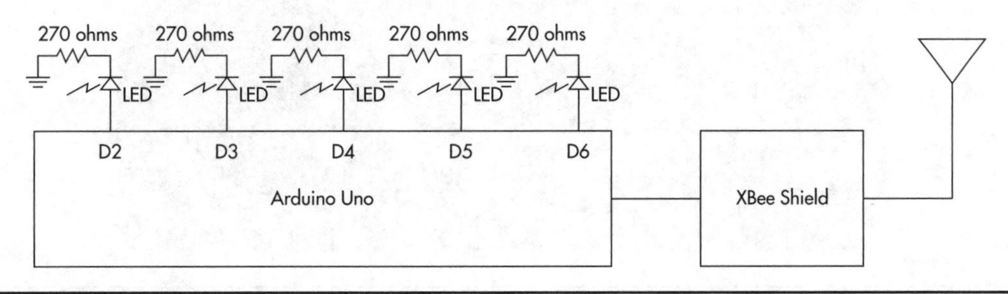

FIGURE 7-25 Receiver calibration test schematic

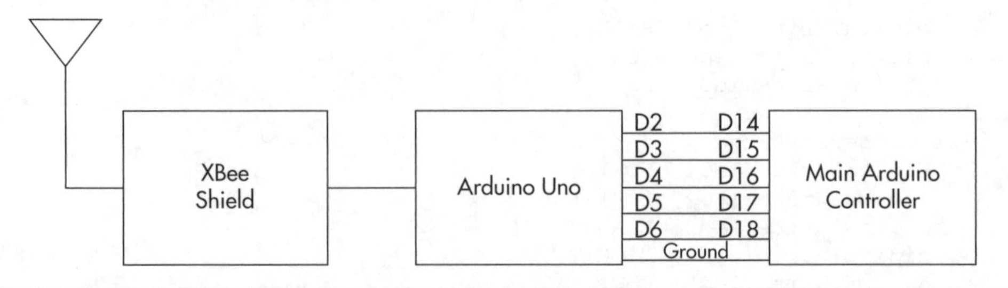

FIGURE 7-26 XBee receiver and Arduino development board schematic

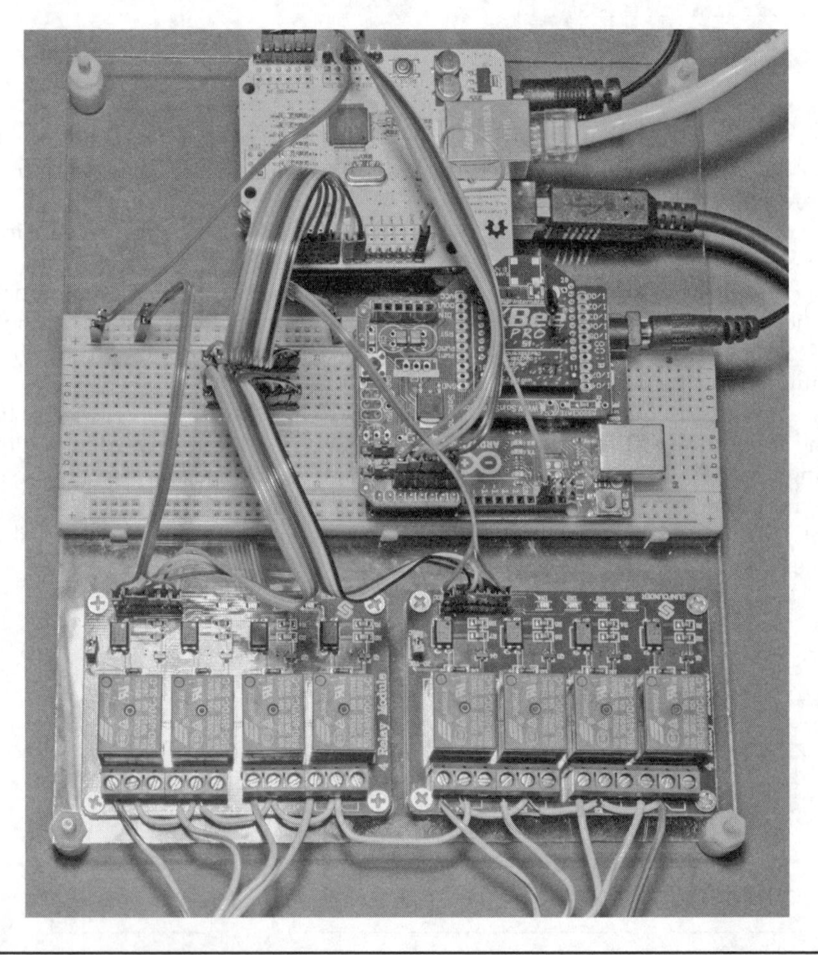

FIGURE 7-27 XBee receiver module connected to Arduino development board

A revised control program named Irrigation_Control_r1 must be uploaded into the Arduino mounted on the development board. This revised code is shown next, and encodes the zone numbers into specific soil condition strings, which are in turn displayed on the browser.

```
/*---------------------------------------------------------------
  Program: Irrigation_Control_r1
 Description: Arduino web server that serves up a web page allowing the user
              to control a residential six-zone irrigation system
 Hardware: Arduino Uno and official Arduino Ethernet shield. Should work
          with other Arduino compatible Ethernet shields.
 Software: Developed using Arduino 1.0.5 software.  Should be compatible
          with Arduino 1.0 +
 References: eth_websrv_LED modified by D. J. Norris
             WebServer example by David A. Mellis and modified by Tom Igoe
             Ethernet library documentation:
             http://arduino.cc/en/Reference/Ethernet
Date: May 2014
 --------------------------------------------------------------*/

#include <SPI.h>
#include <Ethernet.h>

// MAC address from Ethernet shield sticker under board
byte mac[] = { 0xDE, 0xAD, 0xBE, 0xEF, 0xFE, 0xED };
IPAddress ip(192, 168, 1, 177); // IP address, may need to change depending on network
EthernetServer server(80);  // create a server at port 80

String HTTP_req;          // stores the HTTP request

const int level1 = 14;
const int level2 = 15;
const int level3 = 16;
const int level4 = 17;
const int level5 = 18;

void setup()
{
    Ethernet.begin(mac, ip);  // initialize Ethernet device
    server.begin();           // start to listen for clients
    Serial.begin(9600);       // for diagnostics
    pinMode(2, OUTPUT);       // to relay on pin 2
    pinMode(3, OUTPUT);       // to relay on pin 3
    pinMode(4, OUTPUT);       // to relay on pin 4
    pinMode(5, OUTPUT);       // to relay on pin 5
    pinMode(6, OUTPUT);       // to relay on pin 6
    pinMode(7, OUTPUT);       // to relay on pin 7
}

void loop()
{
    EthernetClient client = server.available();  // try to get client
    if (client) {  // got client?
        boolean currentLineIsBlank = true;
```

```
        while (client.connected()) {
            if (client.available()) { // client data available to read
              char c = client.read(); // read 1 byte (character) from client
              HTTP_req += c;        // save the HTTP request 1 char at a time
                                   // last line of client request is blank and


                                   // ends with \n
                                   // respond to client only after last line received
                if (c == '\n' && currentLineIsBlank) {
                    // send a standard http response header
                    client.println("HTTP/1.1 200 OK");
                    client.println("Content-Type: text/html");
                    client.println("Connection: close");
                    client.println();
                    // send web page
                    client.println("<!DOCTYPE html>");
                    client.println("<html>");
                    client.println("<head>");
                    client.println("<title>Arduino Irrigation Controller</title>");
                    client.println("</head>");
                    client.println("<body>");
                    client.println("<h1>Arduino Irrigation Control</h1>");
                    client.println("<p>Click on zone # and then Submit to activate
                    zone</p>");
                    client.println("<form method=\"get\">");
                    ProcessCheckbox(client);
                    GetSensor(client);
                    client.println("</form>");
                    client.println("</body>");
                    client.println("</html>");
                    Serial.print(HTTP_req);
                    HTTP_req = "";    // finished with request, empty string
                    break;
                }
                // every line of text received from the client ends with \r\n
                if (c == '\n') {
                    // last character on line of received text
                    // starting new line with next character read
                    currentLineIsBlank = true;
                }
                else if (c != '\r') {
                    // a text character was received from client
                    currentLineIsBlank = false;
                }
            } // end if (client.available())
        } // end while (client.connected())
        delay(10);      // give the web browser time to receive the data
        client.stop(); // close the connection
    } // end if (client)
}

// turn on selected zone relay
void ProcessCheckbox(EthernetClient cl)
{
```

```
cl.println("<input type=\"radio\" name=\"zones\" value=\"0\" > None   <br>");
cl.println("<input type=\"radio\" name=\"zones\" value=\"1\" > Zone 1 <br>");
cl.println("<input type=\"radio\" name=\"zones\" value=\"2\" > Zone 2 <br>");
cl.println("<input type=\"radio\" name=\"zones\" value=\"3\" > Zone 3 <br>");
cl.println("<input type=\"radio\" name=\"zones\" value=\"4\" > Zone 4 <br>");
cl.println("<input type=\"radio\" name=\"zones\" value=\"5\" > Zone 5 <br>");
cl.println("<input type=\"radio\" name=\"zones\" value=\"6\" > Zone 6 <br>");
cl.println("<input type=\"submit\" value=\"Submit\" >");
cl.print("<br />");

if(HTTP_req.indexOf("zones=0") > -1){
  digitalWrite(2, LOW);
  digitalWrite(3, LOW);
  digitalWrite(4, LOW);
  digitalWrite(5, LOW);
  digitalWrite(6, LOW);
  digitalWrite(7, LOW);
}
else if(HTTP_req.indexOf("zones=1") > -1){
  digitalWrite(2, HIGH);
  digitalWrite(3, LOW);
  digitalWrite(4, LOW);
  digitalWrite(5, LOW);
  digitalWrite(6, LOW);
  digitalWrite(7, LOW);
 }
 else if(HTTP_req.indexOf("zones=2") > -1){
   digitalWrite(3, HIGH);
   digitalWrite(2, LOW);
   digitalWrite(4, LOW);
   digitalWrite(5, LOW);
   digitalWrite(6, LOW);
   digitalWrite(7, LOW);
 }
else if(HTTP_req.indexOf("zones=3") > -1){
  digitalWrite(4, HIGH);
  digitalWrite(2, LOW);
  digitalWrite(3, LOW);
  digitalWrite(5, LOW);
  digitalWrite(6, LOW);
  digitalWrite(7, LOW);
 }
else if(HTTP_req.indexOf("zones=4") > -1){
  digitalWrite(5, HIGH);
  digitalWrite(2, LOW);
  digitalWrite(3, LOW);
  digitalWrite(4, LOW);
  digitalWrite(6, LOW);
  digitalWrite(7, LOW);
 }
else if(HTTP_req.indexOf("zones=5") > -1){
  digitalWrite(6, HIGH);
  digitalWrite(2, LOW);
  digitalWrite(3, LOW);
  digitalWrite(4, LOW);
  digitalWrite(5, LOW);
```

```
        digitalWrite(7, LOW);
       }
      else if(HTTP_req.indexOf("zones=6") > -1){
        digitalWrite(7, HIGH);
        digitalWrite(2, LOW);
        digitalWrite(3, LOW);
        digitalWrite(4, LOW);
        digitalWrite(5, LOW);
        digitalWrite(6, LOW);
      }
}

void GetSensor(EthernetClient cl) {
       if(digitalRead(level1)){
       cl.print("soil is very dry");
       cl.print("<br />");
       }
       else if(digitalRead(level2)){
       cl.print("soil is dry");
       cl.print("<br />");
       }
       else if(digitalRead(level3)){
       cl.print("soil is damp");
       cl.print("<br />");
       }
       else if(digitalRead(level4)){
       cl.print("soil is wet");
       cl.print("<br />");
       }
       else if(digitalRead(level5)){
       cl.print("soil is very wet");
       cl.print("<br />");
       }
}
```

I next used the calibration circuit to send various simulated soil conditions and checked the corresponding browser display. Figure 7-28 shows the result for zone 1 or "Very dry" conditions.

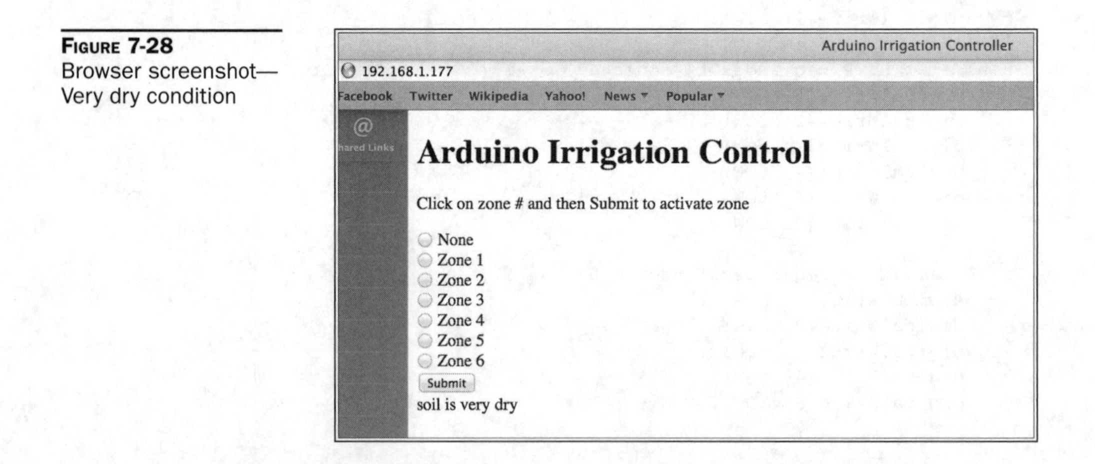

FIGURE 7-28 Browser screenshot—Very dry condition

I then increased the voltage to the mid-point and observed the "Damp" condition displayed on the browser, as shown in Figure 7-29, which is a partial browser screenshot.

I also checked all the other zones and determined that accurate conditions were displayed on the browser corresponding to the calibration voltage.

Submit
soil is damp

Figure 7-29 Browser screenshot—Damp condition

Actual System Operation

I next proceeded to install the moisture sensor subsystem outdoors, as I described in my previous section, after the calibration testing was successfully completed. The outdoor system is pretty much a hands-off box, which means you don't have to fiddle with it after you install it. I did find that my local bird population seemed to like "bombing" the solar panel, but I guess that's part of the price of rural living.

Figure 7-30 shows a browser screenshot of the irrigation system receiving a signal from the outdoor sensor. I knew the soil was very wet as it had been raining for a day prior to the test.

I also cycled all the zones to confirm that the sprinklers did activate when their zone was clicked. This last action completed the basic Arduino irrigation control project.

Enhancements

The most obvious project improvement is to incorporate a timing control element within the Arduino development system. This would allow the user to preset the zone watering start and duration times as well as the days of the week. This additional functionality would completely replace the existing irrigation controller. I purposely did not include timing within this project as the project focus was on implementing a remote control function via the Web.

The moisture sensor capabilities can also be easily extended to include additional remote stations such that multiple areas could be monitored and an overall moisture assessment be developed especially for an extended area. The main irrigation program would have to be modified to accommodate the additional sensor inputs. It also turns out

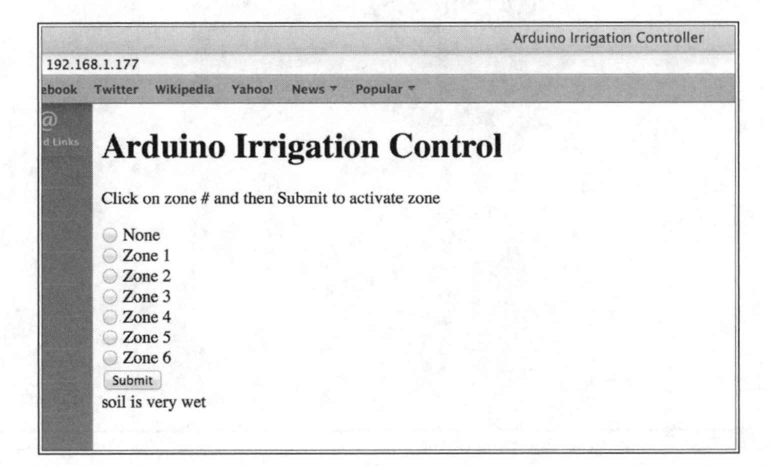

Figure 7-30 Browser screenshot of a system with an operating outdoor moisture sensor

that the XBee protocol easily handles multiple nodes such that each sensor can have its own ID, which enables the main program to identify which sensor is sending the data. It would be entirely feasible to customize the watering based on the sensor inputs to only turn on specific zones for specific durations. This would truly turn the irrigation system into a fully automated watering "robot."

My last suggestion relies on an interesting and very inexpensive soil moisture sensor that I found for sale on eBay (see Figure 7-31). The probe tips are only 1.75 inches in length, which is quite suitable for potted plants.

The electronics portion seen on the left-hand side of the figure outputs a digital high level when the moisture probes' voltage drops below a level set by the board's potentiometer. The analog voltage is also available as an output, which may be connected to an ADC, if desired. This mini probe uses a voltage divider sensor circuit, which is almost identical to my original design. At the simplest level, I can imagine an Arduino connected to this sensor, which would light an LED when the potted plant becomes too dry. You could also make the plant's soil moisture status available on the Web using a modified version of the Irrigation Control

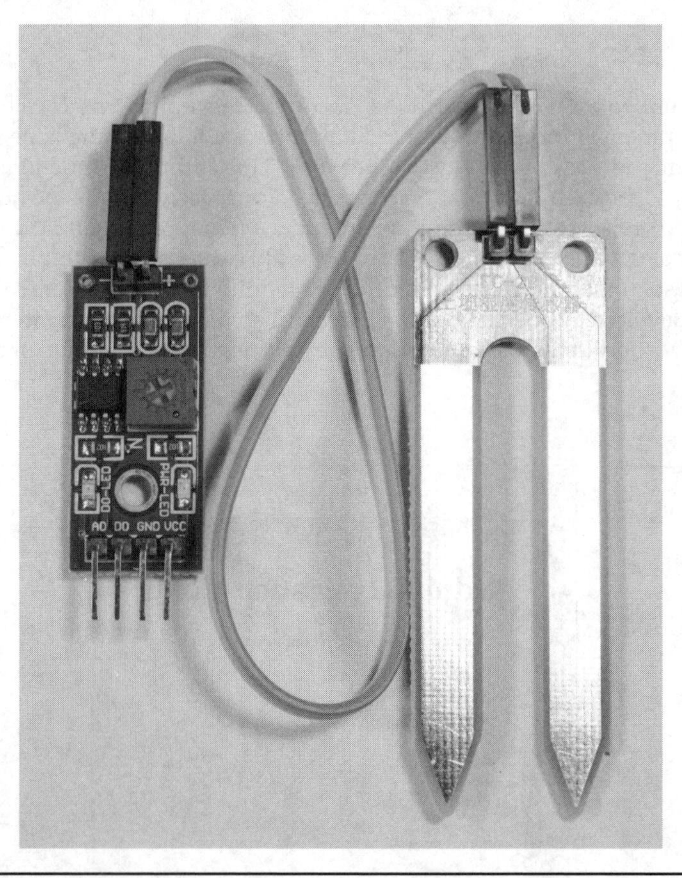

FIGURE 7-31 Mini soil moisture probe

program that only reports the soil moisture level. I do think it would be a bit too much to install a solenoid operated watering system for a single, potted plant, but I will leave that challenge to any ambitious reader. Please send me an e-mail c/o of my publisher if you do accomplish this project. The email-id is listed in the front matter of the book.

Summary

This chapter's project was the design and implementation of a Web-controlled, home irrigation system. It also included a soil moisture sensor, which reported to the user the actual soil conditions and if irrigation was needed.

I discussed the basic design, which included two relay modules controlled by an Arduino Uno's GPIO pins. These modules were necessary because the existing water solenoids required 24VAC to be operated, which far exceeds the Uno's digital output capabilities.

A basic Arduino sketch was created that allowed the user to operate any one of six zones using a web interface. I also showed you how to interconnect the Arduino development board, which held the Uno, Ethernet Shield, relay modules, and solderless breadboard, to the existing irrigation terminal strip.

A brief operational word test was next conducted that confirmed that the new system would control the irrigation zones, as desired, using a web interface.

I next discussed a moisture sensing subsystem that employed XBee wireless technology so that no wires were needed to interconnect between the main board and the remote moisture sensor. A detailed section was also provided for those readers who want to learn more about the XBee technology.

A discussion followed on how the moisture sensor was designed based upon measuring Earth Conductivity (EC) as a moisture content indicator. I also showed the complete design of the outdoor moisture sensor system, which included a solar panel to trickle charge a lead acid battery for maintenance-free operation.

The XBee transmitter and receiver programs were discussed next as well as my concept of using moisture zones to report soil conditions.

I next demonstrated a calibration test, which proved the system accurately reported soil conditions based upon simulated voltage levels. The outdoor moisture sensor was installed after the calibration tests were completed and it reported the correct soil conditions.

The chapter ended with some suggested enhancements, including a possible potted plant monitoring system.

CHAPTER 8

Arduino Lighting Controller

This chapter's project will use the Uno board with an Ethernet Shield connection to remotely operate lights both inside and outside a residence. The XBee wireless technology will also be used to eliminate the need for hard wiring from the controller to the lights that are placed at a distance. Only lights with external power cords will be used in this project to make it reasonable and easy to implement. I will also demonstrate a radio frequency (RF) key fob that will power-on selected lights when the user is within range.

System Design

This system was designed to use both wired and wireless segments to control power controllers, which I discuss in detail in a later section. Using wireless controls will provide a high degree of flexibility in configuring the system to fit a particular residence. Figure 8-1 is a system block diagram, which shows all the principal components.

The controller node is the key system component that supports multiple functions including:

- Web server providing an Internet gateway
- XBee broadcast module
- Digital I/O for wired power controllers

Each of these controller functions will be further discussed in later sections, which will also include appropriate build and configuration instructions.

There are two XBee nodes also shown in the block diagram, which provide the wireless control for both remote power controllers. The software design does provide for a maximum of 18 wireless controllers if such a large amount was required for particular installation. There are a total of three XBee nodes in this design, which at first glance might seem a bit confusing as I used a point-to point in the previous chapter. This works because the XBee module installed in the controller node will be functioning in a point-to-multipoint mode. This means it will be broadcasting to all nearby XBee nodes. However, it does not respond or acknowledge these nodes, thus decreasing the overall system reliability. I do not believe that this lack of positive acknowledgment would be detrimental to overall system operation. That belief was subsequently confirmed in later system tests. I have included the following aside for those readers interested in learning more about the XBee mesh mode, which does provide for full duplex communication between three or more XBee modules.

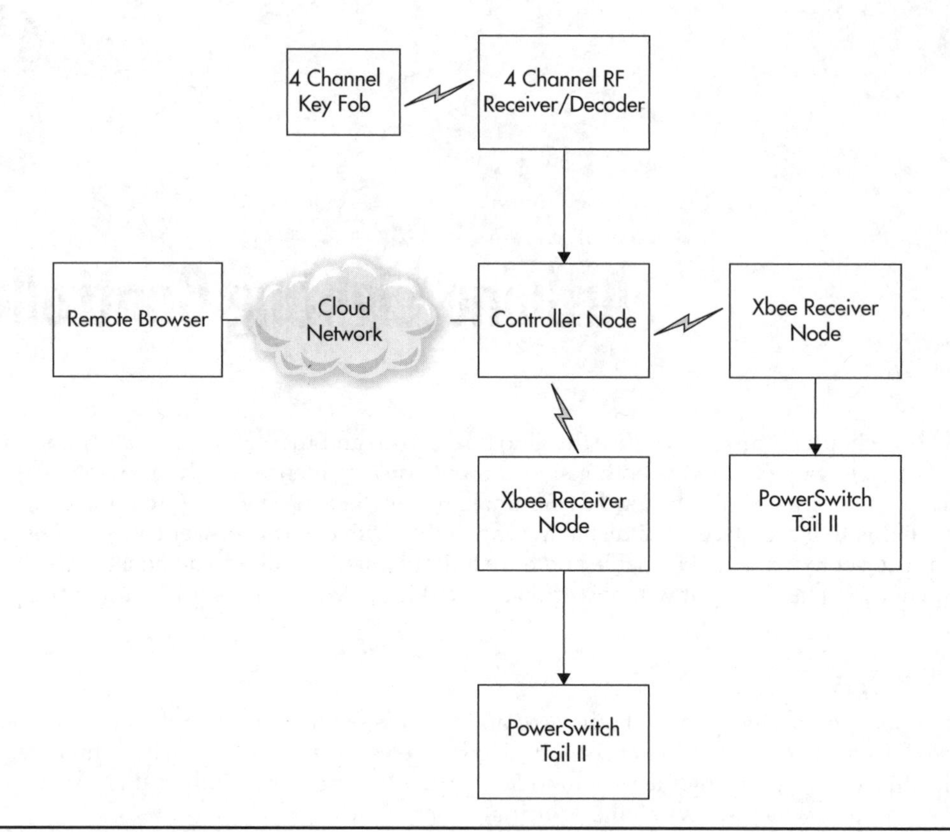

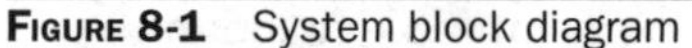
FIGURE 8-1 System block diagram

XBee Mesh

A lot of the discussion that follows is based on information presented in the excellent book *Building Wireless Sensor Networks: with ZigBee, XBee, Arduino, and Processing* by Robert Faludi. Faludi presents a very clear explanation of the basics involved in setting up and running a wireless sensor network using XBee transceivers with Arduino boards.

It should be clear that the name XBee refers only to a line of small, digital, radio transceivers manufactured by Digi International (Digi). The XBee radios follow the ZigBee communications/network protocol just as a multitude of Wi-Fi devices follow the Wi-Fi standard, which formally is designated as IEEE 802.11 a/b/g/n. ZigBee also has the formal standards designation of IEEE 802.15.4.

All XBee radios have been manufactured and programmed with either Series 1 (S1) or Series 2 (S2) firmware. S1 firmware will only support either point-to-point or point-to-multipoint communications. The point-to-point communications link is also known as a peer-to-peer link. S2 firmware supports both point-to-point as well as mesh communications.

XBee radios have two operational modes: Transparent, also known as the AT Modem mode, and application program interface (API). The default mode is Transparent, which allows an XBee user to quickly and automatically set up a peer-to-peer link capable of handling two-way or full-duplex communications using a 9600 baud rate. The Transparent operational mode allows either XBee S1 or S2 radios to seamlessly establish a serial communications link without any user configuration effort. On the other hand, using the API mode does require some user configuration efforts. Digi provides a free software configuration tool named X-CTU, which is available from its website at www.digi.com/support/productdetail?pid=3524&type=utilities. This configuration utility is freely available for Windows, Mac, and Linux platforms. The following configuration discussion will be based on a sample network configuration, as shown in Figure 8-2.

This network configuration is called a star topology because there is a central node with several nodes surrounding it. You should note that the nodes are labeled with these specific ZigBee functions:

- **Coordinator node** This is the master or overall network control node. There is only one coordinator in a ZigBee network. It can transmit and receive data to/from any remote XBee node.
- **Router node** This node type acts as a digital data repeater, receiving and transmitting data between the coordinator and end node or another router.
- **End node** This node is the Arduino Uno's gateway into the ZigBee network. Its purpose is to receive data designated for the Uno's use as well as transmit data that is generated by the Uno.

ZigBee networks will normally have only one coordinator node and two or more end nodes. There could be one or more router nodes in the network depending on

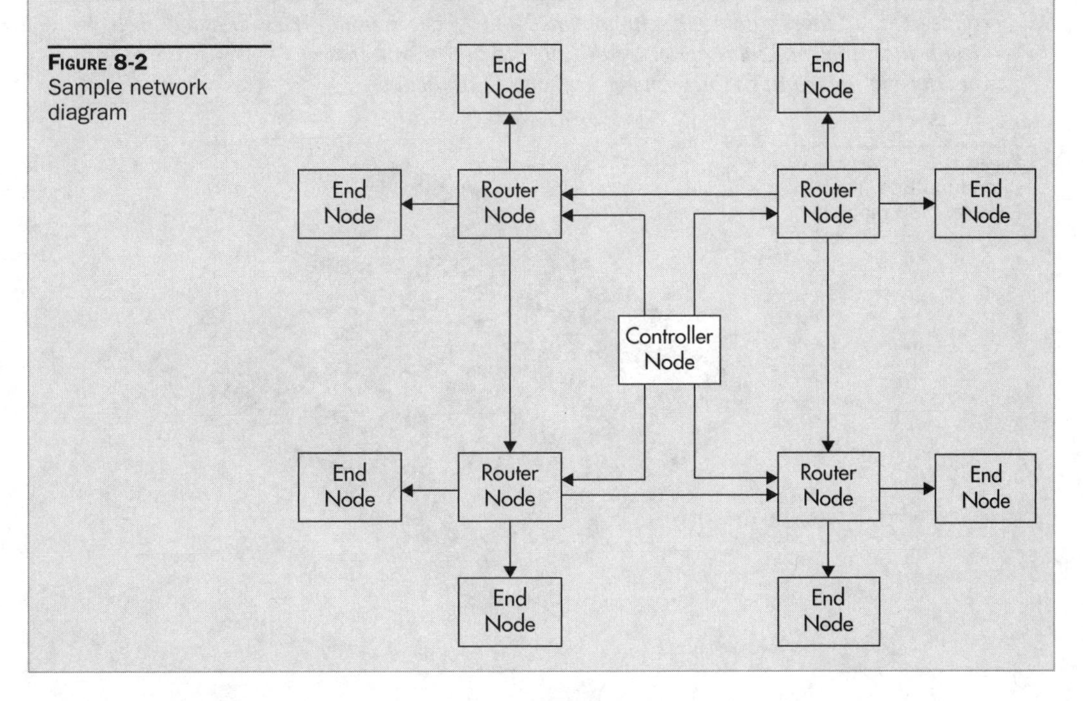

FIGURE 8-2 Sample network diagram

the physical distances between the coordinator and end nodes. Using routers is also dependent upon the transmitter power for the XBee nodes. The XBee Pro series has a maximum 60mW output, which is 30 times greater than the 2mw nodes that I use in this project. The more powerful nodes will mean a greater transmitting range without the need to use a router as a digital repeater.

Every node in an XBee network must be associated with a personal area network (PAN), which is a common identifier shared by all the network nodes. The default PAN ID preprogrammed into all the factory firmware is 3332. There is also a specific node ID associated with each node. This is known as the MY address and is a four-digit number that is user assigned. The PAN ID and MY address are both programmed into a node using the X-CTU utility.

There is another important device that is required before any XBee modules can be programmed. This device is an adapter board that has an onboard USB serial interface, which allows the computer running X-CTU to connect to the XBee transceiver module. The Sparkfun Explorer USB module is the one I would recommend and is shown in Figure 8-3.

NOTE *It is possible to use the XBee Shield and Uno to program the XBee module if your Uno uses a socketed microcontroller. To do so, you must remove the ATMega328P microcontroller from its socket on the Uno board. You must also ensure that the XBee jumpers located on the XBee Shield are set in the USB position. I discussed these jumpers in the previous chapter. However, I do not recommend that you follow this procedure as it is easy to damage the controller chip because it takes considerable force to remove the 28 DIP chip from its socket. There is also the possibility that damage will occur when the chip is reinserted. Of course, doing the chip removal and reinsertion multiple times will likely result in a problem. I have presented this note to inform you that there is an alternative to purchasing the relatively inexpensive Explorer USB module.*

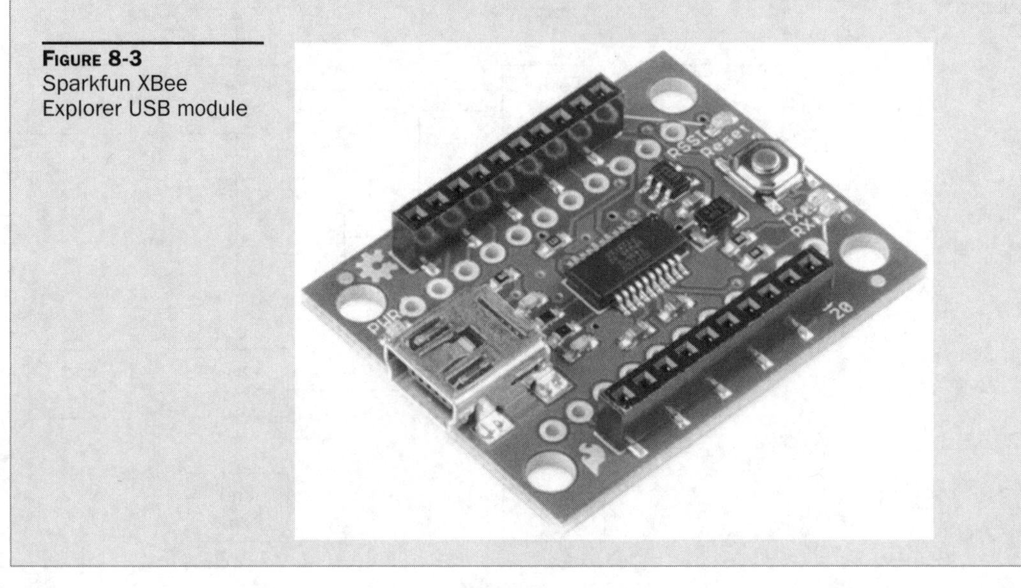

FIGURE 8-3 Sparkfun XBee Explorer USB module

You will also need to download and install the appropriate Virtual Com driver from www.ftdichip.com/Drivers/VCP.htm. This software utility creates a virtual serial port that works with the FTDI chip onboard the XBee Explorer module.

You are ready to program all the XBee modules once you have acquired the XBee adapter and downloaded and installed both the X-CTU utility and the FTDI driver utility. The next section discusses how to use the X-CTU utility to program the XBee modules to meet the network configuration requirements. I will now refer you back to Faludi's book to learn how to actually program the XBee modules.

Finally, there is an analog radio frequency (RF) control segment shown in Figure 8-1, which allows the user to activate lights when the RF key fob is actuated within 10 to 25 meters of the controller node.

Controller Node

The controller node is composed of an Uno stacked with Ethernet and XBee shields. Figure 8-4 shows this stack with the Uno on the bottom, the Ethernet Shield in the middle, and the XBee Shield on top.

However, there is a serious issue that requires a minor modification of the Ethernet Shield before this stack can be successfully interconnected. The XBee Shield is designed to

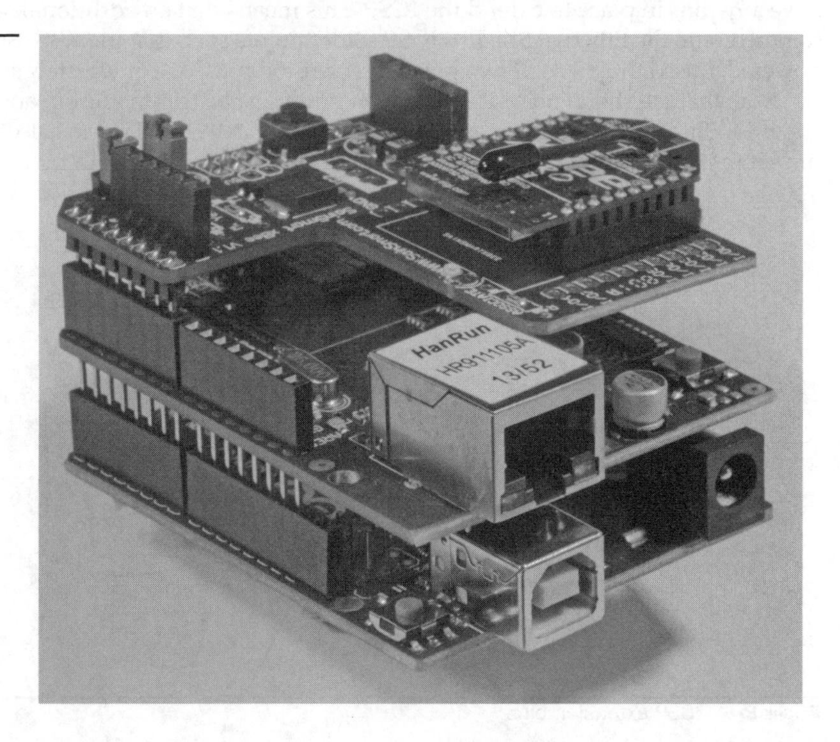

Figure 8-4 Controller node

Figure 8-5 XBee socket

obtain its power using the ICSP socket. This six-pin socket can easily be seen at the bottom of Figure 8-5, which is a photograph of the XBee Shield's back.

It turns out that the Ethernet Shield, which is placed in the middle of the stack, does not have any pins in place to extend the ICSP. This means that an additional set of pins must be soldered onto the Ethernet Shield's ICSP solder points such that the XBee Shield can be powered. I used three sets of two position extension pins, one of which is shown in Figure 8-6.

Note that the short end of the extension pins must be trimmed from about 3 millimeters to about 1 millimeter in order for the pins to fit properly in the final stack. I also show an unaltered pin set in the figure so that you can get an idea of the proper length to trim the pins.

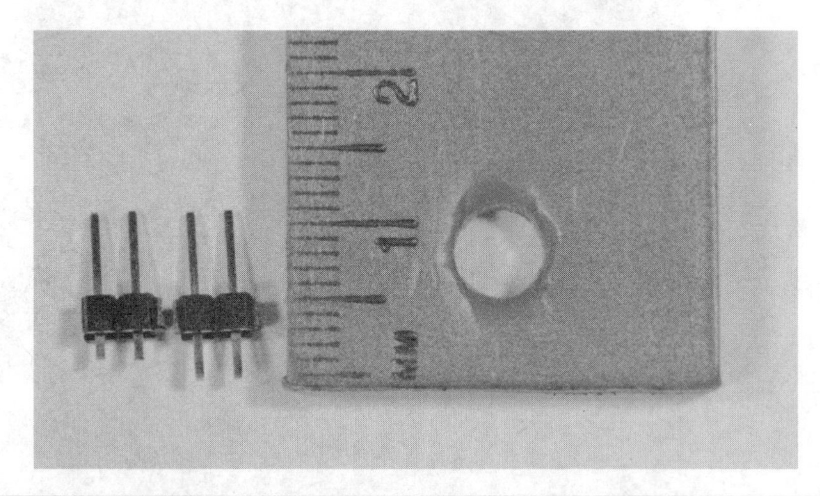

Figure 8-6 ICSP extension pins

FIGURE 8-7 Soldered ICSP extension pins

I found out that the easiest and most effective way to solder the pins to the Ethernet Shield is to first insert them into the XBee Shield's ICSP socket and then mount the XBee Shield to the Ethernet Shield. You then can use a very fine point soldering iron to solder the outermost row to the Ethernet Shield's ICSP solder points. I next carefully removed the XBee Shield and proceeded to solder the innermost set of solder points. Figure 8-7 is a close-up of the ICSP extension pins after they all have been attached.

After soldering these pins, you are all set to finish building the controller node.

Ethernet Shield

Figure 8-8 is a picture of the top of the newly modified Ethernet Shield. Notice the soldered ICSP extension pins at the left-hand side of the board.

FIGURE 8-8
Ethernet Shield board

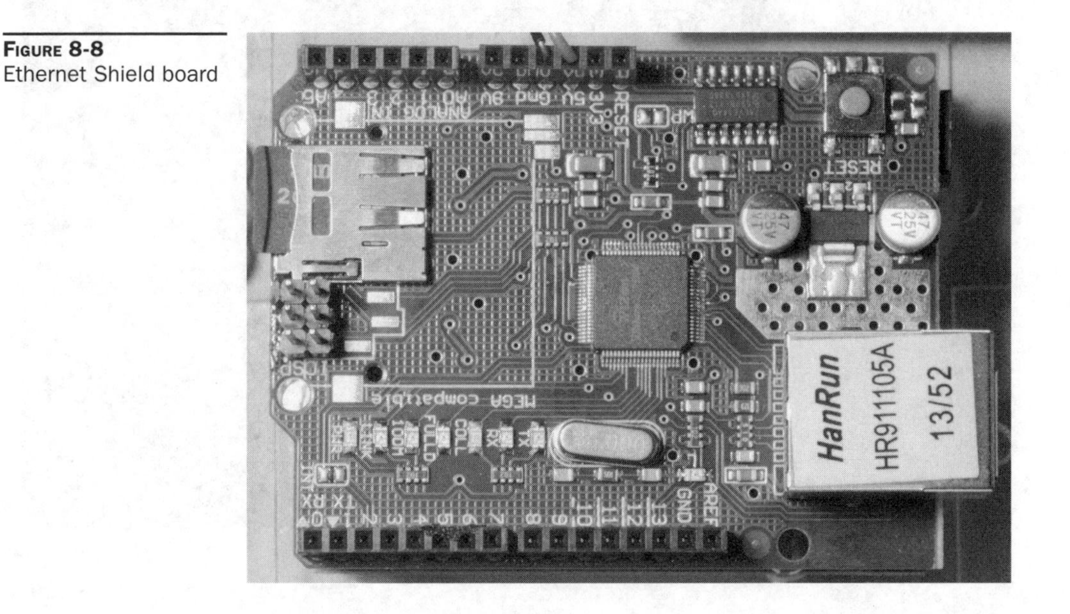

You probably already noticed that this Ethernet Shield board is different than a similar board I used in earlier chapters. I switched the boards for two reasons. First, this new board has a full set of Arduino shield pin sockets and extensions, except for the ICSP pin extensions. Second, the new board has provision to accept a micro SD memory card, which is used to hold the web server HTML code. This new Ethernet Shield is simply named the "Ethernet Shield with Micro SD Slot for Arduino Uno" and is available from a number of online sources. Just be forewarned that I found a wide disparity in prices online for what appears to be the same board. I don't think you should pay more than $20USD for this board. You will need the micro SD card slot installed in order to complete this project.

Controller Node Case and Mounting Arrangement

I decided to use a very nice looking case to enclose the controller node as it would be visible in the home and you would not want a visible "rat's nest" of wires. I purchased a plastic case, part number 905 from Adafruit Industries, which is shown in Figure 8-9.

Its overall dimensions are 125mm × 175mm × 75mm, which is more than ample to contain all the controller node components. It also features a transparent, screw-down case top that makes it more resistant to inadvertent liquid spills, a feature that may prove handy, especially for homes with small children present.

I next mounted the controller stack on a piece of clear Lexan, which in turn is mounted in the case by four 3/8 x 20 flathead Phillips screws. The Lexan sheets and the mounting hole dimensions are shown in Figure 8-10.

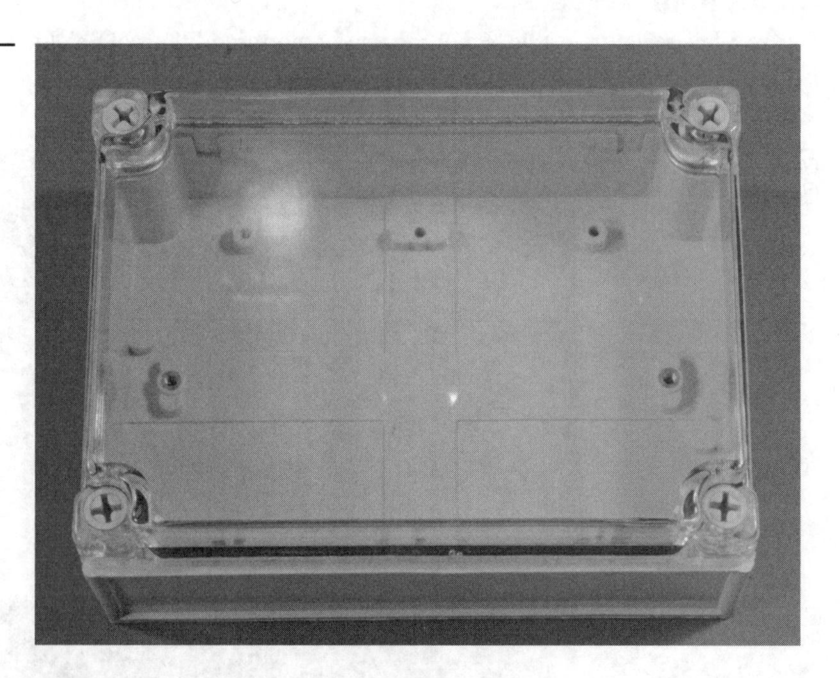

FIGURE 8-9 Controller node case

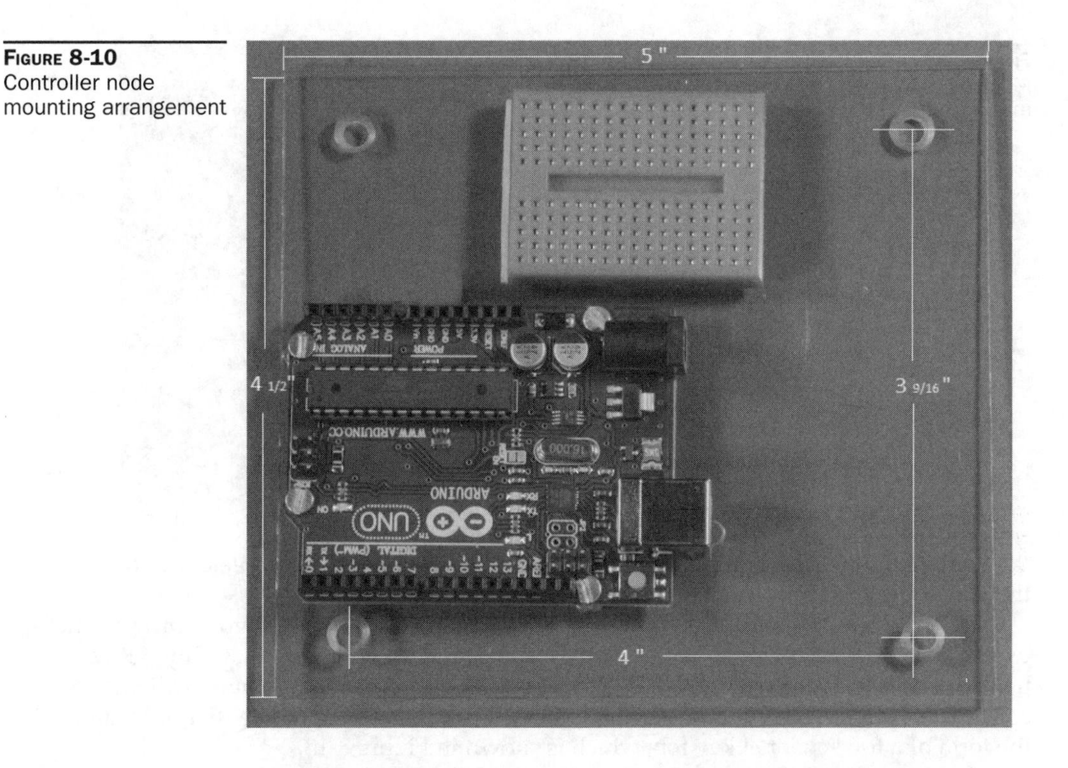

Figure 8-10 Controller node mounting arrangement

The Uno board and a small, solderless breadboard are also shown in the figure. The breadboard is used to hold and wire an RF receiver/decoder that I discuss in the next section. I have shown the Uno board to help illustrate a quirk that slightly complicates how it is mounted to the supporting Lexan board. You may observe that the lower, right-hand mounting hole has been drilled very close to a 10-position socket. It is so close that it prevents the head of a 4-40 mounting screw from sitting flush to the PCB surface. This will cause the screw to be tilted in the hole and generally disturb how the board is mounted to the Lexan sheet. To avoid this issue, I ground down the head of a 4-40 machine screw using a Dremel tool to provide sufficient clearance. Figure 8-11 is a close-up image of this modified screw. Simply slide the modified screw in place and you will be all set for a quality board mount.

Figure 8-12 shows the interior of the controller node with all the components mounted and wired. The external power supply, Ethernet, and power controller cables are also shown in the figure.

Note *I found it convenient to place the Uno board on the Lexan sheet and use a black, fine tip Sharpie marker to locate the board's mounting holes as a drill guide. This technique avoids any problems with transferring measurements to the Lexan sheet and it is a much quicker approach to mounting the board.*

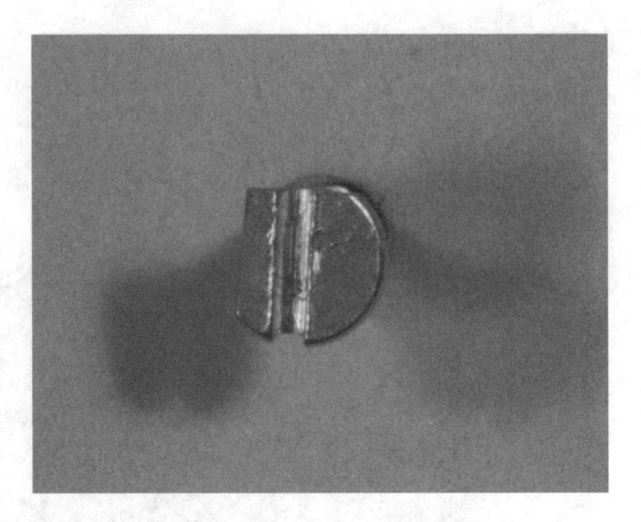

FIGURE 8-11 Modified 4-40 machine screw

I will defer discussing the program that runs this node until I complete my coverage of the remaining system components.

An RF transmitter and receiver was also incorporated into the design to provide the user, who is approaching the home, with a quick and easy way to activate any desired lighting. This activation method does not require any Internet access, but any lights so activated by the RF scheme will also be indicated on the web page. The RF transmitter is in the form of a four-channel key fob, which is shown in Figure 8-13.

The channels are labeled A through D on the fob. The fob RF transmitter operates with about 1mW of power in the 315MHz ISM band. It has a direct line-of-sight range of approximately 25 meters and about 10 to 15 meters indoors with walls and other obstructions. The key fob is part number 1095 and maybe purchased from Adafruit Industries.

The complementary receiver is shown in Figure 8-14.

FIGURE 8-12 Controller node interior

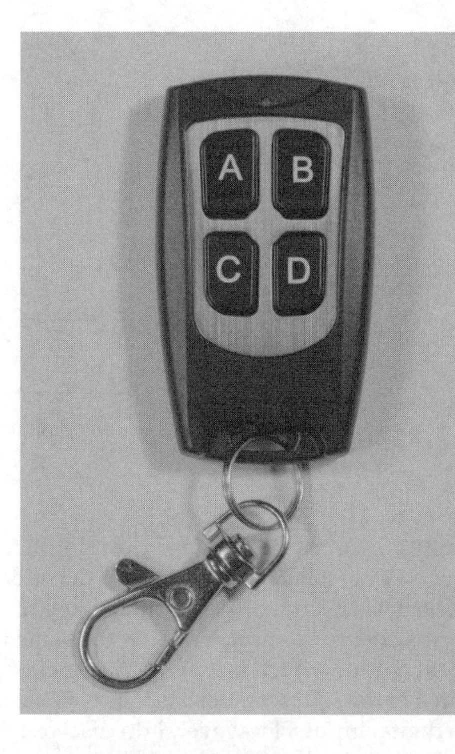

FIGURE 8-13
Four-channel key fob

FIGURE 8-14
RF four-channel latching receiver

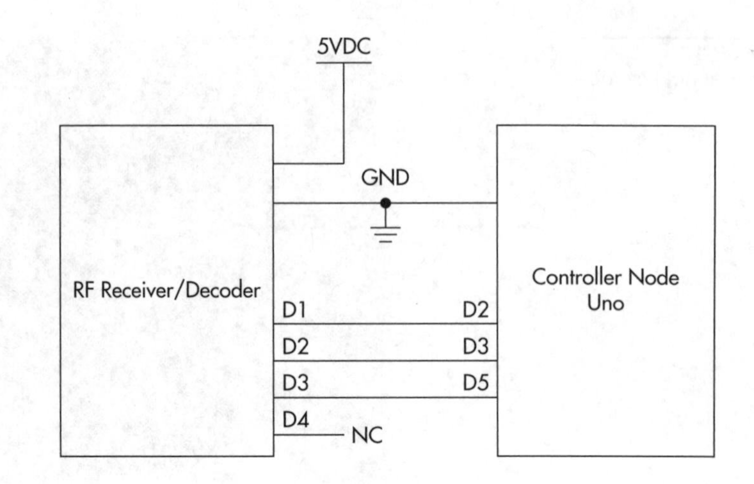

Figure 8-15 RF receiver and Uno interconnection schematic

The receiver may be purchased from Adafruit Industries as part number 1098. The receiver board uses a Princeton Technology Corp decoder chip, model number PT2272, to decode which channel has been received from the key fob. The chip will then latch the corresponding digital channel output. Pressing the same fob channel key again will unlatch the receiver's digital output, which is exactly the functionality needed for this design.

I used only three of the four channels as that was the maximum number supported by the available Uno digital inputs. However, I do discuss in the software section how to alter the number of digital inputs allocated to the RF system, if you later discover that you really need the additional channel.

The receiver board is simply plugged into the small breadboard that is shown in Figures 8-9 and 8-10. Figure 8-15 is the interconnection schematic for the receiver and the Uno board.

I found it very useful to use jumper wires to make all the interconnections between the breadboard and the Uno as that allowed me to quickly change the configuration as I tested the system. You should also note that the receiver is powered from the Uno's 5V supply. I also measured the receiver's digital high-level output and determined it to be 3.84V, which is perfectly acceptable for the Uno's digital input pins.

The RF transmitter key fob and receiver/decoder board do not require any programming or configuration efforts. They simply work out-of-the-box.

XBee Receiver Node

The XBee receiver node is a remotely located module that communicates with the controller node using the XBee network and passes commands over to the installed Uno board, which is directly interfaced to a power control device that I discuss in the next section. The complete node, consisting of an Uno board, an XBee shield, and an XBee Pro transceiver is shown in Figure 8-16.

It is identical to the moisture sensor transmitter node that I discussed in the previous chapter. However, the program controlling this node will either be XBee_Receiver1 or

FIGURE 8-16 Uno XBee receiver node

XBee_Receiver2 depending upon the node. The receiver programs are discussed later in this chapter.

I also used a smaller plastic case, Adafruit part number 903, to enclose the XBee receiver node. This case has overall dimensions of 80mm × 110mm × 45mm and is shown in Figure 8-17.

I mounted the XBee receiver node stacks on Lexan sheets just as I did for the controller node. However, this time the Lexan sheets are attached to the case bottom using two 1/4-inch thick nylon spacers along with a 20mm M4 machine screw and a 3/4-inch sheet metal screw, all of which you can see in Figure 8-18.

I didn't add any measurements to this figure as I would recommend that you make a paper template to help cut and fit the Lexan sheet to the case. The template will also help you to locate the four Uno mounting holes and the two Lexan case mounting holes.

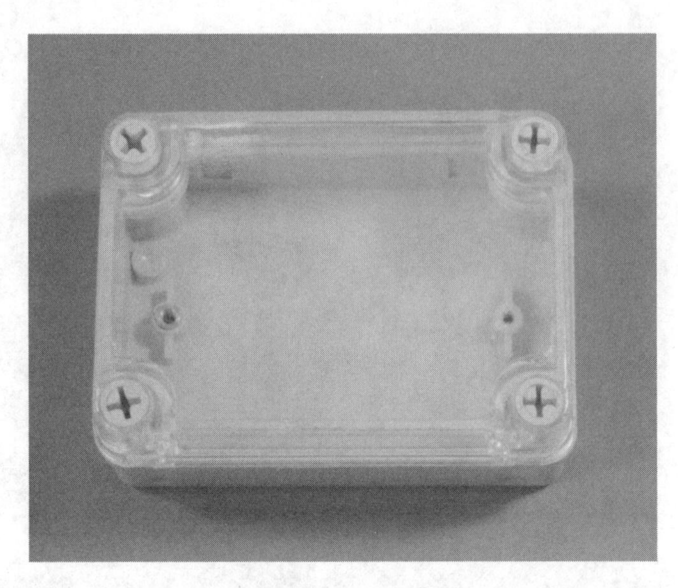

FIGURE 8-17 XBee receiver node case

Figure 8-19 shows the interior of one of the XBee receiver nodes with all the components mounted inside it. I will discuss the programs that control the XBee receiver nodes after the following section.

FIGURE 8-18 XBee receiver node mounting arrangement

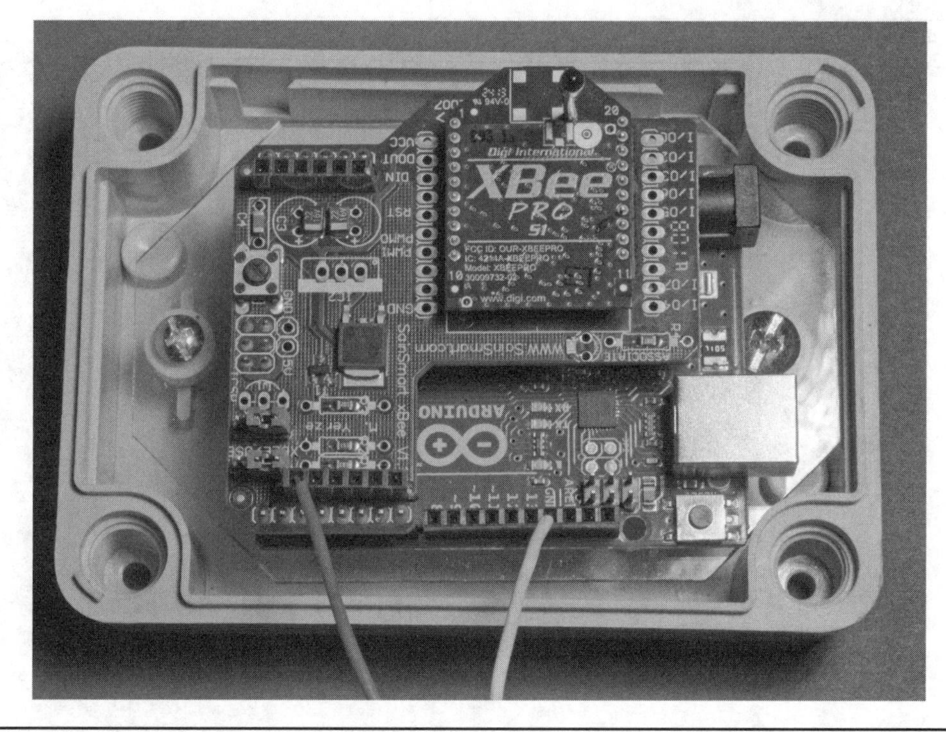

Figure 8-19 XBee receiver node interior

PowerSwitch Tail II

The PowerSwitch Tail II is a power control device and a key component in this system design. Figure 8-20 shows this device.

It is essentially a power cord that is controlled by a low-level digital signal that will be directly connected to an Uno's GPIO pin. The PowerSwitch Tail II, which I will now refer to as the PST2, uses an optically isolated digital input to control a power relay capable of handling up to 15A at 120VAC. The optical isolation eliminates any safety concerns about dealing with mains type power with the Uno board. The PST2 is also ruggedly constructed and very well insulated, making it extremely safe to use in a home environment. This power control device can also handle loads up to 1.5kW, which is well beyond anything I will use in this project.

The PST2 schematic is shown in Figure 8-21, which helps point out the robust and safe design that makes up the PST2.

I highly recommend that you purchase the appropriate number of PST2s to control all the power loads for both safety and convenience. Building your own power controllers is really not a good idea. The PST2 has already passed all UL and other safety certifications and is ready to use.

Figure 8-22 is a simple schematic showing how the Uno Xbee receiver node connects to a PST2.

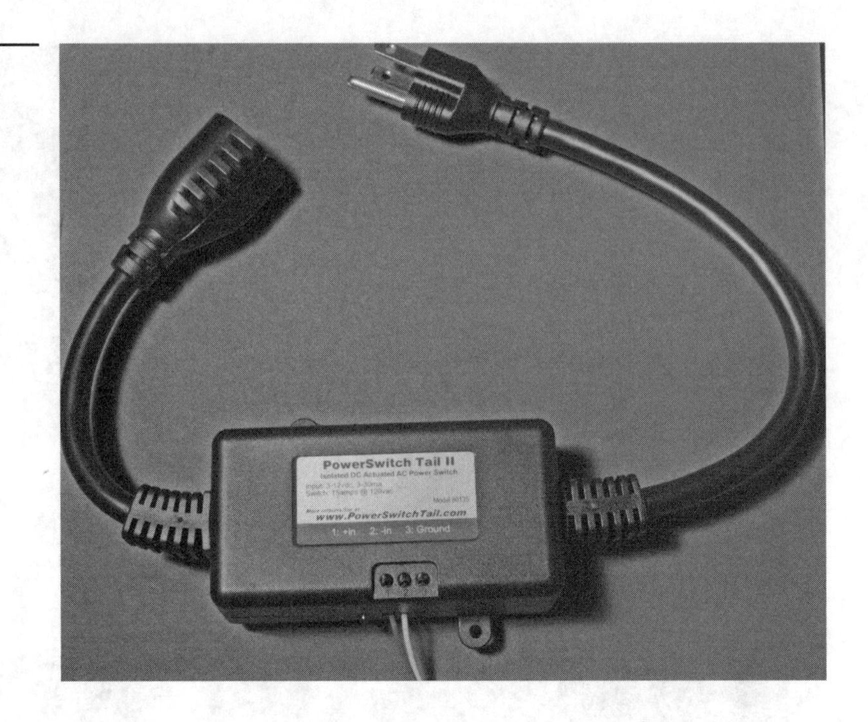

FIGURE 8-20 PowerSwitch Tail II

Not much is required for the connection other than a pair of wires carrying the GPIO signal and ground. I selected the Uno's pin D2 for the control output.

CAUTION *Do not connect the PST2 ground to the XBee receiver node ground. Simply leave it unconnected. It is neither required nor needed and it could possibly be an entry point for mains power if there were some odd and strange failure on the PST2 load side.*

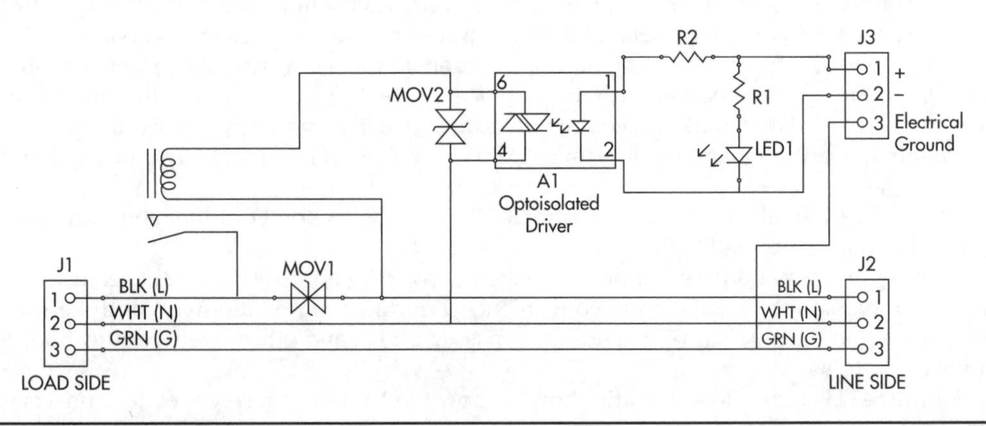

FIGURE 8-21 PST2 schematic

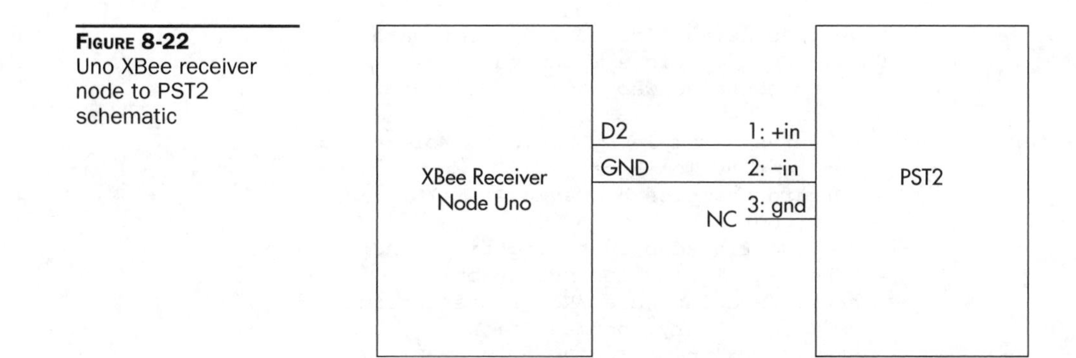

FIGURE 8-22 Uno XBee receiver node to PST2 schematic

This completes the hardware portion of the project. It is now time to discuss the software that brings this system to "life."

System Software

Three programs are required for this project. Two of them are almost identical except for a minor configuration difference. The first I will discuss is the main one that is stored in the controller node.

Controller Node Program

I decided to use an existing program named `eth_websrv_SD_Ajax_in_out_r1` that fortuitously contained all the desired functionality required for this project after I added some modifications. It was originally created to control a group of LEDs via a web interface, but it really makes no difference if the program sends Uno outputs to optoisolators instead of LEDs. I didn't even bother to change the web page descriptions from LEDs to lights as it seemed far more important to test the program's functionality. This program also uses an HTML file that is prestored on a micro SD memory card that is part of the Ethernet Shield. I will discuss this HTML file after I discuss the main program.

```
/*--------------------------------------------------------------
  Program:      eth_websrv_SD_Ajax_in_out

  Description:  Arduino web server that displays 4 analog inputs,
                the state of 3 switches and controls 4 outputs,
                2 using checkboxes and 2 using buttons.
                The web page is stored on the micro SD card.

  Hardware:     Arduino Uno and official Arduino Ethernet
                shield. Should work with other Arduinos and
                compatible Ethernet shields.
                2Gb micro SD card formatted FAT16.
                A2 to A4 analog inputs, pins 2, 3 and 5 for
                the switches, pins 6 to 9 as outputs (LEDs).
```

```
  Software:      Developed using Arduino 1.0.5 software
                 Should be compatible with Arduino 1.0 +
                 SD card contains web page called index.htm

  References:    - WebServer example by David A. Mellis and
                   modified by Tom Igoe
                 - SD card examples by David A. Mellis and
                   Tom Igoe
                 - D.J. Norris adapted to lighting control project
                 - Ethernet library documentation:
                   http://arduino.cc/en/Reference/Ethernet
                 - SD Card library documentation:
                   http://arduino.cc/en/Reference/SD

  Dates:           19 June 2013
                   removed use of the String class
                   21 June 2014
                   modified for the lighting controller project

  Author:        W.A. Smith, http://startingelectronics.com
--------------------------------------------------------------*/

#include <SPI.h>
#include <Ethernet.h>
#include <SD.h>
// size of buffer used to capture HTTP requests
#define REQ_BUF_SZ   60

// MAC address from Ethernet shield sticker under board
byte mac[] = { 0xDE, 0xAD, 0xBE, 0xEF, 0xFE, 0xED };
IPAddress ip(192, 168, 1, 130); // IP address, may need to change


// depending on network
EthernetServer server(80);  // create a server at port 80
File webFile;               // the web page file on the SD card
char HTTP_req[REQ_BUF_SZ] = {0}; // buffered HTTP request stored


// as null terminated string
char req_index = 0;              // index into HTTP_req buffer
boolean LED_state[4] = {0}; // stores the states of the LEDs

void setup()
{
    // disable Ethernet chip
    pinMode(10, OUTPUT);
    digitalWrite(10, HIGH);
    Serial.begin(9600);       // for debugging
    // initialize SD card
```

```
    Serial.println("Initializing SD card...");
    if (!SD.begin(4)) {
        Serial.println("ERROR - SD card initialization failed!");
        return;    // init failed
    }
    Serial.println("SUCCESS - SD card initialized.");
    // check for index.htm file
    if (!SD.exists("index.htm")) {
        Serial.println("ERROR - Can't find index.htm file!");
        return;  // can't find index file
    }
    Serial.println("SUCCESS - Found index.htm file.");
    // switches on pins 2, 3 and 5
    pinMode(2, INPUT);
    pinMode(3, INPUT);
    pinMode(5, INPUT);
    // LEDs
    pinMode(6, OUTPUT);
    pinMode(7, OUTPUT);
    pinMode(8, OUTPUT);
    pinMode(9, OUTPUT);

    Ethernet.begin(mac, ip);  // initialize Ethernet device
    server.begin();           // start to listen for clients
}

void loop()
{
    EthernetClient client = server.available();  // try to get client
    if (client) {  // got client?
        boolean currentLineIsBlank = true;
        while (client.connected()) {
            if (client.available()) {
                // client data available to read
                char c = client.read();
                // read 1 byte (character) from client
                // limit the size of the stored received HTTP request
                // buffer first part of HTTP request in HTTP_req
                // array (string)
                // leave last element in array as 0 to null terminate
                // string (REQ_BUF_SZ - 1)
                if (req_index < (REQ_BUF_SZ - 1)) {
                    HTTP_req[req_index] = c;
                // save HTTP request character
                    req_index++;
                }
                // last line of client request is blank and
                // ends with \n
                // respond to client only after last line received
                if (c == '\n' && currentLineIsBlank) {
```

```
                // send a standard http response header
                client.println("HTTP/1.1 200 OK");
                // remainder of header follows below, depending
                // on if web page or XML page is requested
                // Ajax request - send XML file
                if (StrContains(HTTP_req, "ajax_inputs")) {
                    // send rest of HTTP header
                    client.println("Content-Type: text/xml");
                    client.println("Connection: keep-alive");
                    client.println();
                    SetLEDs();
                    // send XML file containing input states
                    XML_response(client);
                }
                else {  // web page request
                    // send rest of HTTP header
                    client.println("Content-Type: text/html");
                    client.println("Connection: keep-alive");
                    client.println();
                    // send web page
                    webFile = SD.open("index.htm");
                    // open web page file
                    if (webFile) {
                        while(webFile.available()) {
                            client.write(webFile.read());
                            // send web page to client
                        }
                        webFile.close();
                    }
                }
                // display received HTTP request on serial port
                //Serial.print(HTTP_req); // DO NOT USE!
                // This will stop the project code
                // reset buffer index and all buffer elements to 0
                req_index = 0;
                StrClear(HTTP_req, REQ_BUF_SZ);
                break;
            }
            // every line of text received from the client ends
            // with \r\n
            if (c == '\n') {
                // last character on line of received text
                // starting new line with next character read
                currentLineIsBlank = true;
            }
            else if (c != '\r') {
                // a text character was received from client
                currentLineIsBlank = false;
            }
        } // end if (client.available())
```

```
        } // end while (client.connected())
        delay(1);      // give the web browser time to receive
                       // the data
        client.stop(); // close the connection
    } // end if (client)
}

// checks if received HTTP request is switching on/off LEDs
// also saves the state of the LEDs
void SetLEDs(void)
{
    // LED 1 (pin 6)
    if (StrContains(HTTP_req, "LED1=1")) {
        LED_state[0] = 1;  // save LED state
        digitalWrite(6, HIGH);
        Serial.print(1);
    }
    else if (StrContains(HTTP_req, "LED1=0")) {
        LED_state[0] = 0;  // save LED state
        digitalWrite(6, LOW);
        Serial.print(2);
    }
    // LED 2 (pin 7)
    if (StrContains(HTTP_req, "LED2=1")) {
        LED_state[1] = 1;  // save LED state
        digitalWrite(7, HIGH);
        Serial.print(3);
    }
    else if (StrContains(HTTP_req, "LED2=0")) {
        LED_state[1] = 0;  // save LED state
        digitalWrite(7, LOW);
        Serial.print(4);
    }
    // LED 3 (pin 8)
    if (StrContains(HTTP_req, "LED3=1")) {
        LED_state[2] = 1;  // save LED state
        digitalWrite(8, HIGH);
        Serial.print(5);
    }
    else if (StrContains(HTTP_req, "LED3=0")) {
        LED_state[2] = 0;  // save LED state
        digitalWrite(8, LOW);
        Serial.print(6);
    }
    // LED 4 (pin 9)
    if (StrContains(HTTP_req, "LED4=1")) {
        LED_state[3] = 1;  // save LED state
        digitalWrite(9, HIGH);
        Serial.print(7);
    }
```

```
    else if (StrContains(HTTP_req, "LED4=0")) {
        LED_state[3] = 0;  // save LED state
        digitalWrite(9, LOW);
        Serial
    }
}

// send the XML file with analog values, switch status
//  and LED status
void XML_response(EthernetClient cl)
{
    int analog_val;            // stores value read from analog inputs
    int count;                 // used by 'for' loops
    int sw_arr[] = {2, 3, 5};  // pins interfaced to switches

    cl.print("<?xml version = \"1.0\" ?>");
    cl.print("<inputs>");
    // read analog inputs
    for (count = 2; count <= 5; count++) { // A2 to A5
        analog_val = analogRead(count);
        cl.print("<analog>");
        cl.print(analog_val);
        cl.println("</analog>");
    }
    // read switches
    for (count = 0; count < 3; count++) {
        cl.print("<switch>");
        if (digitalRead(sw_arr[count])) {
            cl.print("ON");
        }
        else {
            cl.print("OFF");
        }
        cl.println("</switch>");
    }
    // checkbox LED states
    // LED1
    cl.print("<LED>");
    if (LED_state[0]) {
        cl.print("checked");
    }
    else {
        cl.print("unchecked");
    }
    cl.println("</LED>");
    // LED2
    cl.print("<LED>");
    if (LED_state[1]) {
        cl.print("checked");
    }
    else {
```

```
        cl.print("unchecked");
    }
     cl.println("</LED>");
    // button LED states
    // LED3
    cl.print("<LED>");
    if (LED_state[2]) {
        cl.print("on");
    }
    else {
        cl.print("off");
    }
    cl.println("</LED>");
    // LED4
    cl.print("<LED>");
    if (LED_state[3]) {
        cl.print("on");
    }
    else {
        cl.print("off");
    }
    cl.println("</LED>");

    cl.print("</inputs>");
}
// sets every element of str to 0 (clears array)
void StrClear(char *str, char length)
{
    for (int i = 0; i < length; i++) {
        str[i] = 0;
    }
}

// searches for the string sfind in the string str
// returns 1 if string found
// returns 0 if string not found
char StrContains(char *str, char *sfind)
{
    char found = 0;
    char index = 0;
    char len;
    len = strlen(str);
    if (strlen(sfind) > len) {
        return 0;
    }
    while (index < len) {
        if (str[index] == sfind[found]) {
            found++;
            if (strlen(sfind) == found) {
                return 1;
            }
```

```
        }
        else {
            found = 0;
        }
        index++;
    }

    return 0;
}
```

I know this is a long code listing, but much of it is involved with LED states and status, which is important to track while the program is running. There is also a section of code dealing with real-time reporting of analog value input into pins A2 to A6. I chose to leave this bit of code in place as it didn't slow the program and I have included some project enhancements at the end of the chapter that use this analog value reporting feature.

You will need to perform two tasks before you can successfully configure the controller node:

1. Load this program into the controller node's Uno EEPROM using the Arduino IDE as you have done in past projects. Remember to change the jumpers to the USB position on the XBee Shield in order to program the Uno, as I explained in the previous chapter. Also, restore the jumpers to the XBee position, which enables the XBee communications link once you have completed the initial programming.
2. Format a micro SD card as FAT16 (also referred to as simply FAT). Then copy the index.htm file onto this SD card and insert it into the holder located on the Ethernet Shield.

You should be ready now to program the two XBee receiver nodes. The program stored in each node is almost identical except for two configuration values that specify if the node should respond to a broadcast message from the controller node.

XBee Receiver Node 1 Program

This is one of the XBee receiver node programs and is aptly named Receiver_Test1.

```
int led = 2;

void setup() {
  Serial.begin(9600);
  pinMode(led, OUTPUT);
}

void loop() {
  int c = Serial.read(); // read a byte
  if(c == '1')
    digitalWrite(led, HIGH);
  if(c == '2')
    digitalWrite(led, LOW);
}
```

It is a much shorter code listing compared to the controller node program because it only has to listen for either a 1 or 2 being sent from the controller node and then either turn on or turn off pin 2, which is the one connected to the PST2. Just follow Step 1 shown previously regarding how to program this node. I would also put a piece of tape on the case with the marking "LED 1," which conforms to the web page control that actuates this particular node. Clicking the web page control labeled LED 1 should activate this node.

XBee Receiver Node 2 Program

The other receiver node program is named Receiver_Test2 and is identical to the preceding code except that the values used are 3 and 4 instead of 1 and 2. I suggest identifying this node with a piece of tape as "LED 2" after it is programmed. Clicking the web page control labeled LED 2 should activate this node.

This last step completes all of the required programming, which means the operational test is next.

Operational Test

I used three table-top lamps to conduct this operational test. Two of the lamps were plugged into the PST2s that were connected to the XBee receiver nodes. The third lamp was connected to a PST2 whose control leads were loose, allowing it to be connected to any one of the available controller node digital inputs.

The PST2s must be plugged in in order for their control circuits to be come active. There is also a red LED located next to the PST2 terminals that will light up when a high level is applied between the + and – terminals.

Figure 8-23 shows my browser window after I logged into the web server at my network's URL of 192.168.1.130. Yours will be the same provided you entered those values in the controller node program code.

The test sequence, which I have detailed here, is fairly easy to follow:

1. Ensure that all the nodes are powered on and all XBee jumpers are in the XBee position.
2. Plug the lamps into the respective powered PST2s.

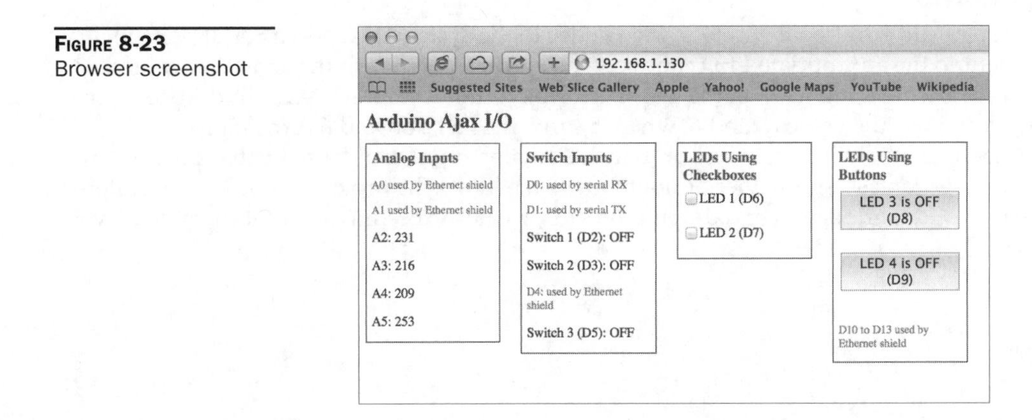

FIGURE 8-23 Browser screenshot

3. The RF receiver should be powered on and its channel one, two, and three connected to the controller node's Uno D2, D3, and D5 inputs, respectively.
4. Connect the third PST2's ground lead to the Uno ground. You should leave the ground connected throughout the remaining steps. Connect the loose PST2 control lead to the D2 input.
5. Select the LED1 checkbox. The lamp connected to the LED1 node should turn on.
6. Clear the LED1 checkbox, and the lamp should go off.
7. Repeat Steps 5 and 6 for the LED2 checkbox.
8. Press the A button on the key fob. The lamp connected to the third PST2 should light up.
9. Press the A button again and the lamp should go off.
10. Shift the loose PST2 control lead to D3 and repeat Steps 8 and 9.
11. Shift the loose PST2 control lead to D5 and repeat Steps 8 and 9.
12. Shift the loose PST2 control lead to D8 and click the web page button named LED3. The lamp should go on and the button indication change from OFF to ON.
13. Press the LED3 web page button again and the lamp should go off and the indication change from ON to OFF.
14. Shift the loose PST2 control lead to D9 and repeat Steps 12 and 13.
15. Shift the loose PST2 control lead to D3 and repeat Steps 8 and 9 using the B key.
16. Shift the loose PST2 control lead to D5 and repeat Steps 8 and 9 using the C key.

After completing this test sequence, it should be readily apparent to you that there is an incredible amount of flexibility designed into this lighting control system. You can also easily change the web page control labels by editing the index.htm file that is stored in the micro SD card. Just change the LED x labels to whatever label suits your home installation. Always ensure you make a backup copy of the original index.htm file, just in case you mess it up and it no longer functions as expected.

Enhancements

It is easy to add additional wireless XBee nodes to the system. Just add a Serial.print(x) statement to the appropriate LED code sequence, where x refers to the actuation values, which are always 1 higher than the last set of programmed values—i.e., a third XBee receiver node would use the values 5 and 6, where 5 turns the PST2 on and 6 turns it off.

Attaching a light intensity sensor to the A2 analog input could also enable a feature by which a user can assess whether or not to turn on a light. The program could also be altered to take an automatic action based upon the sensor value exceeding a preset threshold level.

Summary

This chapter's project showed you how to build an Arduino-controlled home lighting system. Portions of this system used the digital, wireless Xbee technology, while other portions were wired or used an analog RF technology. I also introduced a new Ethernet Shield that incorporates a micro SD card, which holds the web server program code.

I reused the XBee point-to-point network concept that I discussed in the last chapter even though I used three XBee nodes in this project. I explained how this was possible although it somewhat degraded the system reliability. An extensive aside discussing the XBee mesh network technology was included for interested readers who might want to attempt to build a more complex and robust XBee network.

Next, I showed you how to carefully modify the Ethernet Shield such that it could be stacked to create the central controller node.

I discussed the remote XBee receiver nodes next along with the PowerSwitch Tail II, which is the device that actually controls the home lamps.

I also discussed a short-range, four channel, analog RF system, which allows the user to remotely operate selected lamps using a handy key fob when approaching the home. This feature just adds more flexibility regarding how the system can be configured to meet many different home layouts.

The last part of the chapter concerned the software and showed you how to conduct a comprehensive operational test. I showed you a screenshot of the control web page and made some suggestions about how it might be altered to suit your personal preferences. I also included some suggestions regarding system enhancements and feature changes.

CHAPTER 9

BeagleBone Black Message Controller

This chapter's project will use the BeagleBone Black (BBB), which is a single board computer that implements a simple text display system. The text will originate from a remote web browser wherein a user can enter a message and have it appear on a 16 x 2 LCD display. I recognize that the system cannot and will not replace any cell phone text capability, but that is not my intent in presenting this project. I want to introduce you to another highly useful microcontroller board, which can readily be used for IoT projects as well as many other applications. The next section provides an in-depth discussion of the BBB. I will focus on some unique capabilities not supported by either the Raspberry Pi or the Arduino Uno boards, which have been used in the previous chapter projects.

Beagle Boards

The BBB is the latest in a series of single board computers that where designed to host a Linux operating system as well as provide all the usual features expected in a microcontroller, suitable for both home and industrial automation projects. Figure 9-1 shows a BBB top view.

The first board in the series leading up to the BBB was known simply as the BeagleBoard. It was a design project that was begun in 2007 by the non-profit BeagleBoard.org whose intent was to make available a relatively low-cost learning platform that could easily be used by beginners to learn and experiment with computer science technology. This approach was exactly the same as the Raspberry Pi Foundation goals, except that it started several years before the Raspberry Pi came into being. BeagleBoard.org worked with Texas Instruments (TI) to make the early BeagleBoard development kits available to anyone with a desire to learn about how a single board computer could run a full-blown Linux OS and also be used for any one of a myriad of automation projects that extended well beyond blinking an LED. The original BeagleBoard was released as open source using the Creative Commons share-alike license scheme. This meant that all the design documentation was made freely available, including the Cadence OrCAD schematics and the Cadence Allegro PCB design files. Any manufacturer was thus free to start making the BeagleBoards under this arrangement. I believe there are currently several offshore manufacturers that have made and are distributing early BeagleBoard models. I am not aware of any manufacturer other than TI that is currently making the BBB.

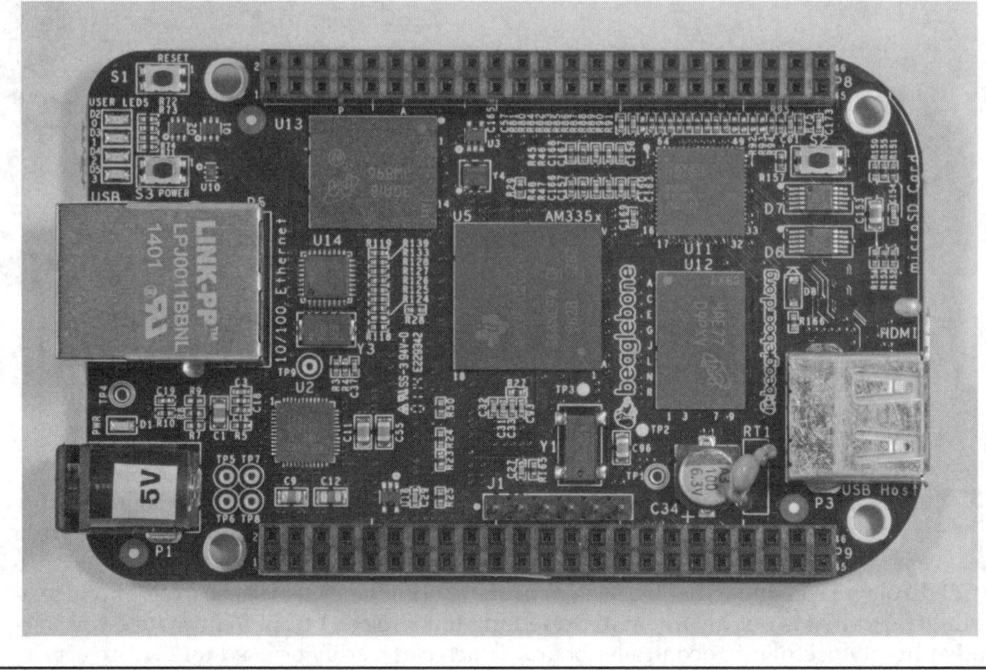

FIGURE 9-1 BBB top view

Table 9-1 lists the lineage of BeagleBoards that have been released and are still manufactured at the time of this writing.

After reviewing Table 9-1, you should recognize that the latest BeagleBones operate at very fast 1 GHz clock speed and have 4GB of onboard flash memory in which the Linux OS is stored. In addition, both the BeagleBone (BB) and the BBB have two 32-bit PRUs (programmable real-time units). The PRUs are 200 MHz coprocessors that can handle intensive computing chores such as pulse width modulation (PWM), thus allowing the

Name	Revision (latest)	Processor (TI)	Speed	Memory	Other
BeagleBoard	D	OMAP 3530 Cortex-A8	720 MHz	512MB RAM	3D Accel
BeagleBoard-xM	C2	ARM AM37x Cortex-A8	1 GHz	512MB RAM	3D Accel
BeagleBone	C	Sitara AM335x	1 GHz	512MB RAM 4GB eMMC	2x 32 bit PRUs 3D Accel
BeagleBone Black	A5A	Sitara AM335x	1 GHz	512MB RAM 4GB eMMC	2x 32 bit PRUs NEON FPU 3D Accel

TABLE 9-1 Beagle Boards

main processor to handle all the other computing requirements in an efficient manner. This project has no need for the PRUs and they will not be discussed. There is a lot of information regarding the PRUs and other BBB areas at eLinux.org

You should also note that the cost of the newest Beagle Boards is much lower than the original boards. The BBB is about one-third the cost of the BeagleBoard-xM and it has more functions.

You should also note that some community members refer to the BB as the BeagleBone White because it has a white PCB with black silkscreen letters as compared to the BBB, which has a black PCB with white silkscreen lettering.

BeagleBone Black

The BBB is an excellent development platform because it is fast, has a lot of memory, and runs various Linux distributions that have been customized for the board. Until very recently, the BBB came with the Angstrom Linux distribution stored in its 4GB onboard flash memory. More recent boards are being delivered with the Wheezy, Debian Linux distribution, which is a fortuitous situation as that is the one I needed to use to implement the web server software. This Wheezy distribution is also the same one that I used in the Raspberry Pi projects discussed in the early chapters of this book. I believe the entire configuration discussions, which I presented earlier, will also be applicable for the BBB running the Wheezy distribution.

The BBB has a micro SD memory cardholder installed, which can be seen in Figure 9-2.

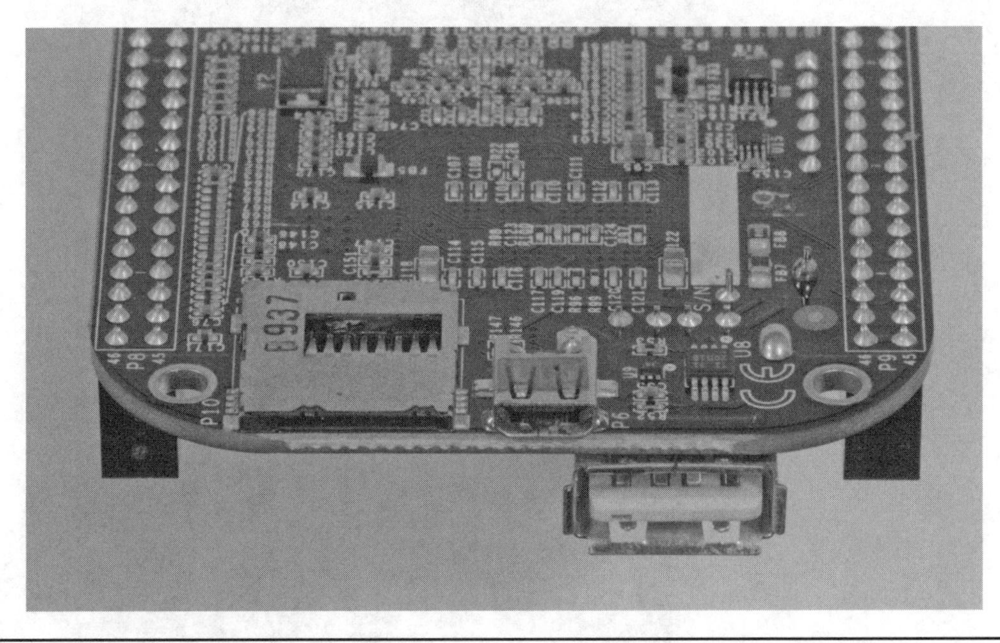

Figure 9-2 Micro SD memory cardholder

A 4GB, FAT-formatted, micro SD card may be inserted into the holder. A new Linux distribution can then be stored on the card and the BBB can then be forced to boot from it. This feature makes it easy to try out new Linux operating systems without disturbing the OS stored in the flash memory. I discuss how to create a bootable SD card later in the chapter for those readers wishing to experiment with a different Linux OS. It is also possible to overwrite the OS stored in the flash memory with a new or different version if that is your preference. I just find that the SD card option is much more convenient and allows me to experiment quite well without the possibility of bricking the board by an improper flash memory overwrite. You can unbrick the board, but it is a bit tedious. Why even get yourself into that situation when you can just as easily use an SD card and accomplish the same function?

I would also like to point out the micro HDMI connector, which is located to the immediate right of the micro SD cardholder, as shown in the figure. The BBB outputs full 1080p, high-definition video through this connector. I use a micro-HDMI-to-HDMI cable to connect the BBB to an HDMI capable monitor.

You can also see a single USB port in Figure 9-2. I used a Pluggable, seven-port, self-powered USB hub, which is shown in Figure 9-3.

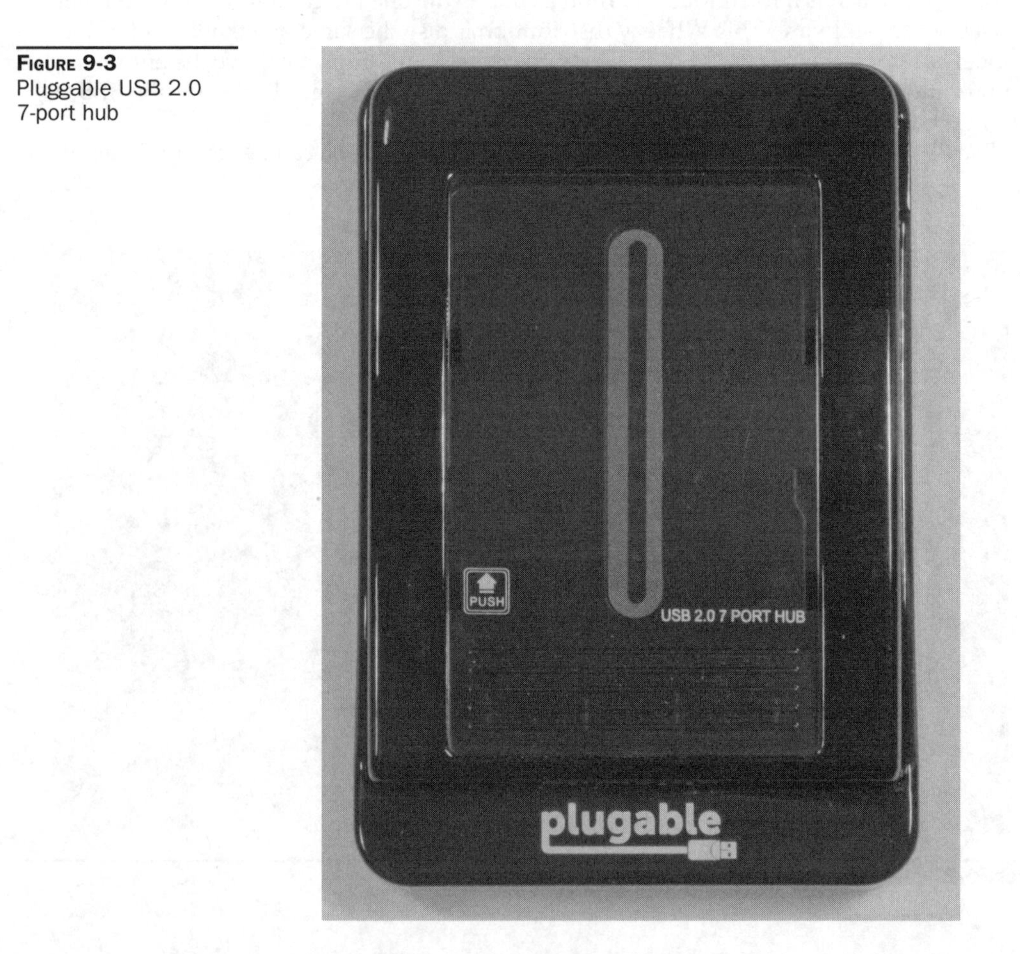

Figure 9-3 Pluggable USB 2.0 7-port hub

I have found that this particular hub model provides sufficient current to power any USB device plugged into it and have it reliably function without any issues. Highly recommended. Just remember to plug the mini USB connector into the hub and the normal-sized USB connector into the BBB.

Connect to and Operate the BBB

The BBB can be operated in several ways. These are listed here in the order I will discuss them:

- Standalone
- USB tether
- SSH over USB
- SSH over Ethernet

The standalone mode is the simplest, but it does require the most peripherals of all of the listed approaches. You will need a powered USB hub, an HDMI compatible monitor, a USB keyboard, and a USB mouse. An Ethernet cable plugged into a wired network port is also strongly recommended to enable you to download any needed software.

You should first connect the USB hub to the BBB and then plug the mouse and keyboard into the hub. Ensure the monitor and hub are both powered on. Finally, connect the 5V power to the BBB. You should next see a lot of text quickly scroll by as the Linux OS boots, and then after 10 to 15 seconds be greeted with the Debian Desktop screen, as shown in Figure 9-4.

NOTE *You should not have the Run dialog box appear. I just needed that to take a screen capture. There also should be an automatic login as no password is required.*

FIGURE 9-4 Initial Debian Desktop screen

The USB tether approach requires a mini-USB-to-USB A cable, which is the same type that was used to connect the BBB with the powered USB hub in the first approach. You will not need an external 5V power supply as the BBB will be powered through the USB cable once it is plugged into the laptop. I used a 15-inch MacBook Pro for this connection and for all of the following connections. First plug the cable into the BBB's mini USB connector and then into the laptop. A new disk icon should appear on the laptop's Desktop screen after a few seconds have elapsed. In my case, the icon was labeled "boot." Double-click on the boot icon and you should see a file listing that includes the entry START.htm. Double-click on the START.htm file and an interactive web page will be displayed. You will now have to install the appropriate FDTI drivers for your OS. Just double-click on the listed drivers corresponding to your laptop's OS. In my case, a .dmg file was downloaded, which I needed to install by doubling-clicking on the dmg package icon. I am sure that the Windows installation will proceed in a similar fashion. Now, open a browser and enter the URL **http://192.168.7.2**. You do not need to connect the BBB to your network in order for this to work. An interactive web page entitled BeagleBone 101 will be displayed (see Figure 9-5).

You should explore this web page to learn about the BBB and how to program it. Also, notice the shaded block near the top of the page, which shows the URL and serial number of the connected BBB.

The next connection approach requires the BBB to be connected in exactly the same way as the previous approach, using only a mini-USB-to-USB A cable. However, this time I will be using an SSH utility to establish the connection. I will not review what SSH is as there is plenty of information readily available explaining this network utility. I need to open a

Figure 9-5 BeagleBone 101

Terminal window because I am using a MacBook. So I enter the following at the command line prompt:

```
ssh root@192.168.7.2
```

That's all you need to do and you should see the result, as shown in Figure 9-6.

The prompt shows that you connect as the root user, which means you now have total control of the BBB OS.

Windows users need to download and install the PuTTY executable, which supports SSH connections in the Windows OS. All that's needed in this case is to enter **192.168.7.2** as the host address and ensure SSH is selected as the connection type. Type in **root** when prompted for a user name. Simply press ENTER when asked for the password. You should see exactly the same prompt as shown in Figure 9-6 when connected.

The last approach I will show is how to connect using SSH but using your network in lieu of a USB cable. You will need to separately power the BBB using a 5V supply and also connect the BBB to the LAN using an Ethernet cable. Next, open an SSH session and enter the following:

```
ssh root@beaglebone.local
```

Figure 9-7 shows the result after the connection is made.

Windows users should use the PuTTY application and enter **beaglebone.local** for the host name, and then click the Connect button. Next enter **root** and press the ENTER key for the user name and password.

NOTE *There is an additional approach to establishing a BBB connection, which uses a special serial connector that attaches to a six-pin header labeled J1 on the BBB and has a USB A connector on the other end that plugs into the laptop. I will not cover this approach as it would only be needed if all the other methods I previously described fail for some reason. For more information on other approaches, see the excellent discussion in Matt Richardson's book* Getting Started with Beaglebone: Linux-Powered Electronic Projects with Python and Javascript.

FIGURE 9-6 SSH over USB connection

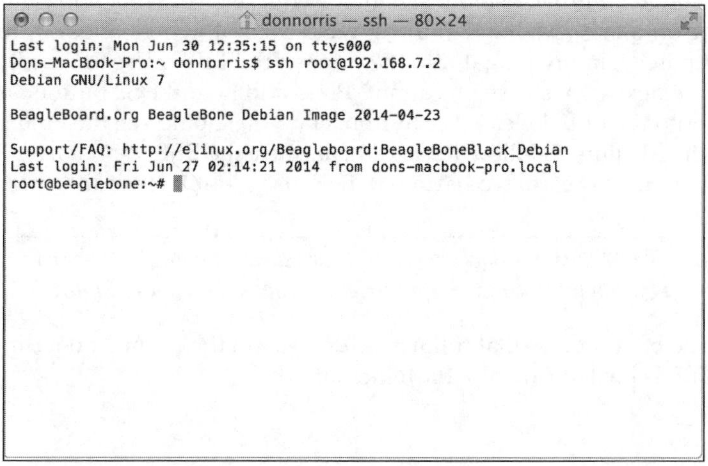

```
donnorris — ssh — 80×24
Last login: Mon Jun 30 12:35:15 on ttys000
Dons-MacBook-Pro:~ donnorris$ ssh root@192.168.7.2
Debian GNU/Linux 7

BeagleBoard.org BeagleBone Debian Image 2014-04-23

Support/FAQ: http://elinux.org/Beagleboard:BeagleBoneBlack_Debian
Last login: Fri Jun 27 02:14:21 2014 from dons-macbook-pro.local
root@beaglebone:~#
```

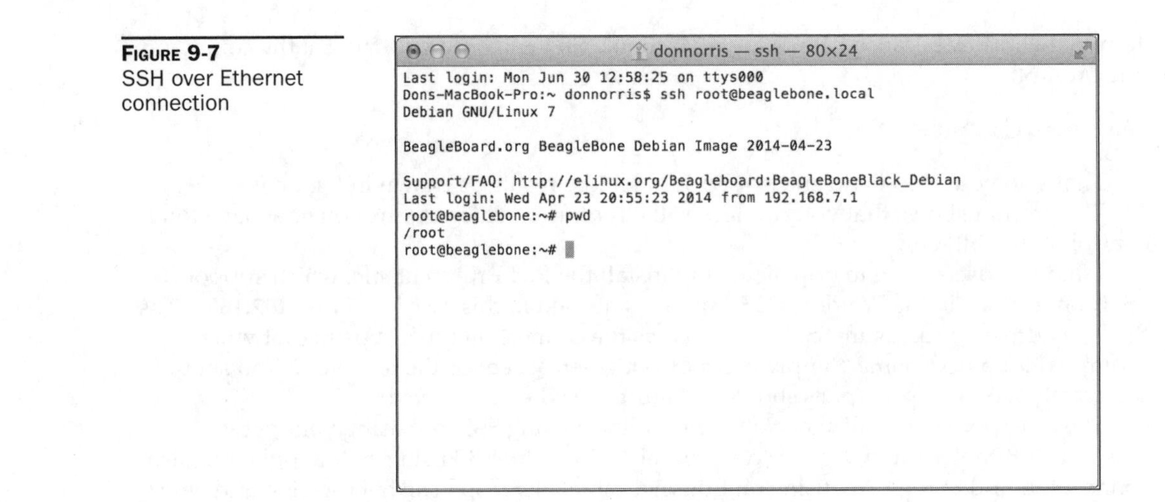

FIGURE 9-7 SSH over Ethernet connection

Downloading, Installing, and Booting a New Linux Distribution

In this section, I take you through the steps to download a new Linux distribution and install it onto a micro SD memory card. I then show you how to boot the BBB from the SD card.

The source for BBB-compatible Linux distributions is http://beagleboard.org/latest-images/. The latest Debian and Angstrom distribution images in a zipped format are normally available for download. Each distribution has two versions, one suitable for an SD card installation and the other designed to be loaded into the BBB eMMC flash memory. You should select the SD card format version as that is what the next procedure will use.

The next step is to extract or unzip the downloaded image such that it can be stored into a micro SD card. You should use whatever favorite extraction program you have installed. Many people use WinZip, while others use 7-Zip. Either one will work just fine.

Once the image is extracted or, more precisely, expanded to its original size, it needs to be loaded onto the micro SD memory card. I have used the Win32DiskImager application to do this step many times without a problem. There are other programs available that function in a similar fashion. Again, it's your choice. What you cannot do is to copy the image onto the card; it will not work and the BBB will not boot from the card. Once the card is prepared, simply plug it into the unpowered BBB micro SD cardholder.

The next step is very important! Press and hold the S2 button and then plug in the 5V power to the BBB. Release the button, and you should see the blue LED activity lights start blinking. Failure to follow this step will cause the BBB to boot from its flash memory and not from the SD card. No harm will be caused, but it is not the outcome you desired.

NOTE *The S2 button is located on the board directly opposite the micro SD cardholder. It is the only push button located in the upper-right corner, as shown in Figure 9-1.*

The boot process that follows is identical to the normal boot process that happens when the BBB is booted from the flash memory.

GPIO Pins

In this section, I discuss some of the general purpose input output (GPIO) that are available to be used by BBB project builders. The BBB top, as shown in Figure 9-1, has two 46-pin sockets aligned at the top and bottom edges. The pins are in groups of two with pin numbers 1, 2 at the left and pins 45, 46 at the right. There is some very tiny silk screen numbers near the socket ends that show these numbers for your reference. Socket P8 is at the top edge while socket P9 is at the bottom edge. I will be using only a dozen of the P9 pins for this project, which is why I will focus on that socket. Figure 9-8 shows all the GPIO pins on both P8 and P9.

P8

	1	2	
	GND	**GND**	
GPIO1_6	P8.3	P8.4	GPIO1_7
GPIO1_2	P8.5	P8.6	GPIO1_3
GPIO2_2	P8.7	P8.8	GPIO2_3
GPIO2_5	P8.9	P8.10	GPIO2_4
GPIO1_13	P8.11	P8.12	GPIO1_12
GPIO0_23	P8.13	P8.14	GPIO0_26
GPIO1_15	P8.15	P8.16	GPIO1_14
GPIO0_27	P8.17	P8.18	GPIO2_1
GPIO0_22	P8.19	P8.20	GPIO1_31
GPIO1_30	P8.21	P8.22	GPIO1_5
GPIO1_4	P8.23	P8.24	GPIO1_1
GPIO1_0	P8.25	P8.26	GPIO1_29
GPIO2_22	P8.27	P8.28	GPIO2_24
GPIO2_23	P8.29	P8.30	GPIO2_25
GPIO0_10	P8.31	P8.32	GPIO0_11
GPIO0_9	P8.33	P8.34	GPIO2_17
GPIO0_8	P8.35	P8.36	GPIO2_16
GPIO2_14	P8.37	P8.38	GPIO2_15
GPIO2_12	P8.39	P8.40	GPIO2_13
GPIO2_10	P8.41	P8.42	GPIO2_11
GPIO2_8	P8.43	P8.44	GPIO2_9
GPIO2_6	P8.45	P8.46	GPIO2_7
	45	46	

P9

	1	2	
	GND	**GND**	
	VDD 3.3V	**VDD 3.3V**	
	VDD 5V	**VDD 5V**	
	SYS 5V	**SYS 5V**	
	PWR_BUT	*SYS_RESETn*	
GPIO0_30	P9.11	P9.12	GPIO1_28
GPIO0_31	P9.13	P9.14	GPIO1_18
GPIO1_16	P9.15	P9.16	GPIO1_19
GPIO0_5	P9.17	P9.18	GPIO0_4
GPIO0_13	P9.19	P9.20	GPIO0_12
GPIO0_3	P9.21	P9.22	GPIO0_2
GPIO1_17	P9.23	P9.24	GPIO0_15
GPIO3_21	P9.25	P9.26	GPIO0_14
GPIO3_19	P9.27	P9.28	GPIO3_17
GPIO3_15	P9.29	P9.30	GPIO3_16
GPIO3_14	P9.31	**VDD_ADC(1.8V)**	
AIN4	P9.33	**GNDA_ADC**	
AIN6	P9.35	P9.36	AIN5
AIN2	P9.37	P9.38	AIN3
AIN0	P9.39	P9.40	AIN1
GPIO0_20	P9.41	P9.42	GPIO0_7
	GND	**GND**	
	GND	**GND**	
	45	46	

FIGURE 9-8 P8 and P9 pin headers

Expansion Header P9 Pinout

PIN	PROC	NAME	MODE0	MODE1	MODE2	MODE3	MODE4	MODE5	MODE6	MODE7
1,2						GND				
3,4						DC_3.3V				
5,6						VDD_5V				
7,8						SYS_5V				
9						PWR_BUT				
10	A10					SYS_RESETn				
11	T17	UART4_RXD	gpmc_wait0	mii2_crs	gpmc_csn4	rmii2_crs_dv	mmc1_sdcd		uart4_rxd_mux2	gpio0[30]
12	U18	GPIO1_28	gpmc_be1n	mii2_col	gpmc_csn6	mmc2_dat3	gpmc_dir		mcasp0_aclkr_mux3	gpio1[28]
13	U17	UART4_TXD	gpmc_wpn	mii2_rxerr	gpmc_csn5	rmii2_rxerr	mmc2_sdcd		uart4_txd_mux2	gpio0[31]
14	U14	EHRPWM1A	gpmc_a2	mii2_txd3	rgmii2_td3	mmc2_dat1	gpmc_a18		ehrpwm1A_mux1	gpio1[18]
15	R13	GPIO1_16	gpmc_a0	gmii2_txen	rmii2_tctl	mii2_txen	gpmc_a16		ehrpwm1_tripzone_input	gpio1[16]
16	T14	EHRPWM1B	gpmc_a3	mii2_txd2	rgmii2_td2	mmc2_dat2	gpmc_a19		ehrpwm1B_mux1	gpio1[19]
17	A16	I2C1_SCL	spi0_cs0	mmc2_sdwp	I2C1_SCL	ehrpwm0_synci	pr1_uart0_txd			gpio0[5]
18	B16	I2C1_SDA	spi0_d1	mmc1_sdwp	I2C1_SDA	ehrpwm0_tripzone	pr1_uart0_rxd			gpio0[4]
19	D17	I2C2_SCL	uart1_rtsn	timer5	dcan0_rx	I2C2_SCL	spi1_cs1	pr1_uart0_rts_n		gpio0[13]
20	D18	I2C2_SDA	uart1_ctsn	timer6	dcan0_tx	I2C2_SDA	spi1_cs0	pr1_uart0_cts_n		gpio0[12]
21	B17	UART2_TXD	spi0_d0	uart2_txd	I2C2_SCL	ehrpwm0B	pr1_uart0_rts_n		EMU3_mux1	gpio0[3]
22	A17	UART2_RXD	spi0_sclk	uart2_rxd	I2C2_SDA	ehrpwm0A	pr1_uart0_cts_n		EMU2_mux1	gpio0[2]
23	V14	GPIO1_17	gpmc_a1	gmii2_rxdv	rgmii2_rxdv	mmc2_dat0	gpmc_a17		ehrpwm0_synco	gpio1[17]
24	D15	UART1_TXD	uart1_txd	mmc2_sdwp	dcan1_rx	I2C1_SCL		pr1_uart0_txd	pr1_pru0_pru_r31_16	gpio0[15]
25	A14	GPIO3_21*	mcasp0_ahclkx	eQEP0_strobe	mcasp0_axr3	mcasp1_axr1	EMU4_mux2	pr1_pru0_pru_r30_7	pr1_pru0_pru_r31_7	gpio3[21]
26	D16	UART1_RXD	uart1_rxd	mmc1_sdwp	dcan1_tx	I2C1_SDA		pr1_uart0_rxd	pr1_pru1_pru_r31_16	gpio0[14]
27	C13	GPIO3_19	mcasp0_fsr	eQEP0B_in	mcasp0_axr3	mcasp1_fsx	EMU2_mux2	pr1_pru0_pru_r30_5	pr1_pru0_pru_r31_5	gpio3[19]
28	C12	SPI1_CS0	mcasp0_ahclkr	ehrpwm0_synci	mcasp0_axr2	spi1_cs0	eCAP2_in_PWM2_out	pr1_pru0_pru_r30_3	pr1_pru0_pru_r31_3	gpio3[17]
29	B13	SPI1_D0	mcasp0_fsx	ehrpwm0B		spi1_d0	mmc1_sdcd_mux1	pr1_pru0_pru_r30_1	pr1_pru0_pru_r31_1	gpio3[15]
30	D12	SPI1_D1	mcasp0_axr0	ehrpwm0_tripzone		spi1_d1	mmc2_sdcd_mux1	pr1_pru0_pru_r30_2	pr1_pru0_pru_r31_2	gpio3[16]
31	A13	SPI1_SCLK	mcasp0_aclkx	ehrpwm0A		spi1_sclk	mmc0_sdcd_mux1	pr1_pru0_pru_r30_0	pr1_pru0_pru_r31_0	gpio3[14]
32						VADC				
33	C8					AIN4				
34						AGND				
35	A8					AIN6				
36	B8					AIN5				
37	B7					AIN2				
38	A7					AIN3				
39	B6					AIN0				
40	C7					AIN1				
41#	D14	CLKOUT2	xdma_event_intr1		tclkin	clkout2	timer7_mux1	pr1_pru0_pru_r31_16	EMU3_mux0	gpio0[20]
	D13	GPIO3_20	mcasp0_axr1	eQEP0_index		Mcasp1_axr0	emu3	pr1_pru0_pru_r30_6	pr1_pru0_pru_r31_6	gpio3[20]
42@	C18	GPIO0_7	eCAP0_in_PWM0_out	uart3_txd	spi1_cs1	pr1_ecap0_ecap_capin_apwm_o	spi1_sclk	mmc0_sdwp	xdma_event_intr2	gpio0[7]
	B12	GPIO3_18	Mcasp0_aclkr	eQEP0A_in	Mcaspo_axr2	Mcasp1_aclkx		pr1_pru0_pru_r30_4	pr1_pru0_pru_r31_4	gpio3[18]
43-46						GND				

FIGURE 9-9 TRM P9 header table

As you can readily see, a significant number of GPIO pins are available for use. There are 66 GPIO pins on the headers, which is a far greater number as compared to what is available on either the Raspberry Pi or the Arduino Uno. You should be aware that not every GPIO pin is available as many pins are already preassigned to functions, including HDMI video output. You should also know that the Linux 3.8 kernel that runs on the BBB has eight different modes that are configured by a software abstraction known as a Device Tree Overlay. I will not go into the Device Tree Overlay as it is somewhat complex and really not needed for this project. If you're interested in learning more about the Device Tree Overlay, visit Derek Molloy's excellent blog at derekmolloy.ie where he has several videos that thoroughly explain the Device Tree Overlay and how to program with it using the C++ language. You should understand that mode 7 is the GPIO mode that provides access to the GPIO pins. This is shown in Figure 9-9, which is the P9 header table excerpted from the BBB Technical Reference Manual (TRM).

I have also used a Python library that preconfigures the GPIO pins so that they can be easily used as inputs or outputs without your having to worry about how to set up a Device Tree Overlay configuration.

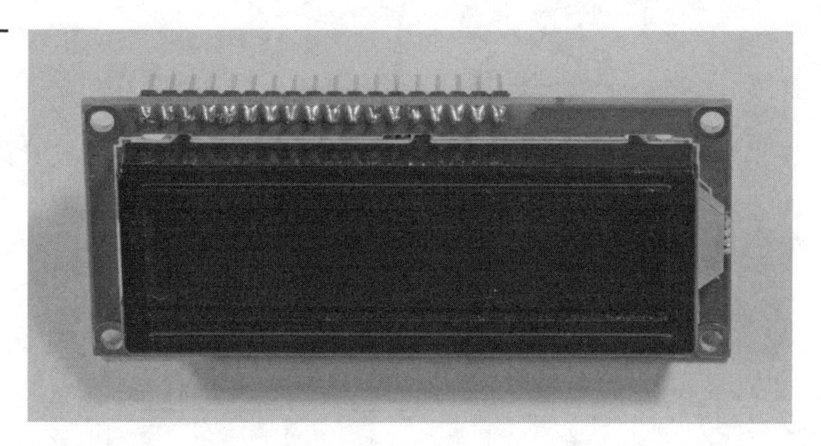

FIGURE 9-10 Generic 16 x 2 LCD display

Setting Up the LCD Display

I used a generic 16 x 2 LCD display wired to the BBB to show a brief text message received from a remote browser. The LCD display can readily be changed to a 16 x 4 or even larger display if additional text display capability is required. Figure 9-10 shows the 16 x 2 LCD display, which I used for this project.

I purchased this display from Adafruit Industries as part number 399. It came with two rows of expansion pins, which permitted me to physically configure the display, as I needed to use it with a solderless breadboard. I soldered 18 pins to the top row of holes as you can see in the figure, which allowed me to easily connect jumper wires between the display and the BBB.

The LCD uses the Hitachi command set designed to function with the Hitachi HD44780 controller chip. This is the reason why I earlier referred to the display as a generic type as it complies with HD44780 command instructions but is likely not manufactured by Hitachi.

The interconnection schematic is shown in Figure 9-11. You should note that all connections were made to the P9 socket.

The next step is to load some test software once all the jumpers are connected.

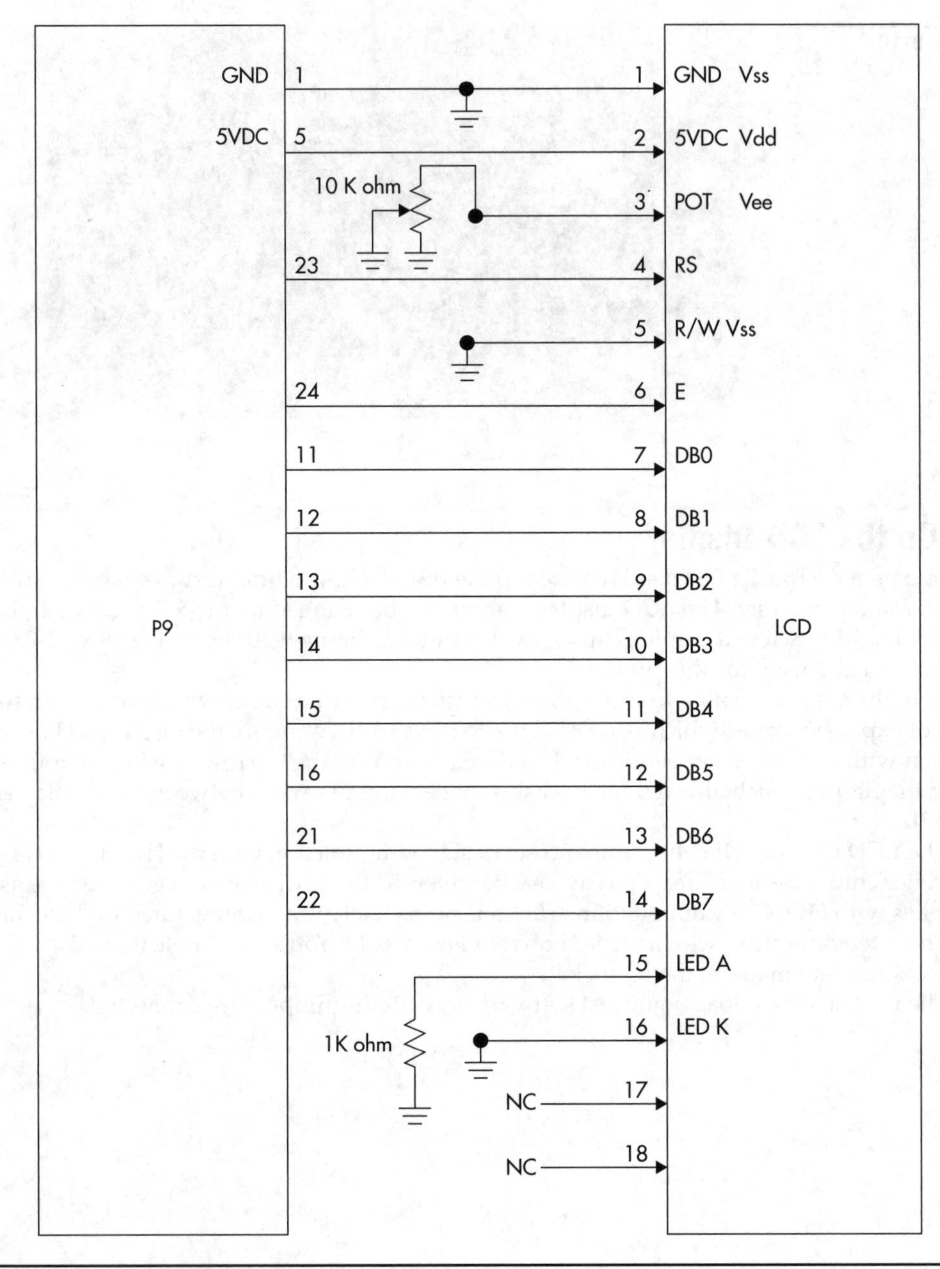

FIGURE 9-11 LCD to BBB interconnection schematic

LCD Operational Test

Some of the test software and procedures used in this section came from Ben Hammel's and Erik McKee's January 12, 2014 blog article "Writing to an LCD screen with the Beaglebone," which is available at TheBrokendesk.com website.

The following steps will test the LCD for proper operation with the BBB:

1. Apply power to the BBB and connect to it using any of the methods previously described. Ensure that the BBB is connected to your LAN as it will need to connect to the Internet in order to download one file. I used an Ethernet SSH connection to conduct this test.
2. Change into the project directory if you already have created one or simply use the home directory.
3. Enter the following command, which will download a required file named lcd.py:
   `sudo wget http://thebrokendesk.s3-us-west-1.amazonaws.com/documents/lcd.py`
4. Start the nano editor with the following command:
   `sudo nano testLCD.py`
5. Enter the following code listing:

```
import os
import sys
import time

ROOT_DIR = os.path.dirname(os.path.abspath(__file__))

if ROOT_DIR not in sys.path:
    sys.path.append(ROOT_DIR)

import lcd

screen = lcd.Screen()
screen.printLine('01234567890123456', 1)
time.sleep(5)
screen.printLine('Hello World!' ,2)
```

6. Save the nano file buffer (CTRL-O) and exit (CTRL-X).
7. Enter the command:
   `sudo python testLCD.py.`

The LCD should now display what is visible in Figure 9-12.

Please recheck your jumper wire interconnections if you do not see the expected display. Also check the 10K ohm potentiometer setting to ensure the contrast is properly adjusted to display the LCD characters. I too have inadvertently turned the potentiometer such that the characters were not visible.

This test completes all the preparatory tasks required before tackling the web server software in the next section.

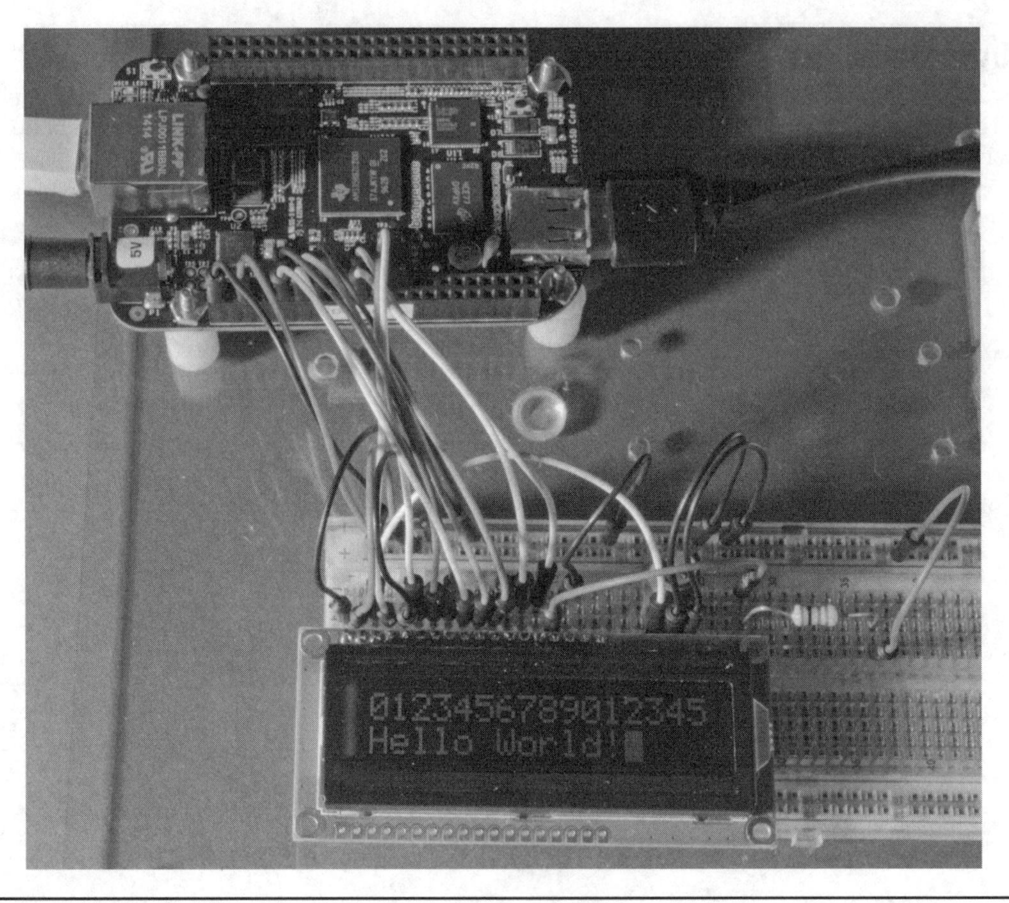

Figure 9-12 Test result

Message Controller Software

The web server portion for this project is based on Flask, which I briefly introduced and discussed in Chapter 2. I chose to use Flask because this particular application does not require all the inherent capabilities and services that are found in the comprehensive Apache web server package. In addition, Flask development is quite easy, once you become accustomed to its layout and how it incorporates HTML, scripts, and templates. I will also try to be as clear as possible regarding how to set up the Flask environment so that you can avoid any potential problems.

Download and Install the Flask Package

For your convenience I will repeat my Chapter 2 instructions regarding how to set up Flask. Flask is available using the pip package manager. It is another package manager similar to apt that I have used up to this point in the book. pip functions with the Python Package

Index (PyPI) repository where the Flask package is stored and available for download. Of course, you must download pip first into the Pi, which is oddly enough done using the apt tool with the following command:

```
sudo apt-get install python-pip
```

After pip is installed, you can use it to install Flask with all its dependencies. Enter this at the command line:

```
sudo pip install flask
```

Flask should be all set up after completing these steps.

Main Application

You will need to create a project directory structure in which all the required files will be located. I decided to create a subdirectory named Sites in the Debian directory, which itself is located in the home directory. In the Sites directory, I can create a specific directory dedicated to the project, which in this case is named webLCD. Finally, Flask requires a templates subdirectory where it can automatically locate any HTML type file it needs. This directory structure is graphically shown in Figure 9-13 to help clarify how it is set up.

The primary or main application is based upon the Python testLCD.py code that I discussed previously. Essentially, all I needed from that code was the LCD object that I referred to as "screen" in the code listing. You should also note that the two files named lcd.py and lcd.pyc are also required to be in the webLCD directory. These two files are located in the same directory as the testLCD.py program. Simply copy them into the webLCD directory.

The following is the complete code listing for the webLCD.py program. As always, it is available from the book's companion website.

```
#!/usr/bin/env python

import lcd

from flask import Flask
from flask import render_template, request

app = Flask(__name__)
```

FIGURE 9-13 Directory structure

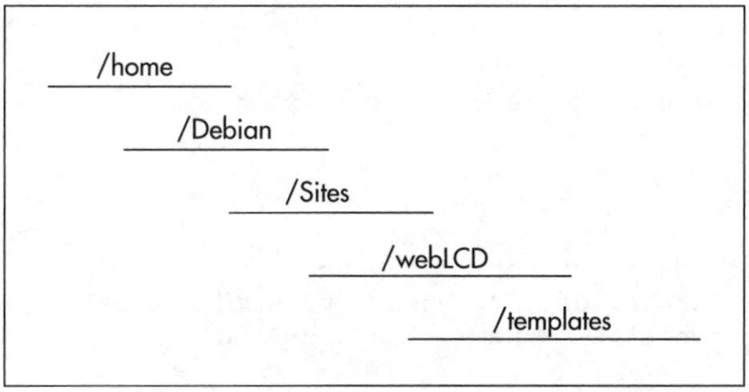

/home
/Debian
/Sites
/webLCD
/templates

```
#initialize an LCD object
screen = lcd.Screen()

@app.route("/")
def index():
    return render_template('index.html')

@app.route("/change", methods=['POST'])
def change():
    if request.method == 'POST':
       # Get the value from the submitted form
       lcdText = request.form['lcd']
       print "---Message is", lcdText

       # Send the message to the LCD
       screen.clear()
       screen.printLine(lcdText, 1)
    else:
       lcdText = None
    return render_template('index.html', value=lcdText)

if __name__ == "__main__":
   app.debug = True
   app.run('0.0.0.0')
```

Much of the code is concerned with interfacing with the HTML code that creates the web page where the user can enter the text message to be displayed. Only one program line, `screen.printLine(lcdText, 1)`, actually causes the text to be displayed on the LCD.

The program line `app.run('0.0.0.0')` causes the Flask web server to listen to all open ports on the LAN for any HTTP request coming from a web client. You should be able to run this application from any computer, tablet, or smartphone that can connect to your network.

HTML Code

Two HTML files are required for this application, index.html and base.html. The index.html file is implemented as a Jinja2 template. Jinja2 is a Python templating language. While it is not critical that you understand Jinja2 for this project, much reference information is available on the Web regarding Jinja2 for those readers interested in further pursuing this topic. The following is the index.html code listing:

```
{% extends "base.html" %}
{% block body %}
<div data-role="header">
  <h1>BeagleBone Message Controller</h1>
</div>

<div data-role="content">

<form method="post" action="change">
  <label for="slider-1">Message to Send:</label>
  <input type="text" name="lcd" id="lcd" />
```

```
  <br />

  <input type="submit" value="Submit" />
</form>

{% if value %}
<p>Message sent: {{ value }}</p>
{% endif %}

</div>
{% endblock %}
```

The index.html code is dependent on additional code that is contained in the base.html file. Readers who are somewhat familiar with object-orientation will recognize this dependency from the first statement in the index.html code: `{% extends "base.html" %}`.

The base.html code listing is shown here:

```
<html>
<head>
<title>BeagleBone Message Controller</title>
<meta name="viewport" content="width=device-width, initial-scale=1" />
<meta name="apple-mobile-web-app-capable" content="yes" />
<meta name="apple-mobile-web-app-status-bar-style" content="black" />

<link rel="stylesheet" href="http://code.jquery.com/mobile/1.2.0


/jquery.mobile-1.2.0.min.css" />
<script src="http://code.jquery.com/jquery-1.8.2.min.js"></script>
<script src="http://code.jquery.com/mobile/1.2.0/jquery.mobile-1.2.0.min.
js"></script>

<script>
$( document ).bind( "mobileinit", function() {
    $.mobile.allowCrossDomainPages = true;
    $.mobile.touchOverflowEnabled = true;
});
</script>
</head>
<body>
<div data-role="page" data-title="Control" id="control">
{% block body %}{% endblock %}
</div>
</body>
</html>
```

This file is created with the Jinja2 script; it also contains references to JQuery Mobile libraries. The JQuery scripts will style the HTTP responses, which accommodate mobile devices and are demonstrated in the following section.

Test Results

The first test demonstration is a result of using a web browser on my MacBook Pro and going to the BeagleBone's local address, which on my LAN was 192.168.1.25:5000.

NOTE *Flask uses port 5000 as the default for the web server. This is fine and should cause no problems. In fact, it alleviates the tedious task of freeing port 80, which the BeagleBone Black 101 website has already claimed in the default configuration.*

Figure 9-14 shows the browser screen after I had entered a short message.

Note that the message that was sent is also shown at the bottom of the screen. Figure 9-15 shows the LCD display after the message was sent.

You can see clearly that some of the trailing characters are not displayed because of the 16-character limit on each LCD line. I didn't really consider this to be much of a limitation as this project is really more for a proof of performance than a finished project. I suggest adding additional code to the main application, webLCD.py, to accommodate any overflow characters so that they can appear on the second line. Another approach might be to put the whole received message into a buffer String and then parse the String such that the first 16 characters are displayed on the first line and the next 16 characters are displayed on the second line. Obviously, any characters beyond 32 cannot be displayed with a 16 x 2 LCD unit. You would need a 16 x 4 or even a 20 x 4 LCD to handle larger messages. Incidentally, the web server software will handle large messages and does not need to be modified. Any needed modifications would be limited to the BBB and LCD interface software. I am sure there are plenty of web references available to help with these modifications.

The next demonstration involved connecting to the BBB using an Android smartphone. Figure 9-16 is a screen capture from my smartphone showing the screen after I sent a text to the BBB.

FIGURE 9-14 Laptop browser screen

Figure 9-15 LCD display for message sent from laptop

This time I intentionally kept the message very short in order to display all the text on the LCD, which is shown in Figure 9-17.

The last demonstration involved sending a message from an iPad connected to the LAN. Figure 9-18 shows the screen capture after I sent a short message to the BBB.

The short message, received and displayed by the BBB, is shown in Figure 9-19.

This last demonstration concludes the BBB message controller project.

Figure 9-16 Android smartphone browser screen

Figure 9-17 LCD display for message sent from smartphone

iPad 6:38 PM 96%
192.168.1.25
Control | Flex-Radio... | CLAYTON M... | Google Earth | The Green H... | Lone Range... | Arduino - So...
BeagleBone Message Controller
Message to Send:
Submit
Message sent: Sent fm iPad

Figure 9-18 iPad browser screenshot

Figure 9-19 LCD display for message sent from an iPad

Summary

I began the chapter with an introduction to the BeagleBone Black (BBB), the latest in a line of single board computers that can host the Linux OS. I described the essential BBB features and also discussed four ways to connect to and operate the BBB.

I next described how to load a new Linux distribution into a micro SD memory card and then described the procedure for how to boot the BBB from the card.

Next came a discussion of the GPIO pins, which is how the LCD display interfaces to the BBB. I demonstrated a simple Python script to test the BBB to LCD display interface to show that it functioned as expected.

I next showed you how to load Flask, a "lightweight" web server application. Flask provides the framework to process messages received from remote browsers, which can then be shown on the local LCD display.

The chapter concluded with a series of demonstrations showing messages being sent to the BBB from a laptop, smartphone, and a tablet. All of these devices were wirelessly connected to the LAN to which the BBB was also connected.

10

CHAPTER

BeagleBone Black with Cloud Service

This chapter's project will demonstrate how to connect a BeagleBone Black (BBB) running the Angstrom Linux distribution to an online web service. The BBB will also have temperature sensors interfaced to it. The BBB will run a program that takes measurements and subsequently sends them to the Xively web service. Users will be able to log into the Xively website and observe both current and past measurements. This project is designed to show you how to set up a cloud-based sensor system, which stores all sensor data remotely, eliminating the need for a local database.

Temperature Sensor

I used a TMP36 as a temperature sensor, which is the same type I used in Chapter 2's home temperature monitoring system project. That project also used the Raspberry Pi as the main control board. I will not repeat the TMP36 sensor description here, but simply refer you back to Chapter 2 for the details. I also reused the handy cat 5 interconnection system, which I also described in detail in Chapter 2. Figure 10-1 is the connection schematic for the BBB and the TMP36 sensor.

Only three connections are required: 5V, output signal, and ground. There is also no need for an external analog-to-digital converter (ADC) chip in this project as the BBB already contains that functionality. Pin P9_40 connects to one of the BBB's ADC channels, but there is one limitation that you should know about. The maximum voltage that a BBB ADC channel can safely accept is 1.8V. Any more can cause damage to the input. The TMP36 sensor can produce a voltage exceeding that level when measuring temperatures at or above 266° F. That's a very high temperature, which should never be encountered when using this project in a typical residence or office environment.

Figure 10-2 shows the physical system with one temperature sensor connected to a BBB using a solderless breadboard. I also used the project BBB in a standalone configuration with a monitor, keyboard, mouse, and powered USB hub. The BBB was also wired to my LAN for the required Internet connectivity.

Nothing else was required from the hardware prospective, so all that's left is to discuss the software, which is where the "magic" happens.

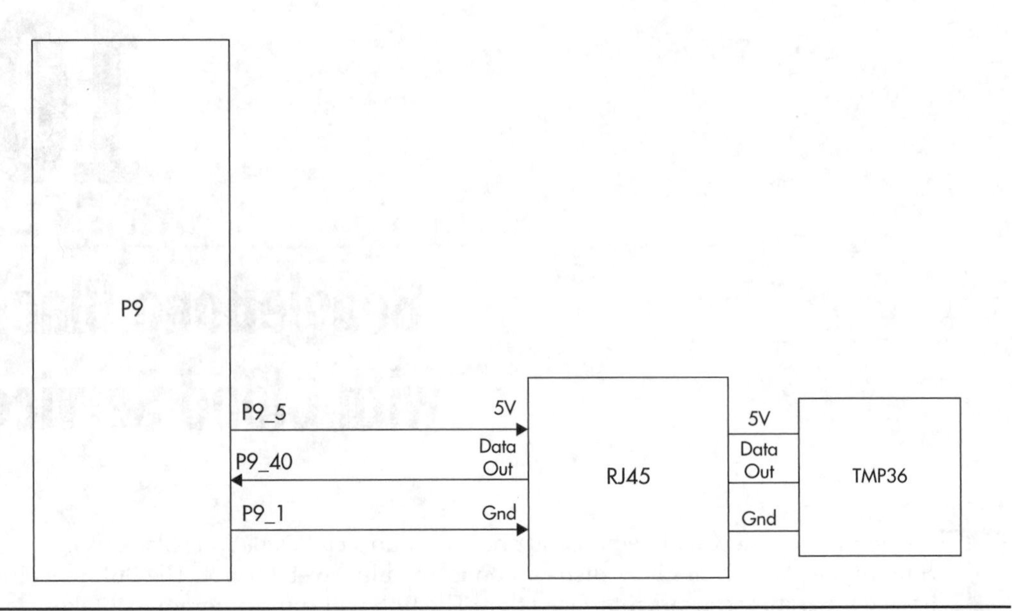

FIGURE 10-1 TMP36 to BBB connection schematic

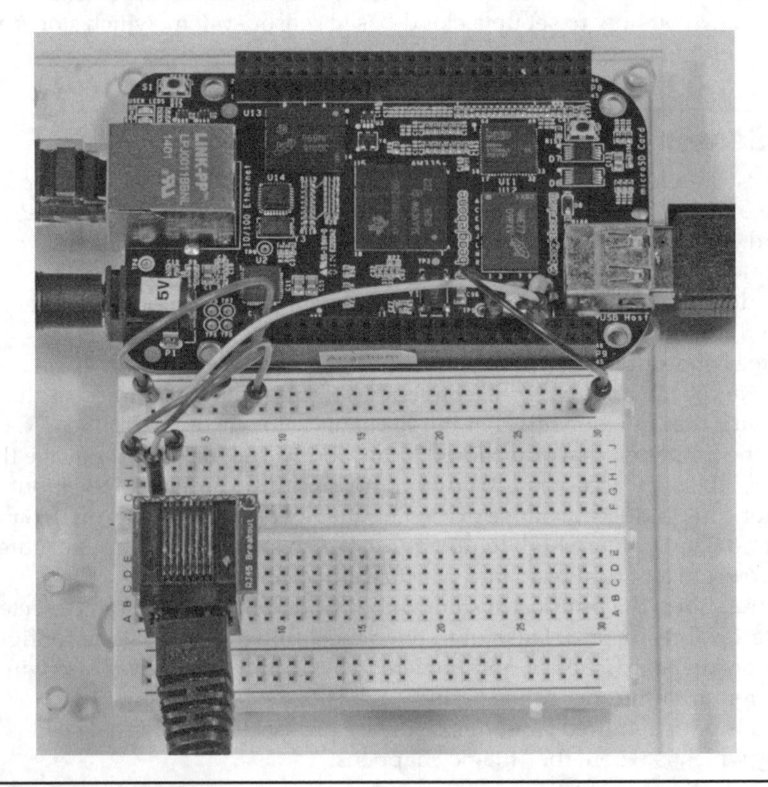

FIGURE 10-2 Physical system

Adafruit_BBIO Library

The very clever folks at Adafruit Industries have made a Python library freely available, which allows you to connect to the BBB GPIO and ADC pins among other things. I also wish to acknowledge Simon Monk's extremely helpful article on how to connect a TMP35 or TMP36 to the BBB. This article is available from Adafruit's Learn website at https://learn.adafruit.com/measuring-temperature-with-a-beaglebone-black.

Note that I am using the Angstrom Linux distribution at the root level. This Linux distribution is running in the BBB's flash eMMC memory. You need to modify the steps somewhat if you are using a different distribution, but that shouldn't be difficult.

1. Set the date and time by entering the following:

```
/usr/bin/ntpdate -b -s -u pool.ntp.org
```

2. Update the Linux distribution by entering the following:

```
opkg update
```

NOTE *You can also upgrade the distribution by entering* `opkg upgrade`*. But be forewarned that the upgrade can take several hours to complete.*

3. Install prerequisite programs by entering the following:

```
opkg install python-pip python-setuptools python-smbus
```

4. Install the Adafruit library by entering the following:

```
pip install Adafruit_BBIO
```

5. Test to see if the installations were completed properly by entering the following:

```
python -c "import Adafruit_BBIO as GPIO; print GPIO"
```

You see something similar to the following displayed on the terminal:

```
<module 'Adafruit_BBIO.GPIO' from '/usr/local/lib/python2.7/
dist-packages/Adafruit_BBIO/GPIO.so'>
```

You are now ready for a test program once the library has been loaded. Ensure that the sensor is correctly wired to the BBB, as shown in Figure 10-1, before proceeding.

Initial Test Program

This test program is from Dr. Monk's article that I previously cited. I have named it TMP36.py, and it is available from this book's companion website.

```
import Adafruit_BBIO.ADC as ADC
import time

sensor_pin = 'P9_40'

ADC.setup()
```

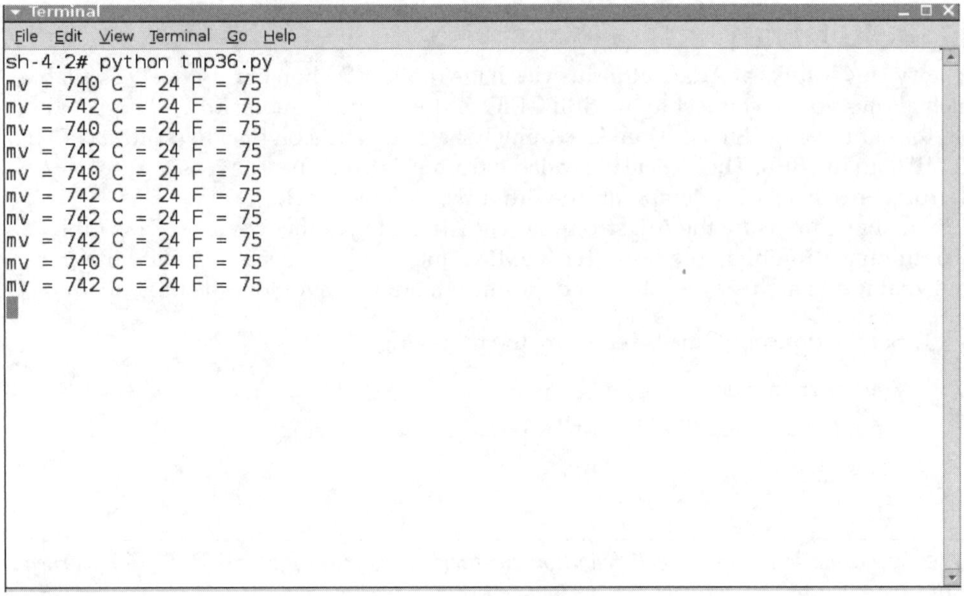

```
Terminal
File Edit View Terminal Go Help
sh-4.2# python tmp36.py
mv = 740 C = 24 F = 75
mv = 742 C = 24 F = 75
mv = 740 C = 24 F = 75
mv = 742 C = 24 F = 75
mv = 740 C = 24 F = 75
mv = 742 C = 24 F = 75
mv = 742 C = 24 F = 75
mv = 742 C = 24 F = 75
mv = 740 C = 24 F = 75
mv = 742 C = 24 F = 75
```

FIGURE 10-3 Test program output

```
while True:
    reading = ADC.read(sensor_pin)
    mv = reading * 1800
    temp_c = (mv - 500) / 10
    temp_f = (temp_c * 9/5) + 32
    print('mv=%d  C=%d  F=%d' % (mv, temp_c, temp_f))
    time.sleep(1)
```

The program is very simple in that it converts a voltage applied to the P9_40 ADC input every second to both °C and °F. The raw reading is first multiplied by an 1800 value to account for the full-scale voltage, which is 1800 millivolts, or 1800 mv. Figure 10-3 shows the program's terminal output.

The next step in this project is to connect a cloud service once the test program proves that the sensor portion works properly.

Xively Cloud Service

Xively is the latest version of a data-based cloud service that began in 2007 as a UK startup known as Pachube (pronounced *patch bay*). Pachube was created to be used as a data infrastructure in support of an early version of the Internet of Things. In 2011, Pachube was acquired by LogMeIn and subsequently rebranded as Cosm; it then began a beta development project. In May 2013, Cosm went from a beta platform into a production version and its name subsequently changed to the current Xively. The Xively website is designed to support automated sensor measurements of which this chapter's project is a simple example. A more complex example would be Xively's support of volunteers using Geiger counters to monitor radioactive fallout from the Yokashima nuclear power plant disaster.

Xively Developer's Account

You will need to establish a free developer's account at www.xively.com before you can use the Xively web service. It is a fairly quick process: You register as a developer and eventually get access to the Xively Developer Workbench, which is the means by which you can attach the BBB system to the Xively website. Simply follow the instructions on the site to register. You can next register your BBB system to Xively using the following steps. Please note that some of these steps are based on the excellent discussion provided by Matt Richardson in his book *Getting Started with BeagleBone: Linux Powered Electronic Projects with Python and Javascript*.

1. Log into your new account and click on the Develop button located near the top of the page.
2. Click the Add Device button.
3. Provide a name for your device. I named mine BBB1, perhaps with the intention of adding some more BBBs at a later time. You will also be asked if you want the submitted data to be public or private. I chose public, but you can change this option at a later date if you want.
4. Click the Add Channel button. This enables you to create a sensor channel or datastream where you can add data to the Xively website. You will also need to provide a name for this new channel. I used room_temp for the channel name. Remember to write it down as you will need it later for the BBB program. You may also optionally add appropriate tags, units, and a symbol to describe the data that will be stored in the channel. Again, you can always edit this data at a later date if you need to do so.
5. In the lower-right section, you will find some additional critical information that you will need for the BBB program. First, write down the 10-digit feed number. There is also a somewhat complex 48-character API key that you will also need. I strongly recommend that you copy and paste this key for entering it into the program. It is almost impossible to manually enter the API key without making a transcription error.
6. Return to the BBB and install the Xively Python library by entering the following:

```
pip install xively-python
```

You are now ready to create the Python program, which will run on the BBB and send temperature data to the Xively website.

BBB to Xively Python Program

Enter the following code into the BBB. This code is also available on the book's companion website, but the feed and API key are not provided as they are unique to each project. The program is named xively-temp.py because it largely mirrors Matt Richardson's code.

```
import Adafruit_BBIO.ADC as ADC
import time
import datetime
import xively
from requests import HTTPError

api = xively.XivelyAPIClient("<your 48 character API key goes here>")
feed = api.feeds.get(<your 10 digit feed number goes here>)

ADC.setup()

while True:
    mv = ADC.read('P9_40') * 1800
    temp_c = mv / 10
    temp_f = (temp_c * 9/5) + 32
    temp_f = round(temp_f, 1)

    now = datetime.datetime.utcnow()
    feed.datastreams = [Xively.Datastream'room_temp',
                        current_value=temp_f, at=now)]

    try:
        feed.update()
        print "Value pushed to Xively: " +str(temp_f)
    except HTTPError e:
        print "Error connecting to Xively: " + str(e)
    time.sleep(20)
```

You may note that I used the same variable names in this program as were used in the test program just to keep the programs somewhat consistent. This program uses the exact same ADC pin as the test program and will not require any physical reconfigurations.

There is a new variable named now, which contains the instantaneous time stamp that is stored along with the °F temperature value. The time is in a UTC format, so you need to know the conversion factor to determine your local time. In my case, the local time zone was EST so I needed to subtract fours hours from the UTC value in order to convert to local time.

If you are the root level, you run the program by entering this:

```
python xively-temp.py
```

Non-root users will need to add sudo before the python command. Figure 10-4 shows the terminal output after the program has run for a while. Remember that readings are taken every 20 seconds due to the `time.sleep(20)` statement in the code listing. You obviously can change that interval to best suit your measurement situation. I would personally not make the interval any time less than 10 seconds as it does take some time to transfer the sensor data and send the appropriate TCP acknowledgements.

```
Terminal
File Edit View Terminal Go Help
Value pushed to Xively: 72.5
Value pushed to Xively: 72.3
Value pushed to Xively: 72.3
Value pushed to Xively: 72.5
Value pushed to Xively: 72.3
Value pushed to Xively: 72.5
Value pushed to Xively: 72.5
Value pushed to Xively: 72.5
Value pushed to Xively: 72.5
Value pushed to Xively: 72.5
Value pushed to Xively: 72.5
Value pushed to Xively: 72.5
Value pushed to Xively: 72.3
Value pushed to Xively: 72.5
Value pushed to Xively: 72.5
Value pushed to Xively: 72.5
Value pushed to Xively: 72.5
Value pushed to Xively: 72.3
Value pushed to Xively: 72.5
Value pushed to Xively: 72.7
Value pushed to Xively: 72.5
Value pushed to Xively: 72.5
Value pushed to Xively: 72.5
```

FIGURE 10-4 Terminal console display for the xively-temp.py program

You should now go to the Xively website and view the results of this program that is running and sending data to it.

Xively Website with an Active Datastream

I used my MacBook Pro browser to go to the Xively website and observe the active data streaming. Figure 10-5 shows the instantaneous data being received at Xively from the BBB.

You can see all the past data received from this datastream by clicking the graphs icon, which is visible in the upper-right side of Figure 10-5. Figure 10-6 shows the graph of temperature data that extends about six hours into the past. It is possible to click on the graph and read specific data and time values, as shown in Figure 10-7.

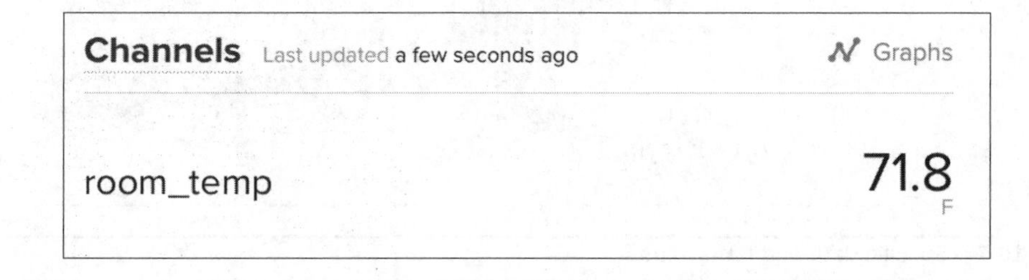

FIGURE 10-5 Instantaneous data at the Xively website

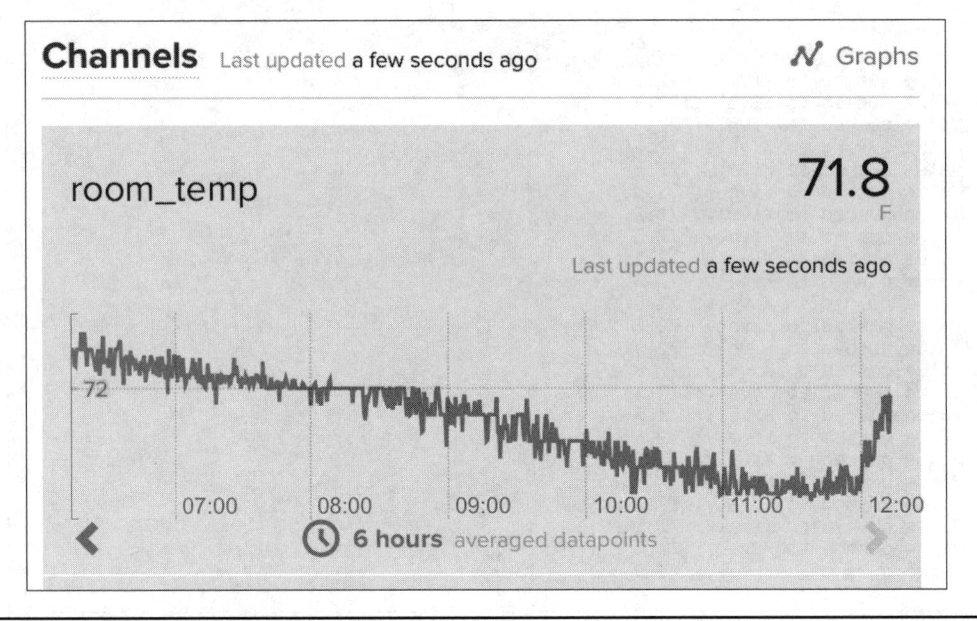

FIGURE 10-6 Temperature vs. time graph

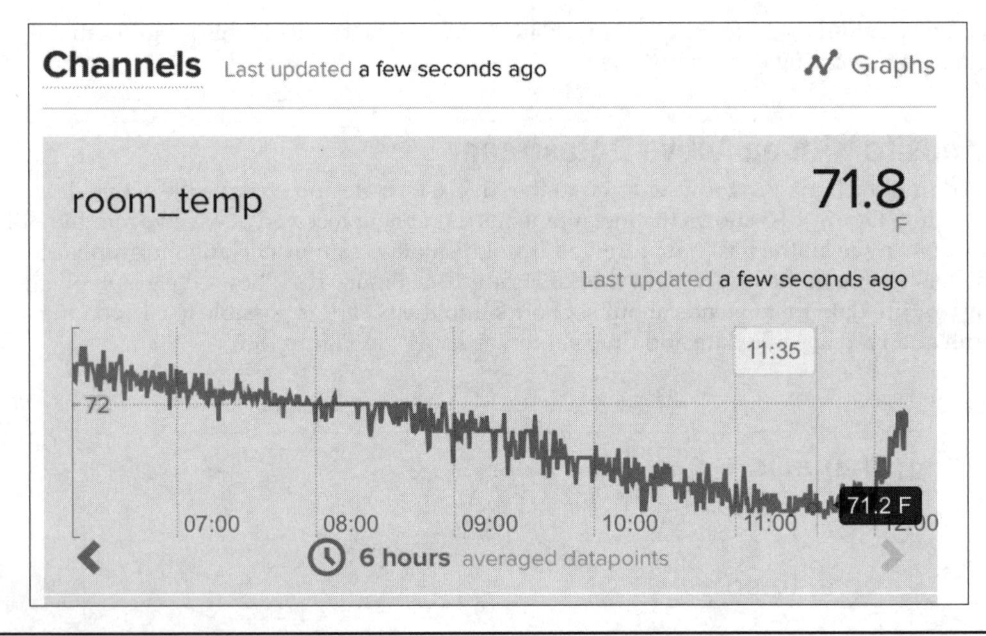

FIGURE 10-7 Specific data and time values

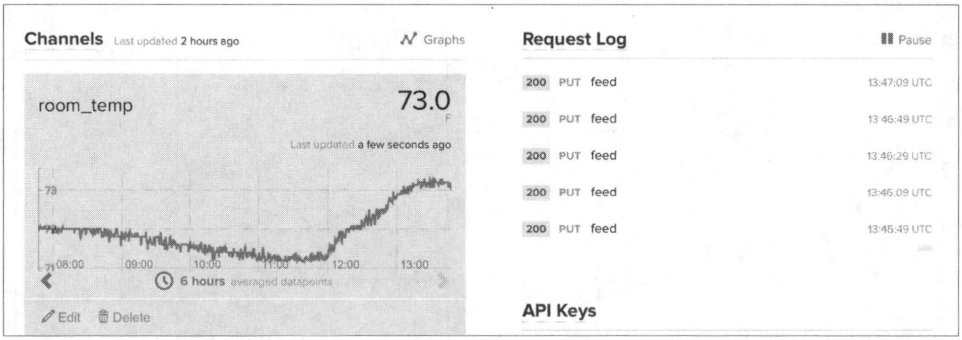

Figure 10-8 Sample HTTP Request Log window

This data set was taken from a sensor located in my office and, as you can see, the temperature slowly dropped during the evening hours and started to rise at sunrise. The data point time shows 1135 UTC, which means the local EST was 0735 in the morning for that specific data point.

There is also an HTTP Request Log window shown on the Workbench just to the right of the graph window. This Log window shows the current and past four HTTP requests that were processed between the BBB and the Xively web server. A sample of this window appears in Figure 10-8.

The Log window would be of great value if communications problems developed between the BBB and the Xively web server. The 200 number that appears before the `PUT` requests indicates a successful TCP transaction. Numbers such as 404 or 500 would indicate that some type of problem existed in the communications link between the BBB and the Xively web server.

An obvious question I have not addressed so far is the issue of how much data can be stored on the Xively website. I could not find a specific limit related to the free developer's account. However, LogMeIn owns the Xively website and it is definitely a profit-minded business. You will have to create and pay for a business account if you choose to go into commercial development with your project. At that point, I am sure that data storage and related services would be based on the fees charged by LogMeIn.

Adding Additional Data Channels

It is a simple matter to add both additional sensors to the BBB and additional channels to your logical Xively device to accommodate those new sensors. I will tackle the physical sensor portion first and then discuss how to expand upon the existing single Xively channel.

Additional TMP36 Sensors

The TMP36 sensor connection system was designed from the beginning to be easily expanded. All you need to do is connect a TMP36 sensor to a new ADC channel and change the Python program to recognize the new sensor. Of course, the data feed will also need to be changed, but I discuss that in the next section.

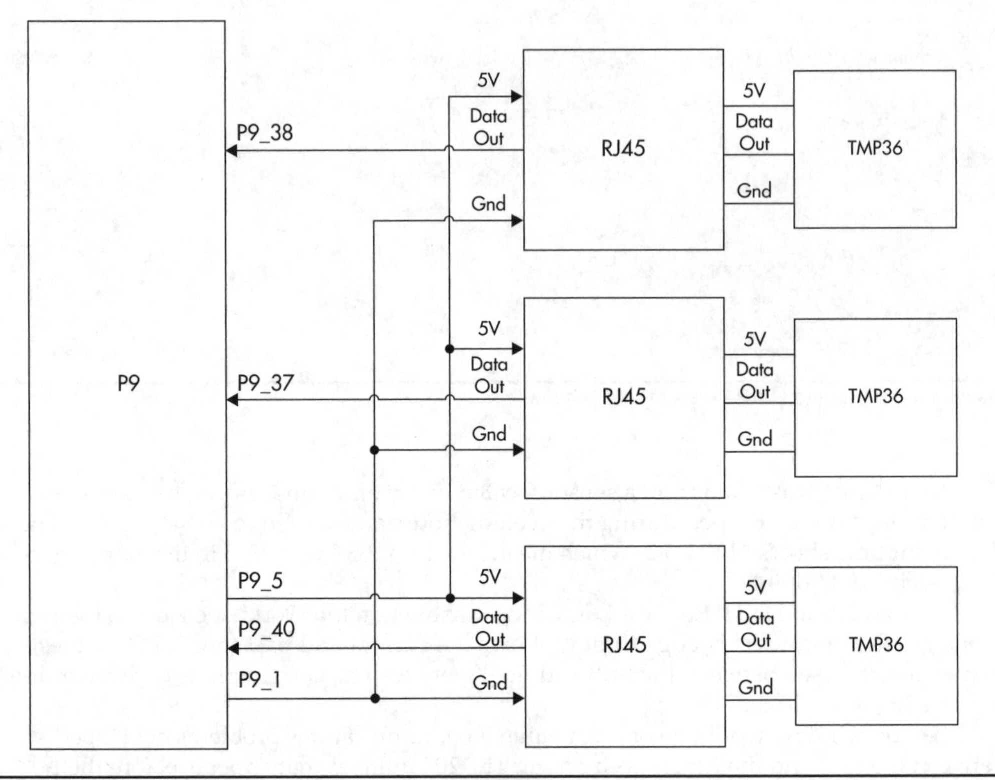

FIGURE 10-9 Expanded TMP36 to BBB connection schematic

Figure 10-9 is an expanded schematic showing two new additional sensors connected to the BBB. I chose P9-37 (A2) and P9-38 (A3) for these new ADC connections.

I also used a longer cat5 cable to connect the outdoor sensor to the BBB. Please refer to the discussion in Chapter 2 regarding some issues that might arise when using long cat5 connecting cables. Figure 10-10 shows the physical setup for the expanded temperature measurement system.

I next created an expanded test program to prove that all the sensor channels functioned as expected. I named this new test program ExpandedTest.py, which is based upon the previous test program. The following is the complete code listing:

```
import Adafruit_BBIO.ADC as ADC
import time

sensor_pin1 = 'P9_40'
sensor_pin2 = 'P9_37'
sensor_pin3 = 'P9_38'

ADC.setup()
```

FIGURE 10-10 Expanded physical system

```
while True:
    reading1 = ADC.read(sensor_pin1) // office room
    reading2 = ADC.read(sensor_pin2) // outdoors
    reading3 = ADC.read(sensor_pin3) // storage room
    mv1 = reading1 * 1800
    mv2 = reading2 * 1800
    mv3 = reading3 * 1800
    temp_c1 = (mv1 - 500) / 10
    temp_c2 = (mv2 - 500) / 10
    temp_c3 = (mv3 - 500) / 10
    temp_f1 = (temp_c1 * 9/5) + 32
    temp_f2 = (temp_c2 * 9/5) + 32
    temp_f3 = (temp_c3 * 9/5) + 32
    print('mv1=%d  C1=%d  F1=%d' % (mv1, temp_c1, temp_f1))
    print('mv2=%d  C2=%d  F2=%d' % (mv2, temp_c2, temp_f2))
    print('mv3=%d  C3=%d  F3=%d' % (mv3, temp_c3, temp_f3))
    print('====================')
    time.sleep(1)
```

Figure 10-11 shows the program's terminal output. The temperature values shown in the figure reflect the actual temperatures present, which is always a good reality check.

It is time to work on an expanded xively-temp program now that the expanded system's hardware has been checked out.

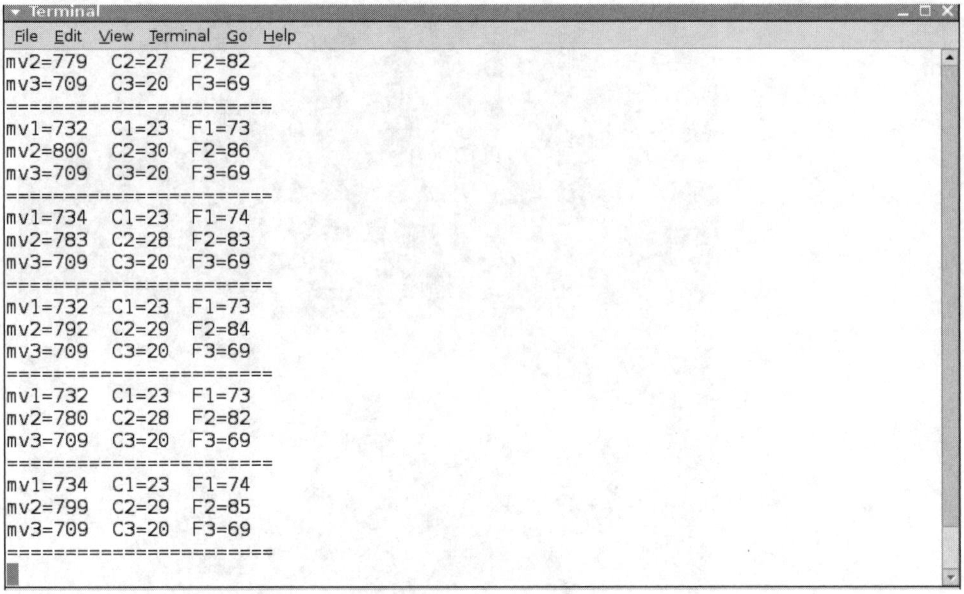

```
Terminal
File Edit View Terminal Go Help
mv2=779  C2=27  F2=82
mv3=709  C3=20  F3=69
======================
mv1=732  C1=23  F1=73
mv2=800  C2=30  F2=86
mv3=709  C3=20  F3=69
======================
mv1=734  C1=23  F1=74
mv2=783  C2=28  F2=83
mv3=709  C3=20  F3=69
======================
mv1=732  C1=23  F1=73
mv2=792  C2=29  F2=84
mv3=709  C3=20  F3=69
======================
mv1=732  C1=23  F1=73
mv2=780  C2=28  F2=82
mv3=709  C3=20  F3=69
======================
mv1=734  C1=23  F1=74
mv2=799  C2=29  F2=85
mv3=709  C3=20  F3=69
======================
```

FIGURE 10-11 Expanded test program output

Expanded xively-temp Program

I named the expanded xively-temp program ExpandedXively.py; it is essentially the same as the original except for the two new temperature channels. The two new channels were named on the Xively website as outdoor_temp and storage_temp, reflecting their actual locations. Adding additional channels to the device is as simple as repeating Step 4, as detailed in the Xively Developer's Account section for every additional channel required. The tags, units, and symbols were the same as I previously set for the room_temp channel. This program's code listing follows and is also available on the book's companion website:

```
import Adafruit_BBIO.ADC as ADC
import time
import datetime
import xively
from requests import HTTPError

api = xively.XivelyAPIClient("<your 48 character API key goes here>")
feed = api.feeds.get(<your 10 digit feed number goes here>)

ADC.setup()

while True:
    mv1 = ADC.read('P9_40') * 1800
    temp_c1 = (mv1 - 500) / 10
    temp_f 1= (temp_c1 * 9/5) + 32
    temp_f 1= round(temp_f1, 1)
```

```
    mv2 = ADC.read('P9_37') * 1800
    temp_c2 = (mv2 - 500) / 10
    temp_f2 = (temp_c2 * 9/5) + 32
    temp_f 2= round(temp_f2, 1)

    mv3 = ADC.read('P9_38') * 1800
    temp_c3 = (mv3 - 500) / 10
    temp_f3 = (temp_c3 * 9/5) + 32
    temp_f3 = round(temp_f3, 1)


    now = datetime.datetime.utcnow()
    feed.datastreams = [
        xively.Datastream(id='room_temp', current_value=temp_f1, at=now),
        xively.Datastream(id='outdoor_temp', current_value=temp_f2, at=now),
        xively.Datastream(id='storage_temp', current_value=temp_f3, at=now)
    ]

    try:
        feed.update()
        print "Value pushed to Xively: " +str(temp_f1)
        print "Value pushed to Xively: " +str(temp_f2)
        print "Value pushed to Xively: " +str(temp_f3)
        print "====================="
    except HTTPError as e:
        print "Error connecting to Xively: " + str(e)
    time.sleep(20)
```

Note that all three Datastream function calls are all contained in the same set of square brackets with commas as delimiters between the first and second as well as between the second and third calls. Failure to follow this format will likely cause at least one datastream to be ignored by the Xively web server. Figure 10-12 shows the current values for all three channels along with the HTTP Request Log.

You should notice from the Request Log that only one HTTP feed is required to push all three channels to the Xively web server. Adding any additional channels would simply follow the same process I have just described. I am also not aware of any particular limit to the number of channels that a given device can push to Xively. I would guess that the device itself would impose a limit rather than the Xively web server. Figure 10-13 shows

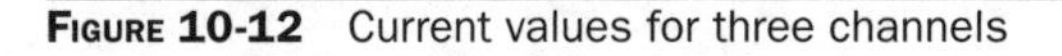

FIGURE 10-12 Current values for three channels

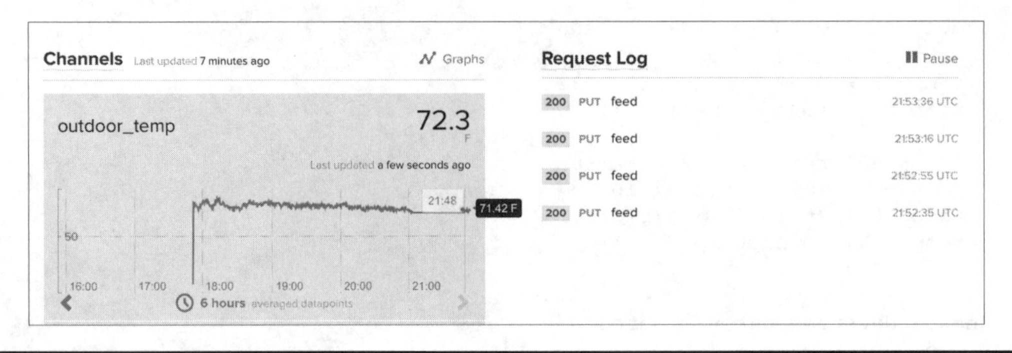

FIGURE 10-13 Outdoor temperature channel graph

a four-hour data graph for the outdoor temperature channel. I also included a specific data point in the figure just to show you that it functions exactly the same as it did for the room temperature plot.

Configuring Angstrom to Auto Start the Application

It would be quite convenient to automatically start this temperature data logging application whenever the BBB is powered on. This is easy to accomplish using a preconfigured service, which you can readily set up in the Angstrom OS to accomplish this task. There are many services listed in the /lib/systemd/system directory. Most are enabled to suit a variety of purposes. Our purpose is to start the ExpandedXively.py file using Python to execute its content. You will create a service file named xively-logger.service in the /lib/systemd/system directory by following these steps:

1. Change into the proper directory by entering the following:
   ```
   cd /lib/systemd/system
   ```
2. Start the nano editor with the service filename given previously by entering the following:
   ```
   nano xively-logger.service
   ```
3. Enter the following content into the nano editor workspace:
   ```
   [Unit]
   Description=Xively client

   [Service]
   WorkingDirectory=/home/root/
   ExecStart=/usr/bin/python ExpandedXively.py
   SyslogIdentifier=xively
   Restart=on-failure
   RestartSec=5

   [Install]
   WantedBy=multi-user.target
   ```

4. Save and exit the nano editor.
5. You use an Angstrom Linux command called `systemctl` to enable, disable, start, stop, and restart service files. Enter the following to enable this new service file:

   ```
   systemctl enable xively-logger
   ```

The ExpandedXively.py will not start after this command but only when the BBB is rebooted. You can force it to start immediately by entering the following:

```
systemctl start xively-logger
```

Similarly you can stop the logger by entering the following:

```
systemctl stop xively-logger
```

You can always force a logger restart by entering the following:

```
systemctl restart xively-logger
```

The status of the logger application can be checked using the `systemctl` status command. The following entry checks the status:

```
systemctl status xively-logger
```

Figure 10-14 shows the result of entering this command.

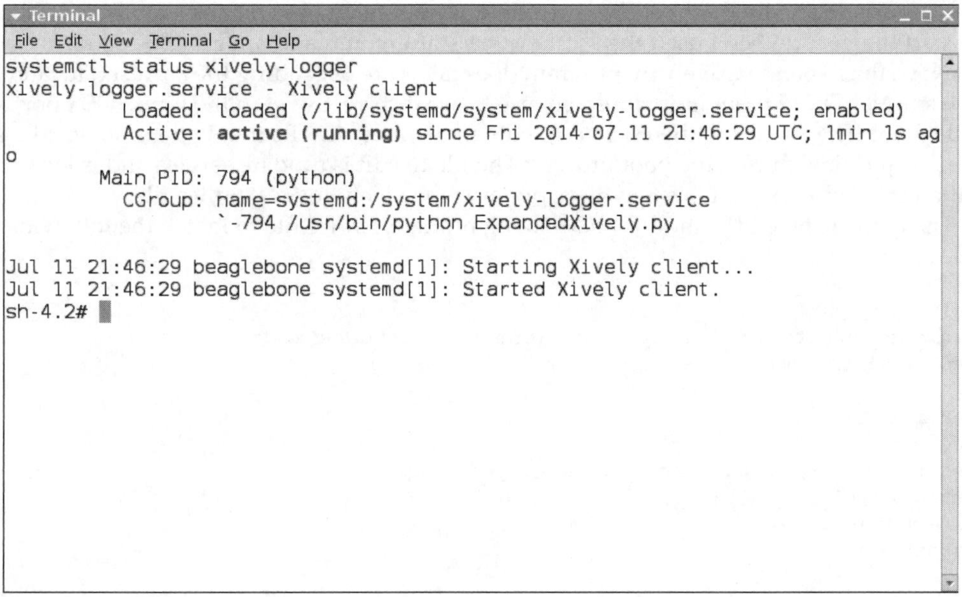

```
Terminal
File Edit View Terminal Go Help
systemctl status xively-logger
xively-logger.service - Xively client
     Loaded: loaded (/lib/systemd/system/xively-logger.service; enabled)
     Active: active (running) since Fri 2014-07-11 21:46:29 UTC; 1min 1s ag
o
   Main PID: 794 (python)
     CGroup: name=systemd:/system/xively-logger.service
             `-794 /usr/bin/python ExpandedXively.py

Jul 11 21:46:29 beaglebone systemd[1]: Starting Xively client...
Jul 11 21:46:29 beaglebone systemd[1]: Started Xively client.
sh-4.2#
```

FIGURE 10-14 xively-logger status

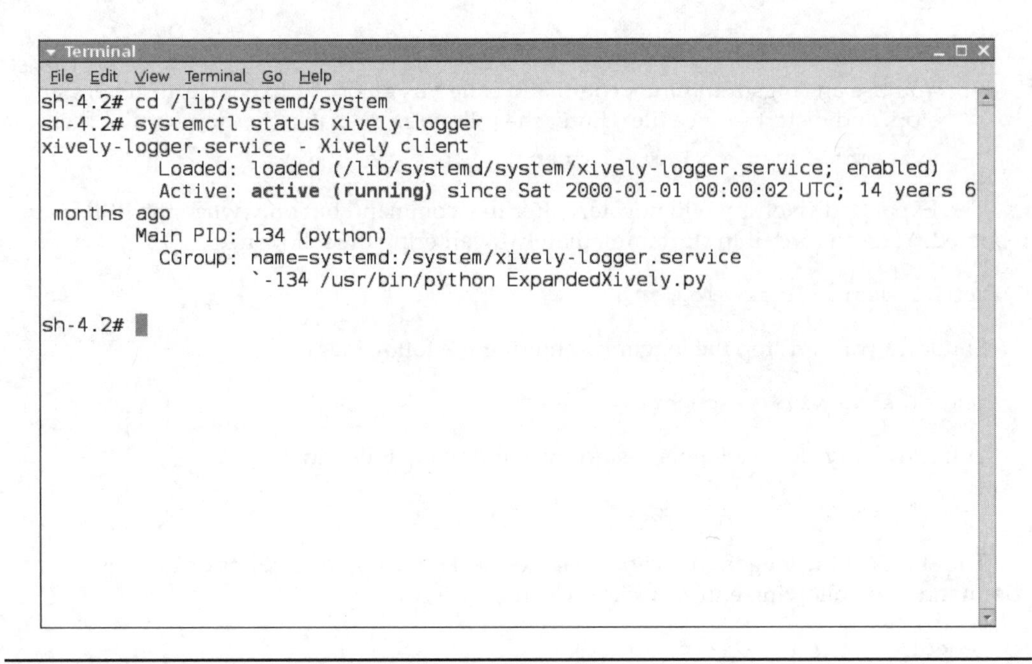

FIGURE 10-15 `Systemctl` status displays immediately after a reboot

One interesting artifact that I discovered while testing this service is that the Linux OS does not automatically discover the correct date and time upon a reboot. Figure 10-15 shows the startling result when I used the `systemctl` status command after a BBB reboot.

Now, the ExpandedXively.py program does take care of sending the correct date and time to Xively, but I thought it would be nice to see it correctly set in the Linux OS upon a BBB reboot. The answer to this issue is to edit another service file that will automatically set the date and time during the boot process. The file to edit is ntpdate.service and is located in the same /lib/systemd/system directory as is the xively-logger.service file.

Using the nano editor, modify the existing ntpdate.service file to match the following code listing:

```
[Unit)
Description=Network Time Service (one-shot ntpdate mode)
Before=ntpd.service

[Service]
Type=oneshot
ExecStart=/usr/bin/ntpdate -b -s -u pool.ntp.org
RemainAfterExit=Yes
Restart=on-failure
RestartSec=5

[Install]
WantedBy=multi-user.target
```

After you save and exit the nano editor, you will need to perform these steps:

1. Enable the modified service by entering the following:

```
systemctl enable ntpdate
```

2. Reload the configuration file by entering the following:

```
systemctl --system daemon-reload
```

3. Restart the service by entering the following:

```
systemctl restart ntpdate
```

That should be all you need to do to now have the correct time and date set for every boot. Just type **date** and check that the correct date and time are displayed. Note that the time will be UTC, but that should not be much of a problem at this stage.

System Case

I considered it important to put the BBB and temperature RJ45 connectors into a case as I will use it as a permanent project. I used the same model of a large plastic case, which I also used to house the remote moisture sensor, part of the Chapter 7 Arduino irrigation controller project.

Figure 10-16 shows the BBB mounted on a piece of clear sheet Lexan along with three RJ45 connectors, which are in turn mounted on a solderless breadboard.

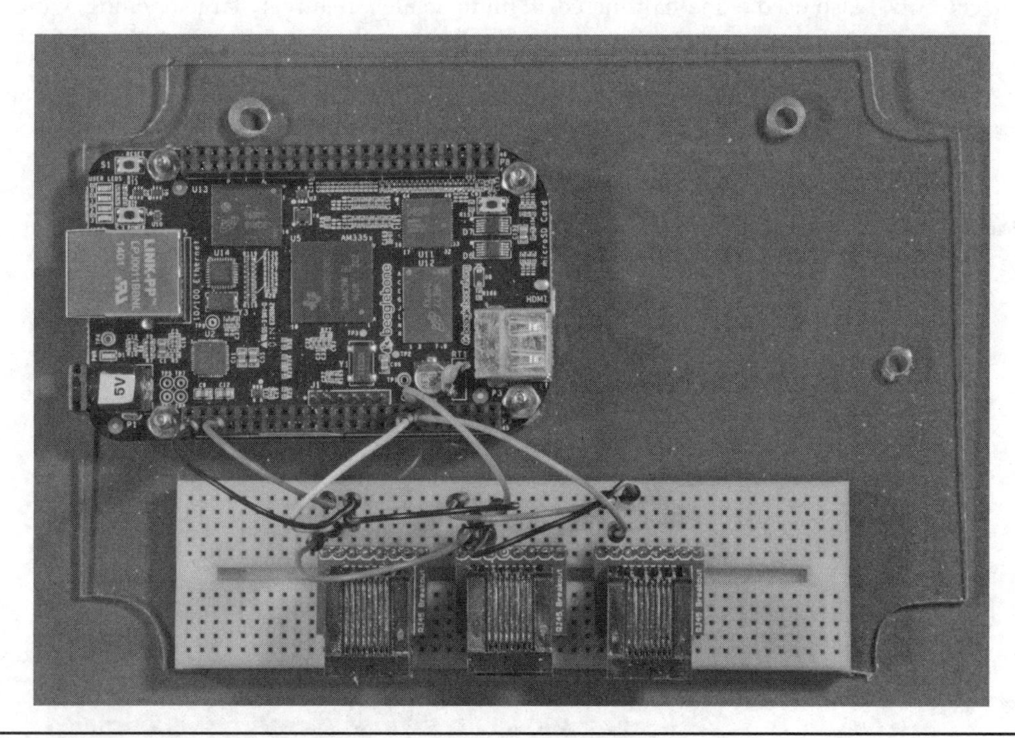

Figure 10-16 BBB and RJ45 connectors mounted on Lexan board

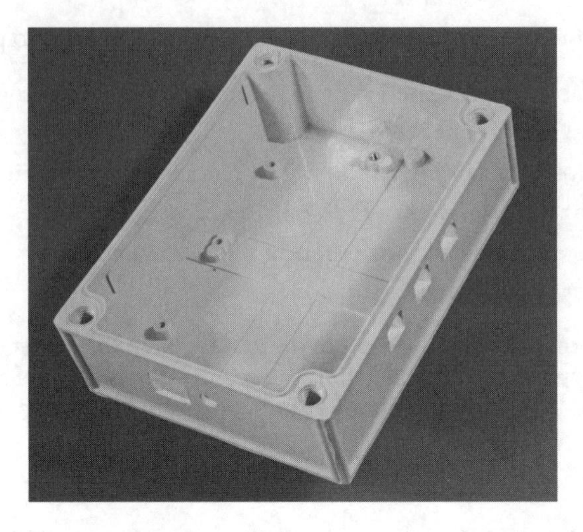

FIGURE 10-17 Modified case bottom

The bottom of the plastic case must have cutouts to allow access to the three RJ45 temperature sensor connectors. It also must have two cutouts on one end for the 5V power plug and the Ethernet cable. Figure 10-17 shows the fully modified case bottom.

I have not included a dimensioned drawing as I found it more convenient to install the mounting plate and use the BBB and RJ45 connectors to mark the inside of the case for the cut-outs. I also used a one-half-inch drill bit to create a relatively large opening, which I subsequently squared off with a file. Figure 10-18 shows the complete assembly without the top cover attached.

This section wraps up this chapter's project and it is time to move on to another interesting project, which will include this BBB measurement system along with a Raspberry Pi controller.

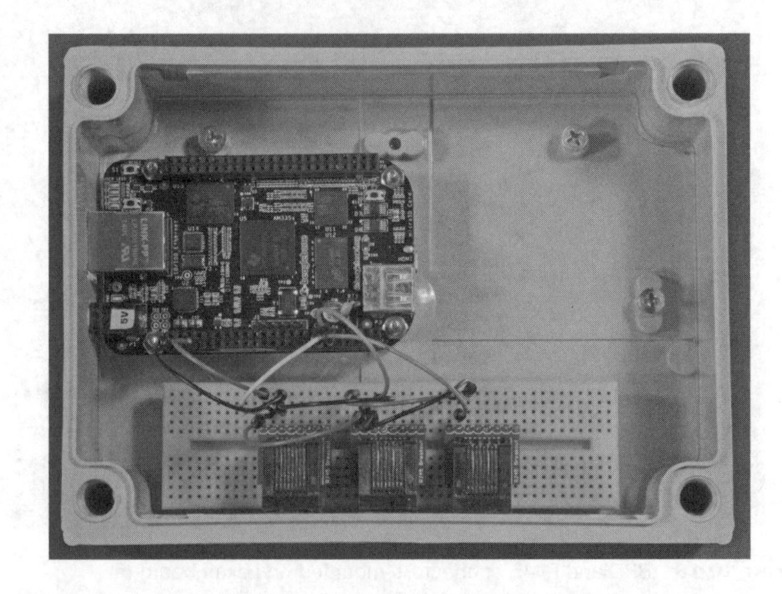

FIGURE 10-18 Complete assembly

Summary

This chapter's focus was to demonstrate how to set up a BeagleBone Black (BBB) with temperature sensors, which automatically transmits measurements to a cloud-based service named Xively.

I reviewed some basic facts about the analog TMP36 temperature sensor, which I first used in the Chapter 2 project, the Raspberry Pi home temperature monitoring system. The BBB is also well suited to interface with the TMP36 as it has an integrated, multiplexed analog-to-digital converter (ADC), which the Raspberry Pi is sorely lacking.

The Adafruit BBIO Python library was discussed next because it allows for a straightforward approach to interface the TMP36 to the BBB using the Python language. I demonstrated a simple test program, which proved that a single TMP36 sensor works fine with the BBB.

A detailed discussion regarding the Xively cloud service followed in which I provided step-by-step instructions on how to set up a free Xively developer's account, which you need for this project.

A single-channel Python program was shown, which both displayed sensor measurements on a terminal screen and sent data to the Xively website. I also used a browser on another computer to go to the Xively website to examine the temperature data being sent or "pushed" to it. Past or historical data is also readily available from the website in the form of a graph.

I next demonstrated how to add two additional sensors to the project. You first need to create named channels at the Xively website to handle the new sensors. Next, you need to change the Python program that is pushing the data to accommodate the new sensors. The Xively website clearly picked up the new channels and displayed the data without any issues.

The BBB's Angstrom Linux OS was next modified to autostart the Python application, which means that you don't need to manually restart it after every BBB reboot. I also showed you how to set up Angstrom so that the correct date and time are set after a reboot.

I finished the chapter by showing you how I modified a nice plastic case to house the project components. I wanted to have them in a presentable enclosure as I plan on indefinitely retaining and using this particular project.

11 CHAPTER

Machine-to-Machine (M2M) Communications

The focus of the chapter, as its name implies, is machine-to-machine (M2M) communications. There is no human intervention in this type of system as all the computers or machines, as they are generically termed, are set up to communicate with one another using an established protocol. I will be using the BeagleBone Black (BBB) as a data source or publisher client and the Raspberry Pi (Pi) as a data sink or subscriber client. The BBB system will also use a single TMP36 temperature sensor—the same type that was used in the previous chapter.

Before starting the detailed discussion, I would like to acknowledge a fine blog article entitled "Using Eclipse Paho's MQTT on BeagleBone Black and Raspberry Pi," which was written by D. J. Walker-Morgan, an extremely talented developer. I can't help but feel an affinity for Mr. Walker-Morgan as my first two initials are also D. J. His article is great and is available at the Eclipse.org's talkingsmall blog at www.eclipse.org/paho/articles/talkingsmall/. I highly recommend you read it when you have an opportunity to do so.

Paho and Eclipse.org

Paho is an open-source project sponsored by the Eclipse.org foundation. This project is dedicated to providing scalable client implementations for both open and standard messaging protocols. The Paho project is designed to provide an exciting infrastructure in support of new M2M and IoT applications. The home website is www.eclipse.org/paho/. At the heart of the Paho project is a lightweight publish/subscribe message protocol named MQTT, which I describe in the next section.

MQTT

MQTT is the current name for this protocol, although it was originally named Message Queuing Telemetry Transport. I guess the project managers felt that was a mouthful, or perhaps the name was changed because there are no actual queues used in the protocol. In any case, it is now simply called MQTT.

MQTT is over 10 years old, having originally been created by the IBM pervasive software group in conjunction with Arcom, which is now called Eurotech. IBM still supports MQTT

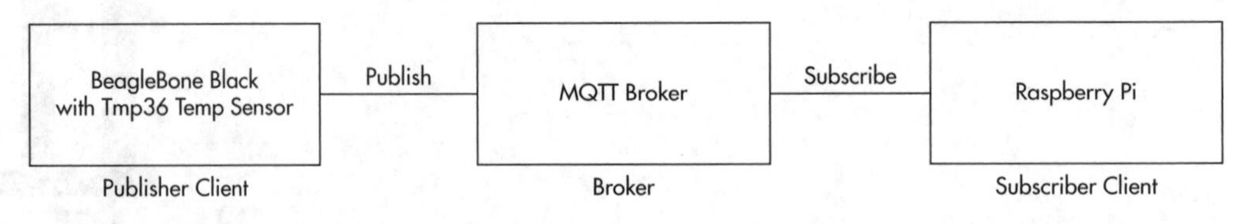

FIGURE 11-1 Project block diagram

with the current 3.1 version specifications available from the IBM developerWorks website at www.ibm.com/developerworks/webservices/library/ws-mqtt/index.html.

MQTT is technically known as a middleware application, as you can see in the Figure 11-1 block diagram for this project. It is important to realize that both publishers and subscribers are treated as client applications in this configuration type.

There is a block named MQTT Broker located between the BeagleBone Black publisher client and the Raspberry Pi subscriber client blocks. This broker may be thought of as a message dispatcher ensuring that the MQTT messages are properly sent from the client publishers to the correct client subscribers. In that way, subscriber clients do not have to constantly monitor all the network traffic looking for messages that are addressed to them. The broker takes over that function and also serves as an acknowledgment intermediary. I discuss this in the section named Quality of Service.

Table 11-1 details some of the MQTT salient features that make it so popular as a messaging protocol.

These features make MQTT very popular for M2M applications, including weather monitoring, stock ticker, smart power grid meters, and even Facebook messaging. It is also a very popular way for cellular services to implement message alerts.

Quality of Service (QoS)

Quality of Service (QoS) refers to the level assurance that MQTT provides regarding message delivery. There are three QoS levels:

- **Level 0** This is also known as "fire and forget." At this level, the publisher sends off messages and there is no attempt to acknowledge their reception by the broker on behalf of the publisher. It is obviously the quickest message delivery method, but it is also the least reliable.
- **Level 1** This is also known as "at least one." Here, messages are sent and resent until the broker receives one acknowledgment from the subscriber. It does provide some assurance that the message did get through to its intended recipient. Level 1 is typically set as the default QoS for a MQTT messaging system.

TABLE 11-1 MQTT Features

Feature	Description
API	Simple, only five methods required.
Packets	Compact binary packets. Capable of up to 250MB payload.
Headers	No compressed headers needed.
Verbose	Minimal text, much less than HTML.

- **Level 2** This is also known as "exactly one." At this level, messages undergo a two-stage process where there is a definitive acknowledgment between the broker and subscriber, ensuring that one and only one message copy was delivered. This QoS level is the slowest among the three levels because of the additional processing overheads required to establish a high reliability level.

Wills

No, this section has nothing to do with legal probate but instead focuses on what happens when a client abnormally loses its connection with the broker. A "will" is both a set of instructions and a prescribed message that is stored by the broker and will only be acted upon if the connection between the broker and a client is unexpectedly broken. Basically, it is a dialog that states, "If you (the broker) cannot connect to me, and I (the client) haven't cleanly disconnected, then carry out the preset instructions and also send out the stored message on my behalf." The "will" concept is implemented in Python by a `setWill` method and in Java by an object of the `MqttConnectionsOptions` class.

Using wills in a MQTT improves both system robustness and reliability and ensures that messages will either be delivered or an error message will be created and distributed describing what went wrong.

Reconnecting

Connections will be broken and MQTT has the inherent ability to reconnect using two system elements. The first is a logical flag known as the "clean flag," which is set at one value for every fresh or new connection. The clean flag informs the client and broker that they must start the messaging process from the beginning as it represents a new connection. However, if the clean flag becomes false or low, a second element comes into play. This is called the client ID and you will see that it plays a key part in establishing an original connection when the test code is discussed. For now, let's assume it had already been set to some String value when the connection broke. Now, assume the connection is restored as might happen when a client briefly loses power. MQTT will attempt to restore the connection to the same precise state because it recognizes that it still has same client ID String stored in its record structure, which existed for this particular connection when it first became disconnected. Note that various MQTT libraries, whether they be Python or Java, have different implementations for storing the client IDs and messages so that the connections can be recovered without any message loss.

It is now time to demonstrate a temperature monitoring application that uses MQTT to distribute single valued data points between one publisher and one client.

BBB MQTT Publisher Client

Initially, I just need to demonstrate the temperature-monitoring program without any messaging features added. The program is named tmp36.py and it is the same one that was used in the previous chapter to interface to a single TMP36 temperature sensor. The code listing is shown here for your convenience:

```
import Adafruit_BBIO.ADC as ADC
import time

sensor_pin = 'P9_40'
```

```
ADC.setup()

while True:
    reading = ADC.read(sensor_pin)
    mv = reading * 1800
    temp_c = (mv - 500) / 10
    temp_f = (temp_c * 9/5) + 32
    print('mv=%d  C=%d  F=%d' % (mv, temp_c, temp_f))
    time.sleep(1)
```

Remember to load the Adafruit BBIO library, following the instructions shown in Chapter 10. I also want to point out that the BBB I used for this project was not the same one I used in Chapter 10. I used a copy of the Debian Wheezy Linux distribution instead of the Angstrom distribution used in Chapter 10. This means that you must substitute the `apt-get` command for the `opkg` command in the instructions if you also use Debian. Everything else should function in exactly the same way.

The hardware setup is shown in Figure 11-2. You can see I reused the RJ45 connector scheme with a cat5 cable to interface the TMP36 sensor with the BBB. Please refer back to Figure 10-1 for the sensor interface wiring schematic.

Figure 11-3 shows the terminal display after I entered the following:

```
sudo python tmp36.py
```

This figure demonstrates that the hardware portion of the project works properly and it is time to incorporate some MQTT features.

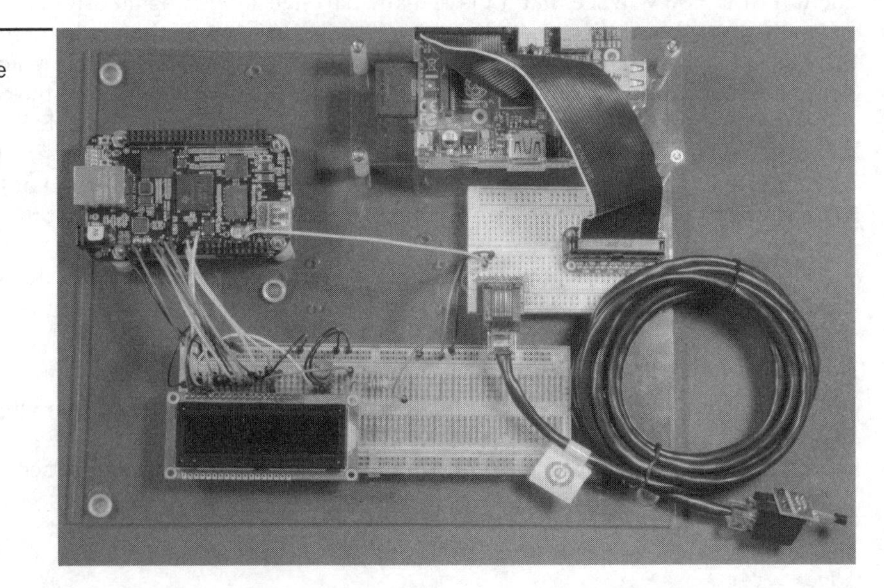

FIGURE 11-2 Physical hardware setup

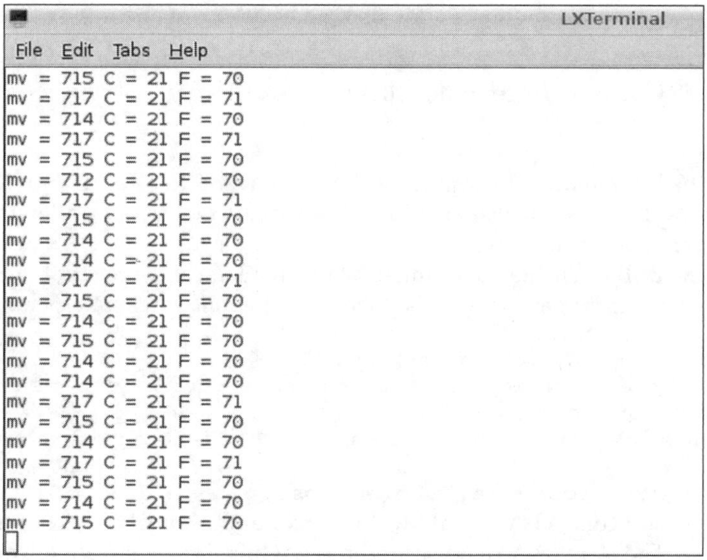

FIGURE 11-3 Terminal screen showing the tmp36.py program output

LXTerminal

File Edit Tabs Help

```
mv = 715 C = 21 F = 70
mv = 717 C = 21 F = 71
mv = 714 C = 21 F = 70
mv = 717 C = 21 F = 71
mv = 715 C = 21 F = 70
mv = 712 C = 21 F = 70
mv = 717 C = 21 F = 71
mv = 715 C = 21 F = 70
mv = 714 C = 21 F = 70
mv = 714 C = 21 F = 70
mv = 717 C = 21 F = 71
mv = 715 C = 21 F = 70
mv = 714 C = 21 F = 70
mv = 715 C = 21 F = 70
mv = 714 C = 21 F = 70
mv = 714 C = 21 F = 70
mv = 717 C = 21 F = 71
mv = 715 C = 21 F = 70
mv = 714 C = 21 F = 70
mv = 717 C = 21 F = 71
mv = 715 C = 21 F = 70
mv = 714 C = 21 F = 70
mv = 715 C = 21 F = 70
```

Adding MQTT Features to the Application

You first need to load the appropriate MQTT client implementation library before adding the messaging features into the application. Please follow these steps to load the Python library that will be used in this project. Also note that I will be using the Debian commands in all of the following instructions:

1. The Linux distribution must first be updated to ensure that all dependencies will be located in the appropriate repositories. Enter the following:

```
sudo apt-get update
```

2. Download the source code from Github using this command:

```
sudo git clone git://git.eclipse.org/gitroot/paho/org.eclipse.
paho.mqtt.python.git
```

NOTE *If you have difficulty in doing a direct git clone, you can also go to http://git.eclipse.org/c/paho/org.eclipse.paho.mqtt.python.git/ and download one of the following compressed files:*

```
org.eclipse.paho.mqtt.python-1.0.zip
org.eclipse.paho.mqtt.python-1.0.tar.gz
org.eclipse.paho.mqtt.python-1.0.tar.bz2
```

Use the extraction application that matches the compressed file extension you downloaded, i.e., winzip or 7zip for the zip file. You should see the same source directory created after extraction as was created for the clone operation.

3. Change into the source directory:

```
cd org.eclipse.paho.mqtt.python/
```

4. Compile the source code using a build script already available in the directory:

```
make
```

5. Install all the compiled files:

```
make install
```

The Python MQTT client should now be ready to be added to the tmp36.py program, but let's first cover some basic concepts, which you should understand before going on to the complete application.

The BBB publisher client must establish a logical connection to the broker before any messages can be passed. This is done with the following statements:

```
import paho.mqtt.client as mqtt
// The preceding import statement sets up a client reference named
mqtt.mqttc = mqtt.Client()
// This statement instantiates a MQTT client object named mqttc.
```

- **`mqttc.connect("m2m.eclipse.org", 1883, 60)`** Goes out to the Internet and connects with an MQTT broker at the website "m2m.eclipse.org" on port 1883. The 60 refers to a 60-second ping, which is a "keep alive," meaning it is sent when no other activity is happening on the connection.
- **`mqttc.loop_start()`** A separate execution thread is started that handles incoming messages from the broker.

The following two statements contain references to what are known as topics and subtopics:

```
mqttc.publish("bbbexample123/tmp36/mv","%.2f" % mv);
mqttc.publish("bbbexample123/tmp36/f",""%.2f" % temp_f);
```

In the preceding statements, `bbbexample123` refers to a root topic created on the broker, which also contains the subtopics, `tmp36`, `mv`, and `f`. Real-time millivolt data is stored in the `mv` subtopic while real-time Fahrenheit temperature data is stored in the `f` subtopic. I will demonstrate how to retrieve this real-time data from the broker shortly. But first, you should enter the following modified tmp36.py program, which I named mqttTMP36.py to reflect the new messaging capabilities:

```
import Adafruit_BBIO.ADC as ADC
import time
import paho.mqtt.client as mqtt

sensor_pin = 'P9_40'

ADC.setup()

mqttc = mqtt.Client()
mqttc.connect("m2m.eclipse.org", 1883, 60)
mqttc.loop_start()


while True:
    reading = ADC.read(sensor_pin)
```

```
    mv = reading * 1800
    temp_c = (mv - 500) / 10
    temp_f = (temp_c * 9/5) + 32
    print('mv=%d  C=%d  F=%d' % (mv, temp_c, temp_f))
    mqttc.publish("bbbexample123/tmp36/mv","%.2f" % mv);
    mqttc.publish("bbbexample123/tmp36/f",""%.2f" % temp_f);
    time.sleep(1)
```

The program is run by entering the following:

```
sudo python mqttTMP36.py
```

You should see exactly the same terminal display that was shown when the tmp36.py program was run in the earlier test except that I added the word `mqtt` to the end of each display line to help me distinguish between the two program outputs. Figure 11-4 shows a terminal screen for this program.

The data in the mqttTMP36.py program is also being sent to the broker located at m2m .eclipse.org and listening on port 1883. I believe that some discussion at this point regarding the broker website would be helpful for your overall understanding of the role that the MQTT broker plays in this messaging scheme.

MQTT Brokers

The web server located at m2m.eclipse.org is a public sandbox hosted by the Eclipse Foundation as part of their open-source IoT project. This web server's software itself is based upon the Mosquito project created and maintained by Roger Light, a highly talented

```
LXTerminal
File  Edit  Tabs  Help
mv = 729 C = 22 F = 73 mqtt
mv = 729 C = 22 F = 73 mqtt
mv = 728 C = 22 F = 73 mqtt
mv = 728 C = 22 F = 73 mqtt
mv = 729 C = 22 F = 73 mqtt
mv = 729 C = 22 F = 73 mqtt
mv = 731 C = 23 F = 73 mqtt
mv = 726 C = 22 F = 72 mqtt
mv = 729 C = 22 F = 73 mqtt
mv = 729 C = 22 F = 73 mqtt
mv = 729 C = 22 F = 73 mqtt
mv = 728 C = 22 F = 73 mqtt
mv = 732 C = 23 F = 73 mqtt
mv = 729 C = 22 F = 73 mqtt
mv = 731 C = 23 F = 73 mqtt
mv = 729 C = 22 F = 73 mqtt
mv = 729 C = 22 F = 73 mqtt
mv = 728 C = 22 F = 73 mqtt
mv = 729 C = 22 F = 73 mqtt
mv = 729 C = 22 F = 73 mqtt
mv = 728 C = 22 F = 73 mqtt
mv = 729 C = 22 F = 73 mqtt
mv = 732 C = 23 F = 73 mqtt
```

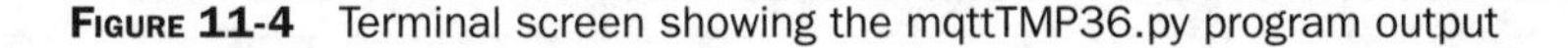

FIGURE 11-4 Terminal screen showing the mqttTMP36.py program output

Address	Ports	Additional Services	Remarks
m2m.eclipse.org	1883, 80	HTTP bridge	Xively stats, topics, mosquito info web page.
test.mosquitto.org	1883, 8883 (SSL), 8884 (SSL), 80	HTTP bridge	Xively stats, topics, mosquito info web page, SSL support.
dev.rabbitmq.com	1883	Dashboard	
broker.mqttdashboard.com	1883, 8000	Dashboard	Stats, HiveMQ info web page, SSL not yet available.
q.m2m.io	1883		Requires registration before use.
www.cloudmqtt.com	18443, 28443 (SSL)		Mosquito info web page, SSL support. Requires registration; paid site but there is a free plan.

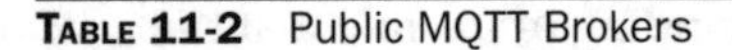

TABLE 11-2 Public MQTT Brokers

UK developer. The sandbox server allows free and public access to an actual MQTT broker where developers may test their software. There are no restrictions at this site and just like an African waterhole, all are welcome to use it, but beware of any predators that might be lurking nearby. This metaphor means that your data, which is being sent to the broker, can be accessed by anyone who is also concurrently on the site. This usually is not a problem, as most developers are typically well behaved.

There are a number of other freely available MQTT brokers in addition to m2m.eclipse .org. Table 11-2 lists all the brokers that were reported as available at the time of this writing. All offer standard MQTT broker support, while some provide additional services such as SSL, a dashboard, or an HTTP bridge, as noted in the Remarks column.

The HTTP bridge is one of the features in the m2m.eclipse.org broker that will allow us to check if the mqttTMP36 application data is actually being received by the broker. To use the HTTP bridge, first ensure that the mqttTMP36 client is running and then, using a browser either on the BBB or another computer, go to http://eclipse.mqttbridge.com/bbbexample123/tmp36/mv.

Once on the website, you should see only a single number such as 734, which represents a millivolt reading taken from the tmp36 sensor. Figure 11-5 shows the HTTP bridge website while I was running the mqttTMP36 application.

FIGURE 11-5 HTTP bridge screen for the mv subtopic

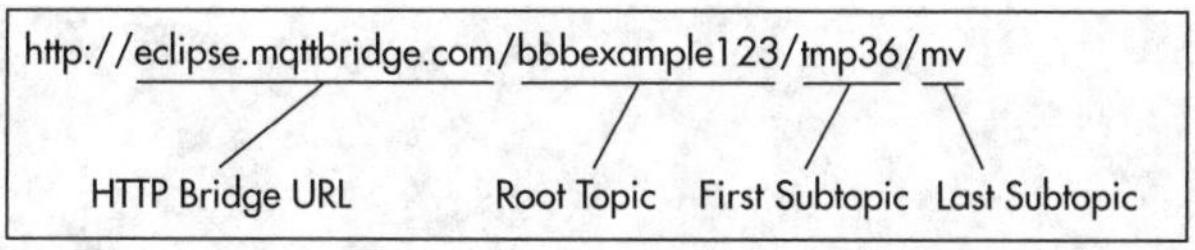

FIGURE 11-6 HTTP bridge URL with topics

http://eclipse.mqttbridge.com/bbbexample123/tmp36/mv

HTTP Bridge URL | Root Topic | First Subtopic | Last Subtopic

You may have noticed the order in which the HTTP bridge URL was constructed specifies the root topic and all the branch subtopics leading to the desired one to be displayed, as shown in Figure 11-6.

Going to the following website will enable you to retrieve the temperature data because the final subtopic is `f`, which matches the published subtopic descriptor. You should examine the published statement in the mqttTMP36.py code listing to confirm this; enter the following URL into a web browser:

http://eclipse.mqttbridge.com/bbbexample123/tmp36/f

Figure 11-7 shows the HTTP bridge website with the `f` subtopic while the mqttTMP36 program was running. There is another MQTT broker display, which I wish to show you in the next section, that provides an interesting insight into the popularity of the MQTT messaging protocol.

MQTT Dashboard

Open the mqttTMP36.py program in the nano editor. Edit the existing client connect statement to the following:

```
mqttc.connect("broker.mqttdashboard.com", 1883, 60)
```

Next, start the mqttTMP36.py program and go to broker.mqttdashboard.com. Figure 11-8 shows an MQTT Dashboard, which is currently processing data from my BBB as well as from some other concurrent users.

In the Recently used Topic box, you can see my `mv` and `f` topics as well as a topic from another user. A bunch of PINGs are also shown, which are probably sent from some connected client that is not sending actual data but only a "keep alive" ping packet without a published topic. The Last Message box on the lower-left side of the figure shows one of my published `f` topic strings along with the temperature value.

You should also examine the tabs on the menu bar, which include PUBLISH and SUBCRIBE functions. These two tabs allow some real-time interaction with the dashboard using a browser instead of a client program. Figure 11-9 is another dashboard screen that shows several other user topics in addition to my topics.

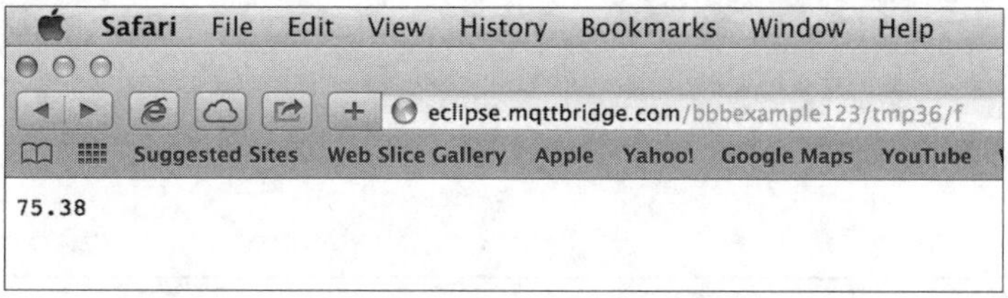

FIGURE 11-7 HTTP bridge website for the f subtopic

Safari File Edit View History Bookmarks Window Help

eclipse.mqttbridge.com/bbbexample123/tmp36/f

Suggested Sites Web Slice Gallery Apple Yahoo! Google Maps YouTube

75.38

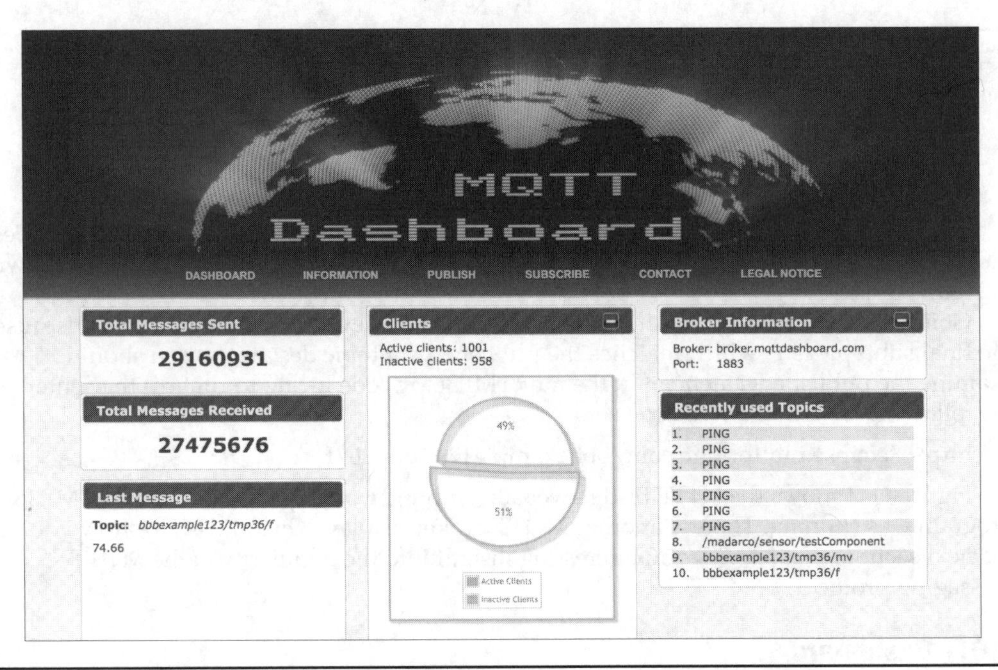

Figure 11-8 MQTT Dashboard

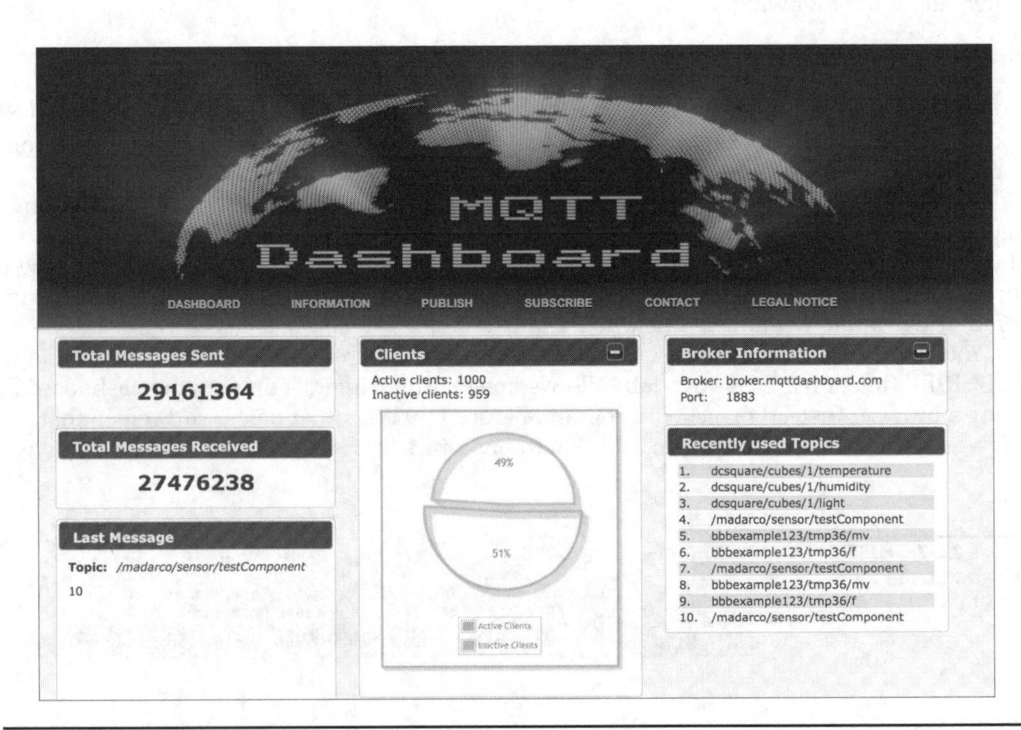

Figure 11-9 MQTT Dashboard screen

I have included this figure to point out the public nature of these MQTT brokers, which allow your data to be examined by anyone who simply browses to the public website. There is a way to safeguard your data and that is to use SSL to encrypt the published topics so only the designated receiver client will be able to decrypt the data. Brokers equipped to handle SSL data use a predesignated port other than 1883. The SSL brokers and their respective ports are described in Table 11-2. I will not be using SSL as my project does not require any data protection and I do not publish any data that I would not want the public to view.

It is time to examine the Raspberry Pi client now that you have had an introduction to the MQTT brokers.

Raspberry Pi Subscriber Client

The MQTT subscriber client will be implemented in Java rather than Python, which emphasizes the MQTT platform agnostic approach. Using Java with the Raspberry Pi also continues the object-oriented theme that was discussed in earlier chapters.

Using a callback method is key to how the MQTT subscriber client functions. A callback method is one that is triggered by an event, which is the arrival of a message at the broker for this situation. Callback methods are specified in the MQTTCallback interface, which is implemented by this subscriber client class named `PahoMqttSubscribe`. The next statement shows the class declaration along with the interface implementation.

```
public class PahoMqttSubscribe implements MqttCallback
```

The client class also requires a supporting library, which is in the form of a Java archive file named mqtt-client-0.4.0.jar. This jar file will need to be downloaded from the Eclipse.org website. Instructions on how to download it will be discussed shortly. This statement is the import for the mqtt client library:

```
import org.eclipse.paho.client.mqttv3.*;
```

An empty client constructor and a reference to the client are created by these statements:

```
public PahoMqttSubscribe() { }
MqttClient client;
```

The following `main` method contains only one method call in this minimal demo project. This method call also incorporates an instantiation of the `PahoMqttSubscribe` class.

```
public static void main(String[] args) {
   new PahoMqttSubscribe().doDemo();
}
```

The `doDemo()` method call made in the `main` method is where the application's forever loop is located. The first action that occurs when this method is first entered is an instantiation of the client object, which is directed to the desired broker website. A `connect` command follows the instantiation.

```
client = new MqttClient("tcp://m2m.eclipse.org:1883, MqttClient.generateClientId());
client.connect();
```

Notice that one of the arguments in the instantiation statement sets up a unique client ID. All clients connecting to a broker require a unique ID, which is typically constructed from metadata elements that the broker can discern from the initial TCP connection.

The next step in the doDemo method is to establish the callback method, which I previously mentioned. This method will be called when a message to which the client is subscribed is received by the broker.

```
client.setCallback(this);
```

The client must next inform the broker which topic it desires to subscribe to:

```
client.subscribe("bbbexample123/tmp36/f");
```

Additional actions are normally added after the subscribe statement. In this minimal demo, there is only a forever loop containing a 1-second sleep statement. The sleep statement is contained by try/catch statements, which are needed for this execution sequence. Obviously, real-time control application statements would also be placed here as desired.

```
while(true) {
   try {
      Thread.sleep(1000):
   }
   catch(InterruptedException e) { }
}
```

The remaining item that is missing in this class definition is the actual callback method. This method is named messageArrived and it takes two arguments, a String for the topic and an MqttMessage type representing the subtopic value. The MqttMessage type value is also known as the payload. The only action that the callback method will perform in this demo is to print the topic and the payload.

```
public void messageArrived(String topic, MqttMessage message) throws Exception {
 System.out.println(topic + " " + new String(message.getPayload()));
}
```

There is a pro-forma action that is also required for this class definition to be complete and able to be compiled. Because the PahoMqttSubscribe class implements the MqttCallback interface, it is required to provide an implementation for all of the methods specified by the interface. One method, messageArrived, has already been implemented. There are two other methods that must be implemented. These are shown here as empty or null implementations:

```
public void connectionLost(Throwable cause) { }
public void deliveryComplete(IMqttDeliveryToken token) { }
```

The Java MQTT API does contain applications that will provide real implementations for the preceding callback methods. They are not needed in this demo, but you should know that they are available.

All of the code discussed in the preceding text is shown next as a complete class definition named `PahoMqttSubscribe.java`. You should use the nano editor to enter it or download it from the book's companion website.

```
import org.eclipse.paho.client.mqttv3.*;

public class PahoMqttSubscribe implements MqttCallback {

 MqttClient client;

public PahoMqttSubscribe() {}

public void messageArrived(String topic, MqttMessage message) throws
Exception
{
 System.out.println( topic + " " + new String(message.getPayload()));
}

public void connectionLost(Throwable cause) {}
public void deliveryComplete(IMqttDeliveryToken token) {}

public static void main(String[] args) {
  new PahoMqttSubscribe().doDemo();
}

public void doDemo() {
  try {
    client = new MqttClient("tcp://m2m.eclipse.org:1883", MqttClient
.generateClientId());
    client.connect();
    client.setCallback(this);

    client.subscribe("bbbexample123/tmp36/f");

    while(true) {
       try { Thread.sleep(1000); }
       catch(InterruptedException e) {}
    }
  }
  catch(MqttException e) { e.printStackTrace(); }
} // end of doDemo()
} // end of class def
```

Do not compile the code at this point because it will not work due to the missing jar file. Please follow these steps to download the jar, compile the source file, and execute the class file.

1. Enter the following while in a Raspberry terminal window:

   ```
   curl -O https://repo.eclipse.org/content/repositories/paho-releases/
   org/eclipse/paho/
   mqtt-client/0.4.0/mqtt-client-0.4.0.jar
   ```

2. Ensure that the jar file is in the same directory as the class source file PahoMqttSubscribe.java. Enter the following to compile the source file:

```
javac -cp mqtt-client-0.4.0.jar PahoMqttSubscribe.java
```

3. Ensure that the BBB is running the mqttTMP36.py program. Also check that you are using the m2m.eclipse.org broker. Enter the following command to execute the class file:

```
java -cp mqtt-client-0.4.0.jar;.  PahoMqttSubscribe
```

NOTE *Don't forget to enter the semicolon and period that follow the* `.jar` *extension. The program will not run unless you have those in the command.*

Figure 11-10 shows the Raspberry Pi terminal window with data streaming from the MQTT broker.

This last step completes the initial M2M demonstration project. To recap, I showed you how to first set up a BBB as a publisher client, which streamed temperature data to a MQTT broker. The BBB was running a Python program for this function. I next showed you how to set up a Raspberry Pi as a subscriber client using a Java program. The Pi was connected to the same MQTT broker as the BBB and was thus able to receive the data messages from the BBB via the broker. This was made possible by an MQTT callback method named `messageArrived`. The next part of this M2M demonstration project is to slightly expand the subscriber Java class such that it can undertake some automatic actions based on the received data messages.

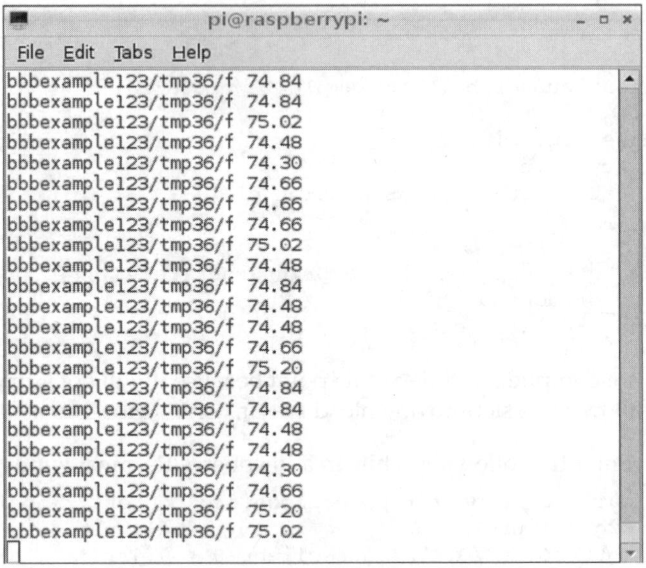

FIGURE 11-10 Raspberry Pi subscriber client terminal screen

```
pi@raspberrypi: ~
File  Edit  Tabs  Help
bbbexample123/tmp36/f 74.84
bbbexample123/tmp36/f 74.84
bbbexample123/tmp36/f 75.02
bbbexample123/tmp36/f 74.48
bbbexample123/tmp36/f 74.30
bbbexample123/tmp36/f 74.66
bbbexample123/tmp36/f 74.66
bbbexample123/tmp36/f 74.66
bbbexample123/tmp36/f 75.02
bbbexample123/tmp36/f 74.48
bbbexample123/tmp36/f 74.84
bbbexample123/tmp36/f 74.48
bbbexample123/tmp36/f 74.48
bbbexample123/tmp36/f 74.66
bbbexample123/tmp36/f 75.20
bbbexample123/tmp36/f 74.84
bbbexample123/tmp36/f 74.84
bbbexample123/tmp36/f 74.48
bbbexample123/tmp36/f 74.48
bbbexample123/tmp36/f 74.30
bbbexample123/tmp36/f 74.66
bbbexample123/tmp36/f 75.20
bbbexample123/tmp36/f 75.02
```

MQTT Two-Phase Thermostat

The two-phase thermostat in the section title refers to a unit that can either start heating or cooling depending upon the measured temperature in the monitored space. In this section, I will show you how to establish two set points, which will cause cooling, heating, or no action based upon the received MQTT temperature data. The `PahoMqttSubscribe` Java class will be modified to incorporate this new controller application. I also renamed the class to `PahoMqttSubscribe1` to distinguish it from the original, non-controller version. The major change to the original class was to import the pi4j library, which provides the capability to control the Pi's GPIO pins. I have already shown you a project in Chapter 4, which also used the pi4j library. In that chapter, I provided an overview of the pi4j library and some of the important features it provides for Java project development. I would urge you to go back and review that chapter's content to refresh your knowledge of this important library.

You will need to download and install the pi4j library in order to compile the modified Java class. You can either follow my original instructions given in Chapter 4 or use the following guidance, which accomplishes the same goal but uses a slightly different approach.

1. Enter the following command in the Pi terminal window:

   ```
   sudo wget http://pi4j.googlecode.com/files/pi4j-0.0.5.deb
   ```

2. Install the newly downloaded code by entering the following:

   ```
   sudo dpkg -i pi4j-0.0.5.deb
   ```

The modified Java class is now named `PahoMqttSubscribe1` and is listed here:

```
import org.eclipse.paho.client.mqttv3.*;
import com.pi4j.io.gpio.GpioController;
import com.pi4j.io.gpio.GpioFactory;
import com.pi4j.io.gpio.GpioPinDigitalOutput;
import com.pi4j.io.gpio.RaspiPin;

public class PahoMqttSubscribe1 implements MqttCallback
{

   MqttClient client;

   final static GpioController gpio = GpioFactory.getInstance();
   final static GpioPinDigitalOutput pinH =
   gpio.provisionDigitalOutputPin(RaspiPin.GPIO_17,"PinH");
   final static GpioPinDigitalOutput pinL =
   gpio.provisionDigitalOutputPin(RaspiPin.GPIO_18,"PinL");


public PahoMqttSubscribe1() { }

public void messageArrived(String topic, MqttMessage message) throws Exception
{
   String msg = new String(message.getPayload());
   System.out.println( topic + " " + msg );
```

```
        Double dValue = Double.parseDouble(msg);
        int iValue = dValue.intValue();
        if(iValue >= 80) {
            pinH.high();
            pinL.low();
            System.out.println("Above 80 F");
        }
        if(iValue <= 60) {
            pinH.low();
            pinL.high();
            System.out.println("Below 60 F");
        }
        if(iValue > 60 && iValue < 80) {
            pinH.low();
            pinL.low();
            System.out.println("Between 60 and 80 F");
        }
    }

    public void connectionLost(Throwable cause) {}
    public void deliveryComplete(IMqttDeliveryToken token) {}

    public static void main(String[] args) {
      new PahoMqttSubscribe1().doDemo();
    }

    public void doDemo() {

      try {
        client = new MqttClient("tcp://m2m.eclipse.org:1883", MqttClient
    .generateClientId());
        client.connect();
        client.setCallback(this);

        client.subscribe("bbbexample123/tmp36/f");

        while(true) {
           try { Thread.sleep(1000); }
           catch(InterruptedException e) {}
        }
      }
      catch(MqttException e) { e.printStackTrace(); }
    } // end   of doDemo()
    } // end of class def
```

I have really only added some new functionality to the messageArrived method where the payload value is compared to two preset values in order to determine which GPIO pins are set to a HIGH value. The logic is simple: If the payload value is 80 or higher, turn on GPIO_17, which is theoretically connected to a relay module controlling an air conditioner for cooling purposes. If the payload value is 60 or lower, turn on GPIO_18, which likewise is theoretically connected to a relay module controlling a heater. Of course, if the temperature is between 60 and 80, do nothing as that is the desired comfort zone.

The set points of 60 and 80 are purely arbitrary but I did need some concrete values to test the system.

If you examine the code listing, you will see that I added some `println` statements in the control logic, which will allow me to display the GPIO control states on the Pi's terminal window. Figure 11-11 shows the system in operation and the state changes as I either added some heat to the tmp36 sensor by holding it or cooled it by touching an ice cube to the sensor. I also changed the measurement interval to 5 seconds in order to give myself enough time to change the sensor temperature without the intentionally changed temperatures scrolling off-screen.

This last figure concludes this M2M demonstration project where there were only computers "talking" to computers without any human intervention. This was a simple example of two computers communicating with each other using the standardized messaging MQTT protocol. There was also an intermediate message broker involved, which received data messages from a publisher client and then passed them on to a subscriber client. Many clients can subscribe to a broker, but only the messages they are interested in are sent to them. They show their interest by subscribing to a specific set of topics and subtopics.

This messaging project is only one of many M2M projects that have been developed to date. It is an exciting area, which promises to have many new and interesting projects available for developers and experimenters now and in the not-too-distant future.

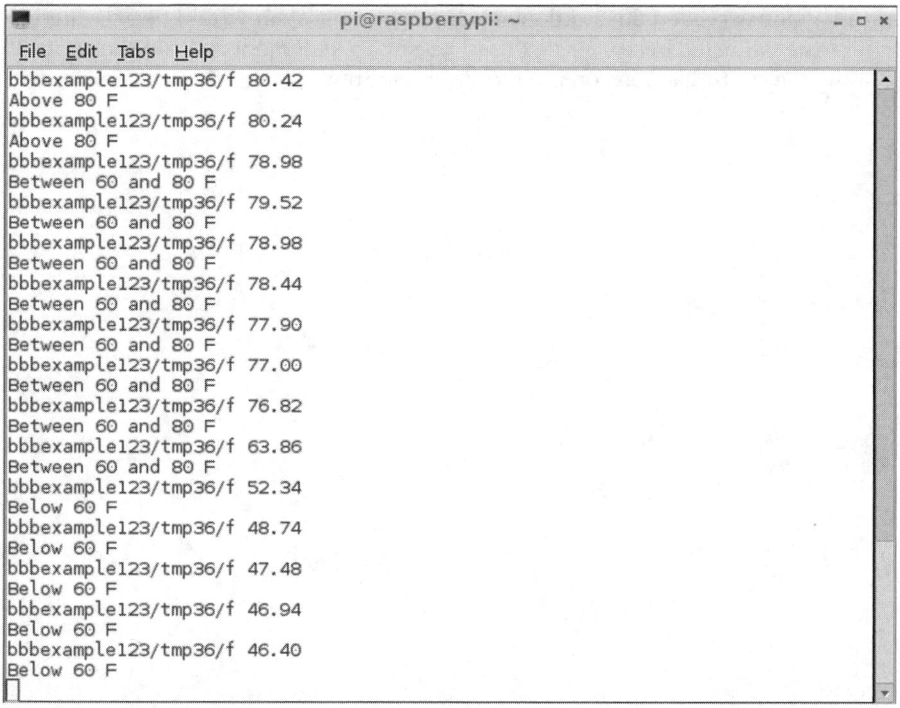

```
pi@raspberrypi: ~
File Edit Tabs Help
bbbexample123/tmp36/f 80.42
Above 80 F
bbbexample123/tmp36/f 80.24
Above 80 F
bbbexample123/tmp36/f 78.98
Between 60 and 80 F
bbbexample123/tmp36/f 79.52
Between 60 and 80 F
bbbexample123/tmp36/f 78.98
Between 60 and 80 F
bbbexample123/tmp36/f 78.44
Between 60 and 80 F
bbbexample123/tmp36/f 77.90
Between 60 and 80 F
bbbexample123/tmp36/f 77.00
Between 60 and 80 F
bbbexample123/tmp36/f 76.82
Between 60 and 80 F
bbbexample123/tmp36/f 63.86
Between 60 and 80 F
bbbexample123/tmp36/f 52.34
Below 60 F
bbbexample123/tmp36/f 48.74
Below 60 F
bbbexample123/tmp36/f 47.48
Below 60 F
bbbexample123/tmp36/f 46.94
Below 60 F
bbbexample123/tmp36/f 46.40
Below 60 F
```

FIGURE 11-11 Two-phase thermostat terminal display

Summary

The overall concept of machine-to-machine (M2M) communications was initially introduced along with a standardized message protocol named MQTT. I explained that I would use a simple temperature sensor connected to a BBB to send data to a Raspberry Pi via a MQTT broker. The first part of the demonstration would have the Pi only displaying the temperature data sent to it. In the second part of the demonstration, the Pi would execute some control action based upon a received data value.

I first demonstrated how the BBB functioned with the tmp36 temperature sensor before going into the actual MQTT portion of the project. The tmp36 would be the data source for the MQTT messages.

A section on MQTT brokers followed in which I explained their purpose and also provided a table of all the public brokers that were currently operating. I also explained the very public nature of the non-SSL MQTT messaging scheme and how your data is readily available to be accessed by anyone using the broker website.

The Python temperature sensor program was next modified to incorporate MQTT messaging. I demonstrated how to use an HTTP bridge website to view the published data without the need for an actual operating client subscriber.

A Raspberry Pi was next used to implement a Java-based subscriber client. This client displayed the temperature being sent from the BBB on a terminal window.

The Raspberry Pi client was then modified to trigger some GPIO pins, which simulated a two-phase thermostat. A cooling system would be activated when the received temperature value exceeded 80 °F. Likewise, the heating system would be activated if the temperature value fell below 60 °F. I used `println` statements within the control logic to show the control actions on the terminal window.

Index

E

F

G

H

I

M

N

O

P

Q

R